Chromatographic Analysis of Pesticide in Environmental and Food

Chromatographic Analysis of Pesticide in Environmental and Food

Editor

Miguel Ángel González-Curbelo

MDPI

Basel • Beijing • Wuhan • Barcelona • Belgrade • Novi Sad • Cluj • Manchester

Editor
Miguel Ángel
González-Curbelo
Universidad EAN
Bogotá
Colombia

Editorial Office
MDPI
St. Alban-Anlage 66
4052 Basel, Switzerland

This is a reprint of articles from the Special Issue published online in the open access journal *Molecules* (ISSN 1420-3049) (available at: https://www.mdpi.com/journal/molecules/special_issues/chromato_pesticide).

For citation purposes, cite each article independently as indicated on the article page online and as indicated below:

Lastname, A.A.; Lastname, B.B. Article Title. *Journal Name* **Year**, *Volume Number*, Page Range.

ISBN 978-3-0365-9462-0 (Hbk)
ISBN 978-3-0365-9463-7 (PDF)
doi.org/10.3390/books978-3-0365-9463-7

Contents

About the Editor

Miguel Ángel González-Curbelo

Dr. Miguel Ángel González-Curbelo holds the position of full professor in the Department of Basic Sciences at Ean University in Bogotá, Colombia. He earned his Ph.D. in Chemistry from the University of La Laguna (ULL) in the Canary Islands, Spain, in 2014. During his doctoral studies, he completed a pre-doctoral research stay at the United States Department of Agriculture (USDA) under the supervision of Dr. Steven J. Lehotay, the father of the QuEChERS method. Dr. González-Curbelo's scholarly contributions comprise approximately 50 publications in international journals and books, with his work focused on the development of environmentally friendly analytical methodologies for the determination of various contaminants in environmental and food samples through the employment of chromatographic techniques.

molecules

Review

Pesticide-Residue Analysis in Soils by the QuEChERS Method: A Review

Miguel Ángel González-Curbelo [1,*], Diana Angélica Varela-Martínez [1] and Diego Alejandro Riaño-Herrera [2,*]

[1] Departamento de Ciencias Básicas, Facultad de Ingeniería, Universidad EAN, Calle 79 nº 11-45, Bogotá 110221, Colombia; davarela@universidadean.edu.co
[2] Departamento de Ingeniería Ambiental y Energías, Facultad de Ingeniería, Universidad EAN, Calle 79 nº 11-45, Bogotá 110221, Colombia
* Correspondence: magonzalez@universidadean.edu.co (M.Á.G.-C.); drianohe7000@universidadean.edu.co (D.A.R.-H.)

Abstract: Pesticides are among the most important contaminants worldwide due to their wide use, persistence, and toxicity. Their presence in soils is not only important from an environmental point of view, but also for food safety issues, since such residues can migrate from soils to food. However, soils are extremely complex matrices, which present a challenge to any analytical chemist, since the extraction of a wide range of compounds with diverse physicochemical properties, such as pesticides, at trace levels is not an easy task. In this context, the QuEChERS method (standing for quick, easy, cheap, effective, rugged, and safe) has become one of the most green and sustainable alternatives in this field due to its inherent advantages, such as fast sample preparation, the minimal use of hazardous reagents and solvents, simplicity, and low cost. This review is aimed at providing a critical revision of the most relevant modifications of the QuEChERS method (including the extraction and clean-up steps of the method) for pesticide-residue analysis in soils.

Keywords: green extraction techniques; sample preparation; clean-up; multiresidue analysis; environmentally friendly methods

Citation: González-Curbelo, M.Á.; Varela-Martínez, D.A.; Riaño-Herrera, D.A. Pesticide-Residue Analysis in Soils by the QuEChERS Method: A Review. *Molecules* **2022**, *27*, 4323. https://doi.org/10.3390/molecules27134323

Academic Editor: Jesus Simal-Gandara

Received: 9 June 2022
Accepted: 30 June 2022
Published: 5 July 2022

1. Introduction

The current widespread use of herbicides, insecticides, fungicides, or other types of pesticides to effectively protect crops from pests and increase agricultural productivity results in unintended negative environmental effects, especially when good agricultural practices are breached. Soils are directly sprayed with pesticides before sowing and at the stage of germination; pesticides can also reach the soil after their application onto crops, even from long distances, through atmospheric volatilization and deposition processes. Consequently, soils contaminated by pesticide residues can be found, even in remote areas where they have never been used [1]. Because most pesticides do not easily dissipate or are biologically or chemically decomposed, their residues can persist in soils, which places soils among the environmental systems most affected by pollution. In fact, this is one of the most significant significant with converting conventional crops to organic, since conventional agriculture depends on the use of pesticides. In this sense, one of the most complete studies to have been carried out recently found that 83% of the 317 agricultural soils analyzed contained one or more residues and 58% contained a mixture [2]. Since pesticide residues have high levels of acute toxicity and endocrine disruptor effects, even at low concentrations, as well as long half-lives, they can affect soil functions, as well as the safety of subsequent productions. In addition, depending on the absorption capacity of the soil materials (pesticides are more strongly absorbed in soils with high clay or organic matter content than in sandy soils) and other environmental conditions, such as temperature, humidity, and pH, pesticide residues can migrate to other environmental compartments, such as ground or surface water. For this reason, water quality is normally monitored near to agricultural areas. Regarding the potential health risks to humans, they are not only exposed indirectly to pesticide residues through food grown in contaminated soil or products derived from grazing animals, but also by the ingestion/inhalation of soil and dust particles, as well as by dermal contact [3]. Hence, current legislation is increasingly restrictive to protect ecological sustainability and human health, even in developing countries, where there is an increase in the application of methods of food production that adopt the maximum residue levels established by international institutions, such as Codex Alimentarius. This implies the need to continuously develop new methods of analysis to assess these residues at trace levels in a fast, economical, and reliable way. In this sense, the improvement of sample pretreatments to extract these multiresidues strongly adsorbed into complex and heterogeneous soils is a fundamental aspect that requires constant revision.

The standard sample-extraction methods routinely employed for pesticide residues from soil include the Soxhlet extraction (Environmental Protection Agency (EPA) method 3540), automated Soxhlet extraction (EPA method 3541), pressurized liquid extraction (PLE) (EPA method 3545), microwave-assisted extraction (MAE) (EPA method 3546), ultrasonic solvent extraction (USE) (EPA method 3550), and supercritical fluid extraction (SFE) (EPA method 3562), as well as solid-phase extraction (SPE), solid-phase microextraction (SPME), matrix solid-phase dispersion (MSPD), and accelerated solvent extraction (ASE). Because soil matrices usually have a high content of natural organic components, mainly composed of humic substances, lipids, pigments, and fulvic acids, the matrix effects from the presence of interfering substances in the injection vial that are coextracted with pesticides should be minimized. Therefore, different clean-up steps, such as SPE using alumina (EPA method 3610), Florisil (EPA method 3620), or silica gel (EPA method 3630), as well as gel permeation chromatography (EPA method 3640) and sulfur (EPA method 3660), have also been incorporated into the analytical methods. However, any combination results in multi-stage procedures that use large amounts of toxic organic solvents and time require a large working place, are very tedious, and can discharge substantial waste. Therefore, increasingly environmentally friendly, fast, and simple alternatives are currently being developed to meet new analysis needs and to observe the principles of green chemistry.

Alternatively, attention has recently been drawn towards the use of a quick, easy, cheap, effective, rugged, and safe (QuEChERS) method to replace previous, less efficient extraction methods for pesticide determination [4]. It was first presented at the Fourth European Pesticide Residue Workshop (EPRW 2002), published in 2003 by Anastassiades et al. [5], and validated by Lehotay et al. [6]. The QuEChERS method has made it possible to quantify a much broader spectrum of pesticides (even hundreds) from different chemical classes simultaneously in a fast, simple, and cost-effective way while minimizing the amounts of sample and organic solvent used. This environmentally friendly and multiresidue method for the high-throughput routine analysis of pesticides involves only two steps, which minimizes errors: (i) a microscale extraction step with acetonitrile (ACN) based on partitioning via salting-out combined with (ii) a dispersive SPE (d-SPE) using a mix of clean-up sorbents composed of anhydrous $MgSO_4$, together with primary secondary amine (PSA) to remove traces of water and matrix interferences (organic acids, fatty acids and sugars), respectively, without large volume transfers or exchanges of solvents, blending, filtration, or evaporation [4]. This allows a single operator to perform multiple extractions simultaneously within a short period of time. The general scheme of the original (unbuffered) QuEChERS method can be observed in Figure 1, which includes the two official buffering-salt methods to increase the recovery of pH-dependent analytes, called the AOAC Official Method 2007.01 [7] and the CEN Standard Method EN 15662 [8]. These methods are called official methods because they were published by the AOAC (Association of Official Analytical Chemists) and the European Committee for Standardization, respectively. Moreover, the QuEChERS method stands out for removing matrix interferences and achieving very accurate results and high sensitivity. Due to all these features, it has evolved into the most popular method for the determination of pesticide residues in soil and related applications [4].

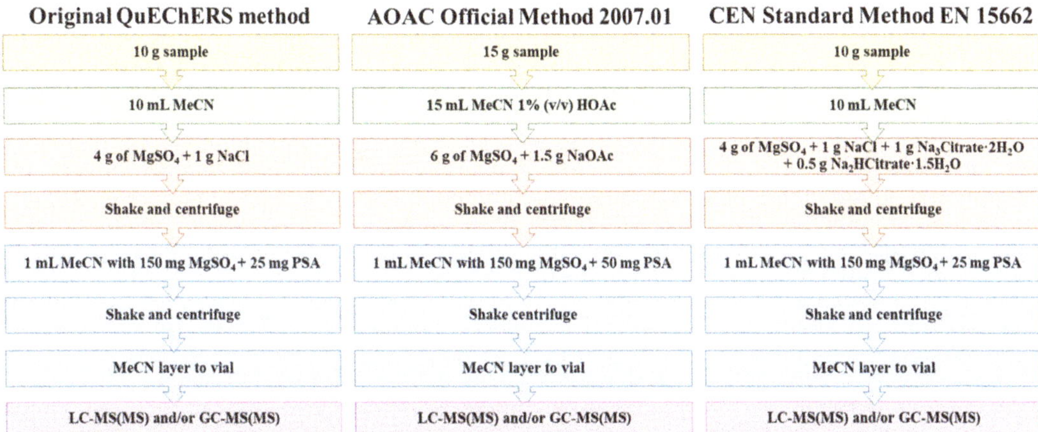

Figure 1. Diagram of the three primary QuEChERS methods based on [5,7,8], respectively.

Several reviews have been published in recent years focused on sample preparation procedures for the determination of pesticides in soils, including the QuEChERS method [9–13]. However, to the best of our knowledge, only one of these review articles was critically focused on modifications involving the use of the QuEChERS method for pesticide-residue analysis in soils, and it was published several years ago [11]. Therefore, the aim of this review is to provide an up-to-date critical assessment of the QuEChERS-based methods that have been employed for the analysis of pesticide residues in soils. On this basis, the modifications to the QuEChERS method are thoroughly described as a reference for researchers interested in this subject and in other types of organic contaminant or similar matrices, as well as for private laboratories and state agencies that seek to apply new and cost-effective methods.

2. QuEChERS Applications to Pesticide-Residue Analysis in Soils

According to Web of Science, 726 articles have been published featuring the terms "soil", "pesticide", and "QuEChERS", of which only 212 directly focus on the determination of pesticide residues in soils at trace levels by the QuEChERS method and chromatographic techniques coupled to mass spectrometry or other detectors, including the development and validation of analytical methods, as well as monitoring studies. The first application of the QuEChERS method for pesticide-residue analysis in soils was published in 2008 by Lesueur et al. [14]. In that study, the authors compared the QuEChERS method with a new USE, the European Norm DIN 12393, and a PLE method combined with gas chromatography–mass spectrometry (GC–MS) and high-performance liquid chromatography–tandem mass spectrometry (HPLC–MS/MS) in three different types of soil. The QuEChERS method was the most efficient extraction procedure: around 50% of the 24 multiclass pesticides analyzed had recoveries satisfying the 70–120% recovery range and a median recovery of 72.7%. Table 1 [14–56] summarizes a representative sample of the subsequent studies regarding the QuEChERS extraction approaches for the analysis of a wide range of pesticide residues belonging to different chemical families, such as organochlorine pesticides (OCPs) [15,16,20,25,29,48], organophosphorus pesticides (OPPs) [25], pyrethroid pesticides (PYPs) [25,47], neonicotinoids [35,57], carbamates [53], and triazole [18] and urea [22] derivatives, among others. Depending on the country, the types of pesticides vary due to the characteristic crops of each geographical and climatic zone. In most of these works, HPLC coupled with MS or MS/MS was the technique adopted for the determination of the pesticide residues, followed by GC-MS(MS), because it shows limitations for volatile pesticides, while HPLC allows the separation of the thermolabile and polar residues, as well as showing higher sensitivity. In some cases, ultra-high-performance liquid chromatography (UHPLC) coupled with MS/MS was employed for high throughput, especially when hundreds of pesticides were analyzed simultaneously [49,50]. Less sensitive techniques for pesticide residue analysis in soil samples include HPLC with traditional detectors, such as diode array detectors (DADs) [17,35], fluorescence detectors (FLDs) [53], ultraviolet (UV) [52] and GC with electron-capture detectors (ECDs) [16,25,29,36], and nitrogen phosphorous detectors (NPDs) [36] or flame photometric detectors (FPDs) [25] for OCPs and OPPs, respectively. In this context, Łozowicka et al. [36] studied the extent and variability of the matrix effects of pesticides using GC with different types of detectors (MS/MS and μECD/NPD). In the case of MS/MS detection, the recoveries for almost all the pesticides were in the range of 70–120% with an acceptable relative standard deviation (RSD) of less than 17% while μECD/NPD detection gave recoveries in the range 60–69% with similar RSD values. Unfortunately, the results for both systems of detection remained poor for captan, dichlofluanid, folpet, thiabendazole, and tolylfluanid, with recoveries between 63 and 69%. Nevertheless, it is well known that captan and folpet tend to degrade when they are pesticides are dissolved in ACN solutions [58], which was the extraction solvent. Analogously, Yang et al. [43] clearly observed some interfering compounds in the chromatograms of GC–ECD for the assessment of chloroacetanilide herbicides, which may cause overestimations or even false-positives. Therefore, the GC–MS/MS was more suitable for the analysis of those herbicides. Even though most MS/MS techniques provide high selectivity and sensitivity [59], sample preparation is still crucial. In this sense, the original version [5] and the two official versions [7,8] of the QuEChERS method were developed for the determination of pesticides in fruits and vegetables. This is why different modifications of the QuEChERS method for the extraction of pesticides from soils have been developed. Many of them focus on optimizing the parameters of both the extraction step and the subsequent clean-up step. These improvements have been made with the aim of obtaining better extraction efficiency and providing greater reliability and robustness to the chromatographic system, which is usually sensitive to matrix effects [36].

Table 1. Evolution of the QuEChERS method for the analysis of pesticide residues in soils.

Pesticides	Sample Amount	Water Added	Extraction		Sorbents in the dSPE Step per mL of Extract	Analytical Technique	Recoveries	LOQs	Comments	Reference
			Solvents	Salts						
24 multiclass pesticides	10 g	-	20 mL ACN	4 g $MgSO_4$, 1 g NaCl, 1 g sodium citrate tribasic dihydrate and 0.5 g sodium citrate dibasic sesquihydrate	150 mg $MgSO_4$ and 25 mg PSA	HPLC-MS/MS and GC-MS	27–121%	0.3–125 µg/kg	The QuEChERS method showed better performance than USE, the European Norm DIN 12393 and PLE	[14]
19 OCPs	5 g	10 mL	10 mL ACN (1% HAc)	4 g $MgSO_4$ and 1.7 g NaOAc	-	GC-MS/MS	70–100%	0.1–1.6 µg/kg	The clean-up step was performed by liquid-liquid partitioning with n-hexane	[15]
Chloroform, 1,2-dichlorobenze and HCB	2.5 g	1.5 mL	10 mL EtOAc	4 g $MgSO_4$	-	GC-µECD	62–93%	0.4–7.2 µg/kg	EtOAc showed higher extraction efficiency than ACN	[16]
Pyrimorphos	15 g	9 mL	15 mL ACN	6 g $MgSO_4$ and 1.5 g NaOAc	150 mg $MgSO_4$ and 50 mg PSA	HPLC-DAD	86–96%	50 µg/kg	The clean-up step of the extract was optimized by vortex	[17]
Clomazone, fipronil, tebuconazole, propiconazole and azoxystrobin	10 g	-	10 mL ACN (1% HAc)	4 g $MgSO_4$ and 1 g NaCl	-	HPLC-MS/MS	70–118%	10–50 µg/kg	PSA, C_{18} and $MgSO_4$ in the d-SPE step did not improve recoveries	[18]
Trifluralin	10 g	-	20 mL ACN	4 g $MgSO_4$, 1 g NaCl, 1g sodium citrate tribasic dihydrate and 0.5 g sodium citrate dibasic sesquihydrate	150 mg $MgSO_4$ and 25 mg PSA	GC-ECD	87–93%	11 µg/kg	Clean-up and preconcentration steps to change the injection solvent from ACN to EtOAc were incorporated	[19]
34 OCPs	2 g	-	15 mL DCM	4 g $MgSO_4$, 1g NaCl, 1g sodium citrate tribasic dihydrate and 0.5 g sodium citrate dibasic sesquihydrate	-	GC-MS	60–100% for almost all pesticides	58–2708 µg/kg	The QuEChERS method showed better performance than ASE. DCM showed higher extraction efficiency than ACN	[20]
Nicotine, sabadine, veratridine, rotenone, azadirachtin, cevadine, deguelin, spynosad D, pyrethrins and piperonyl butoxide	5 g	2.5 mL	5 mL ACN (1% HAc)	4 g $MgSO_4$, 4 g NaCl, 1g sodium citrate dihydrate and 0.5 g sodium citrate dibasic sesquihydrate	-	UHPLC-MS/MS	70–120% for almost all pesticides	4–10 µg/kg	The QuEChERS method showed better performance than SLE, SLE-USE and PLE	[21]
Diafenthiuron	10 g	2 mL	10 mL ACN	4 g $MgSO_4$ and 1 g NaCl	150 mg $MgSO_4$ and 50 mg PSA	HPLC-MS	74–100%	1 µg/kg	USE improved extraction efficiency	[22]
Benazolin-ethyl and quizalofop-p-ethyl	10 g	5 mL	10 mL ACN	3 g NaCl	200 mg PSA and 50 mg C_{18}	HPLC-MS/MS	74–110%	5 µg/kg	GCB gave lower recoveries for quizalofop-p-ethyl and benazolin-ethyl	[23]
36 multiclass pesticides	10 g	3 mL	10 mL ACN	4 g $MgSO_4$, 1 g NaCl, 1g sodium citrate tribasic dihydrate and 0.5 g sodium citrate dibasic sesquihydrate	d-SPE: 150 mg $MgSO_4$, 150 mg PSA and 50 mg C_{18}; DPX: 150 mg $MgSO_4$, 50 mg PSA and 50 mg	GC-MS/MS	70–120% for almost all pesticides	10 µg/kg	There was no significant difference between d-SPE and DPX in term of recoveries	[24]

Table 1. *Cont.*

Pesticides	Sample Amount	Water Added	Extraction		Sorbents in the dSPE Step per mL of Extract	Analytical Technique	Recoveries	LOQs	Comments	Reference
			Solvents	Salts						
10 OPPs, 8 OCPs and 6 PYPs	10 g		15 mL ACN (1% HAc)	6 g $MgSO_4$ and 1.5 g NaCl	-	GC-FPD and GC-ECD	80–120%	2–5 µg/kg	0.2 g PSA for OPPs and 0.2 g silica gel format for OCPs and PYPs, both in DPX format using Pasteur pipettes	[25]
17 multiclass pesticides	10 g	-	10 mL ACN (1% HAc)	4 g $MgSO_4$ and 1 g NaCl	75 mg $MgSO_4$ and 25 mg chitosan	HPLC-MS/MS	70–120% for almost all pesticides	0.1–100 µg/kg	Chitosan was more efficient than PSA, Chitin, and diatomaceous earth for clean-up purposes	[26]
25 multiclass pesticides	5 g	5 mL	10 mL ACN	4 g $MgSO_4$, 1g NaCl, 1g sodium citrate tribasic dihydrate and 0.5 g sodium citrate dibasic sesquihydrate	180 mg $MgSO_4$, 30 mg PSA and 30 mg C_{18}	UHPLC-MS/MS	74–111% for almost all pesticides	0.2–2.5 µg/kg	The QuEChERS method showed better performance than PLE	[27]
Bentazone, atrazine, carbamazepine, phenytoin, and its metabolites 5-(p-hydroxyphenyl-) and 5-phenylhydantoin	5 g	-	10 mL ACN: H_2O (70:30, *v*/*v*) (5% HAc)	4 g $MgSO_4$ and 1 g NaCl	12.5 mg $MgSO_4$ and 6.25 mg C_{18}	HPLC-UV	83–113%	10 µg/kg	C_{18} showed higher clean-up performance than PSA	[28]
10 OCPs	5 g	3 mL	7 mL ACN	6 g $MgSO_4$, 1.5 g NaCl, 1.5 g sodium citrate tribasic dihydrate and 0.75 g sodium citrate dibasic sesquihydrate	-	GC-ECD and GC-MS/MS	57–124%	1–3.6 µg/kg	The QuEChERS and MAE methods showed better performance than ASE and USE, but QuEChERS yielded slightly higher RSD values compared to MAE. Florisil in SPE format showed better clean-up efficiency than a mix of $MgSO_4$, PSA and C_{18} in d-SPE format	[29]
26 multiclass pesticides	5 g	10 mL	10 mL ACN (1% HAc)	4 g $MgSO_4$, 1 g NaCl, 1 g sodium citrate tribasic dihydrate and 0.5 g sodium citrate dibasic sesquihydrate	100 mg acidic alumina	UHPLC-MS/MS	70–114%	1 µg/kg	Acidic alumina showed better performance compared to 14 combinations of sorbents including PSA, GCB, C_{18}, Florisil, silica gel, Z-SEP, and Z-SEP+	[30]
58 multiclass pesticides	5 g	10 mL	10 mL ACN (1% HAc)	4 g $MgSO_4$ and 1 g NaOAc	150 mg $MgSO_4$, 25 mg PSA and 25 mg C_{18}	GC-MS/MS	69–119%	0.1–5 µg/kg	The AOAC QuEChERS version showed better performance than the EN QuEChERS version	[31]
Florasulam, carfentrazone-ethyl, fluroxypyr-meptyl and fluroxypyr	5 g	2 mL	10 mL ACN (1% HAc)	2 g $MgSO_4$ and 1 g NaCl	10 mg GCN	HPLC-MS/MS	80–110%	2.4–6 µg/kg	GCN showed higher clean-up performance than C_{18}	[32]
Benzobicyclon	20 g	20 mL	40 mL ACN (1% FA)	8 g $MgSO_4$ and 2 g NaCl	-	UPLC-MS/MS	64–76%	0.3–2.2 µg/kg	ACN showed higher extraction efficiency than EtOAc. HLB showed higher clean-up performance than C_{18} in SPE format	[33]

Table 1. *Cont.*

Pesticides	Sample Amount	Water Added	Extraction		Sorbents in the dSPE Step per mL of Extract	Analytical Technique	Recoveries	LOQs	Comments	Reference
			Solvents	Salts						
Furon, mesotrione, fluroxypyr-mepty and fluroxypyr	5 g	2 mL	10 mL ACN (1% HAc)	2 g $MgSO_4$ and 1 g NaCl	200 mg $MgSO_4$ and 25 mg C_{18}	**1 HPLC-MS/MS**	80–110%	2.4–6 μg/kg	GCB and PSA were not necessary because the soil had no pigments	[34]
Acetamiprid, imidacloprid, nitenpyram, flonicamid thiacloprid and 6-chloronicotinic acid	10 g	-	25 mL ACN: DCM (1:2, *v*/*v*)	5 g NaCl	400 mg C_{18} for the upper supernatant layer	HPLC-DAD	65–100%	48–246 μg/kg	ACN: DCM (1:2, *v*/*v*) showed higher extraction efficiency than ACN, acetone, EtOAc and ACN: DCM (2:1, *v*/*v*). C_{18} showed higher clean-up performance than PSA	[35]
216 multiclass pesticides	5 g	10 mL	10 mL ACN (1% FA)	4 g $MgSO_4$, 1 g NaCl, 1 g sodium citrate tribasic dihydrate and 0.5 g sodium citrate dibasic sesquihydrate	-	GC-MS/MS and GC-μECD/NPD	71–120%	5–10 μg/kg	A clean-up step with different combinations of $MgSO_4$, PSA, C_{18} and GCB gave lower recoveries	[36]
Spirotetramat and its four metabolites (β-enol, β-keto, β-mono and β-glu)	5 g	-	10 mL ACN (1% FA)	4 g $MgSO_4$ and 1 g NaCl	33 mg Florisil	HPLC-MS/MS	76–94%	1 μg/kg	Florisil showed higher clean-up efficiency than neutral alumina, GCB, PSA, C_{18}, diatomaceous earth, VERDE, ChloroFiltr and Chitosan	[37]
Metaldehyde and niclosamide ethanolamine	5 g	-	10 mL ACN	4 g $MgSO_4$ and 1 g NaCl	150 mg $MgSO_4$ and 50 mg Florisil	HPLC-MS/MS	90–101%	10–200 μg/kg	ACN showed higher extraction efficiency than DCM and EtOAc. Florisil showed higher clean-up efficiency than PSA, GCB, and MWCNTs	[38]
Dioctyl diethylen-etriamine acetate	10 g	5 mL	20 mL ACN	**2→5 g $MgSO_4$ and 5 g NaCl**	-	HPLC-MS/MS	86–97%	10 μg/kg	ACN: H_2O (4:1, *v*/*v*) showed higher extraction efficiency than MeOH: H_2O (4:1, *v*/*v*)	[39]
Fluopicolide, cyazofamid and their metabolites (M-01, M-02 and 4-chloro-5-p-tolylimidazole-2-carbonitrile)	10 g	10 mL	ACN 10 mL (2.5% FA)	6 g NaCl	100 mg $MgSO_4$	HPLC-MS/MS	71–107%	50 μg/kg	ACN (2.5% FA) showed higher extraction efficiency than ACN	[40]
Hexaconazole, flutriafol, triadimenol, tebuconazole, diniconazole, fipronil and picoxys-trobin	5 g	-	20 mL ACN	-	60 mg $MgSO_4$, 10 mg PSA, 10 mg C_{18} and 40 mg GCB	UHPLC-MS/MS	69–106%	0.03–0.25 μg/kg	USE for 20 min improved the extraction efficiency	[41]
Polyoxin B	5 g	-	5 mL H_2O (1% FA)	-	13 mg C_{18}	HPLC-MS/MS	83–112%	3 μg/kg	H_2O (1% FA) showed higher extraction efficiency than H_2O: MeOH (1:1), H_2O and H_2O (1% NH_3)	[42]

Table 1. *Cont.*

Pesticides	Sample Amount	Water Added	Extraction		Sorbents in the dSPE Step per mL of Extract	Analytical Technique	Recoveries	LOQs	Comments	Reference
			Solvents	Salts						
Acetochlor, alachlor, metolachlor, metazachlor, butachlor and pretilachlor	5 g	10 mL	**3→10 mL ACN**	4 g NaCl	50 mg $MgSO_4$, 25 mg PSA, 25 mg C_{18} and 5 mg GCB	GC-MS/MS	87–108%	0.8–2.2 µg/kg	There was no significant difference between ACN and ACN (1% disodium hydrogen citrate sesquihydrate) in terms of recoveries	[43]
12 multiclass pesticides	10 g	20 mL	10 mL ACN	4 g $MgSO_4$, 1 g NaCl, 1g sodium citrate tribasic dihydrate and 0.5 g sodium citrate dibasic sesquihydrate	130 mg $MgSO_4$, 21 mg PSA and 21 mg C_{18}	GC-MS	54–103%	6–21 µg/kg	The QuEChERS method showed better performance than SLE and Soxhlet extraction	[44]
25 multiclass pesticides	5 g	-	20 mL ACN	2 g NaCl	50 mg C_{18}	HPLC-MS/MS	72–108%	80–400 µg/kg	ACN showed higher extraction efficiency than MeOH and DCM. C_{18} showed higher clean-up efficiency than PSA and GCB	[45]
Dimethyl disulfide	10 g	5 mL	15 mL MeOH	-	-	GC-MS	85–98%	1 µg/kg	A simplified QuEChERS method without extraction salts showed better performance than the original QuEChERS extraction	[46]
Bifenthrin, chlorfenapyr, λ-cyhalothrin, pyridaben, pyrimethanil, and pyriproxyfen	5 g	-	10 mL ACN (1% HAc)	-	-	GC-MS	86–100%	0.5–2.4 µg/kg	Florisil showed higher clean-up efficiency than a mix of MgSO4, PSA and GCB both in SPE format.	[47]
16 OCPs	5 g	8 mL	8 mL ACN	4 g $MgSO_4$ and 1 g NaCl	25 mg Fe_3O_4@Triton	GC-MS	65–103%	0.3–5.5 µg/kg	Fe_3O_4@Triton showed higher clean-up efficiency than C_{18}, GCB and Fe_3O_4 in avocado and strawberry, and later this was then validated in soil	[48]
225 multiclass pesticides	10 g	-	10 mL ACN (2.5% FA)	6 g $MgSO_4$ and 1.5 g NaOAc	-	UHPLC-MS/MS and GC-MS/MS	70–120 for more than 87% pesticides	1–5 µg/kg	-	[49]
218 multiclass pesticides	10 g	-	10 mL ACN (2.5% FA)	6 g $MgSO_4$ and 1.5 g NaOAc	-	UHPLC-MS/MS and GC-MS/MS	70–120%	5–20 µg/kg	The AOAC QuEChERS method showed better performance than the EN QuEChERS version. PSA, C_{18}, GCB or EMR- lipid in d-SPE did not improve recoveries	[50]
13 multiclass pesticides	10 g	5 mL	10 mL ACN (0.5% FA)	4 g $MgSO_4$, 1 g NaCl, 1g sodium citrate tribasic dihydrate and 0.5 g sodium citrate dibasic sesquihydrate	150 mg $MgSO_4$ and 25 mg C_{18}	HPLC-MS/MS	70–93%	0.05 µg/kg	The QuEChERS method showed better performance than ultrasonic cylindrical probe and PLE. ACN (0.5% FA) showed higher extraction efficiency than MeOH. PSA did not improve recoveries	[51]

Table 1. *Cont.*

Pesticides	Sample Amount	Water Added	Extraction		Sorbents in the dSPE Step per mL of Extract	Analytical Technique	Recoveries	LOQs	Comments	Reference
			Solvents	Salts						
Isocycloseram	5 g	5 mL	10 mL ACN	1 g $MgSO_4$ and 0.5 g NaCl	-	HPLC-UV	91–109%	7.3–24 μg/kg	ACN showed higher extraction efficiency than DCM, MeOH, EtOAc and petroleum eter	[52]
Atrazine, desethylatrazin, desisopropyla-trazine, carbaryl, carbendazim and diuron	1 g	4 mL	2 mL ACN	1 g $MgSO_4$ and 0.5 g NaCl	66 mg $MgSO_4$ and 16 mg PSA	HPLC-DAD/FLD	74–108%	5–15 μg/kg	PSA showed higher d-SPE efficiency than Florisil	[53]
94 multiclass pesticides	5 g	10 mL	10 mL acetone:n-hexano (1:4, *v*/*v*)	4 g $MgSO_4$, 1 g NaCl, 1g sodium citrate tribasic dihydrate and 0.5 g sodium citrate dibasic sesquihydrate	180 mg $MgSO_4$ and 30 mg PSA	GC-MS/MS	70–117%	5–14 μg/kg	Acetone:n-hexane (1:4, *v*/*v*) showed higher extraction efficiency than ACN	[54]
31 multiclass pesticides	2.5 g	6 mL (EDTA 0.1 M)	5 mL ACN	4 g $MgSO_4$, 1 g NaCl, 1g sodium citrate tribasic dihydrate and 0.5 g sodium citrate dibasic sesquihydrate	150 mg $MgSO_4$, 25 mg C_{18} and 25 mg PSA	UPLC MS/MS	55–118%	0.01–5.5 μg/kg	The QuEChERS method showed better performance than PLE	[55]
Pyraclostrobin	5 g		10 mL ACN	-	150 mg $MgSO_4$ and 50 mg PSA	HPLC-MS/MS	97–102%	0.2 μg/kg	ACN showed higher extraction efficiency than ACN (1% HAc), ACN (0.1% HAc), ACN (1% FA), ACN (0.1% FA) and ACN (1% NH_3). PSA showed higher clean-up efficiency than C_{18}, Florisil, PSA+C_{18} and GCB	[56]

ACN: Acetonitrile; AOAC: Association of Official Analytical Chemists; ASE: Accelerated solvent extraction; C_{18}: Octadecylsilane; DAD: Diode array detector; DCM: Dichloromethane; DPX: Disposable pipette extraction; d-SPE: Dispersive solid-phase extraction; ECD: Electron capture detector; EMR-lipid: Enhanced Matrix Removal-lipid; EtOAc: Ethyl acetate; FA: Formic acid; GC: Gas chromatography; FLD: Fluorescence detector; FPD: Flame photometric detector; GCB: Graphitized carbon black; GCN: Graphitic carbon nitride; HAc: Acetic acid; HCB: Hexachlorobenzene; HPLC: High-performance liquid chromatography; LOQ: Limit of quantification; MAE: Microwave-assisted extraction; MeOH: Methanol; MS: Mass spectrometry; MS/MS: Tandem mass spectrometry; MWCNTs: Multi-walled carbon nanotubes; NaOAc: Sodium acetate; NPD: Nitrogen phosphorous detector; SLE: Solid-liquid extraction; SPE: Solid-phase extraction; OCPs: Organochlorine pesticides; OPPs: Organophosphorus pesticides; PLE: Pressurized liquid extraction; PSA: Primary secondary amine; PYPs: Pyrethroid pesticides; RSD: relative standard deviation; UHPLC: Ultra high-performance liquid chromatography; USE: Ultrasonic solvent extraction; UV: Ultraviolet.

3. The Extraction Step

The original approach, which involves adding anhydrous magnesium sulphate and sodium chloride in the extraction step, has found several applications for the analysis of pesticide residues in soils [18,22,26,28,37,38,48]. Furthermore, many researchers have used extraction liquid–liquid partitioning based on the AOAC Official Method 2007.01, which involves the use of acetic acid (HAc) in can, plus anhydrous $MgSO_4$ and NaOAc (relatively strong buffering capacity) [31], and the CEN Standard Method EN 15662 approach, which uses ACN followed by anhydrous $MgSO_4$ and NaCl, as well as sodium citrate tribasic dihydrate and sodium citrate dibasic sesquihydrate as the buffer (with a relatively low buffering capacity) [14,19,20,24,27,30,36,44,51]. In an interesting example, Yu et al. [31] compared the original method with the AOAC 2007.01 and EN 15662 official methods for the extraction of 58 multiclass pesticides from soil samples. Concretely, the no-buffer method contained 4 g of $MgSO_4$ and 1 g of NaCl, the acetate buffer contained 4 g of $MgSO_4$ and 1 g of NaOAc, and the citrate buffer contained 4 g of $MgSO_4$, 1 g of NaCl, 1 g of sodium citrate tribasic dihydrate, and 0.5 g of sodium citrate dibasic sesquihydrate. The AOAC QuEChERS version gave the higher average recoveries, between 72% and 121% (RSD < 19%), while the EN buffer method gave slightly lower recoveries (67–123%, RSD < 15%). The recoveries for the no-buffer method were lower than 70% for approximately 30% of all the pesticides. It should be noted that although the original works established specific amounts of reagent, many of these studies were slightly modified to obtain increasingly effective methods for high-organic-matter-content and low-humidity-content soils. Based on the analysis of the works under study, it was determined that the factors with the greatest impact on the extraction of pesticides from soils by the QuEChERS are: (i) the sample mass, (ii) the type and volume of solvent, and iii) the type and amount of extraction salt.

3.1. Modifications of the Sample Amount

The amount of sample and even sample size selection play an important role in obtaining the most accurate possible analytical results and high sensitivity. In this context, different sample amounts have been extracted after proper homogenization through mechanical processes, such as grinding and sieving. Methods involving 1 g [53], 2 g [20], 2.5 g [16], 5 g [52], 10 g [51], 15 g [17], and 20 g [33] of soil sample have been developed for pesticide-residue analysis, although most authors opted for 5 or 10 g. In all cases, the amount of sample used can be considered relatively low, which in turn is one of the great inherent benefits of the QuEChERS method. However, it must be considered that the extraction is normally carried out in 50-milliliter centrifuge tubes. Consequently, smaller sample amounts allow good homogenization and better separation of the supernatant because there is more free volume. Unfortunately, the lower the amount of sample, the lower the amount of analyte injected in the chromatographic system, so a proper balance must be found between the amount of sample that provides acceptable recoveries and the required sensitivity. Fernández et al. [24] reduced the sample amount from 10 g to 5 g, achieving a higher mean recovery (104% versus 68%) for the 36 multiclass pesticides analyzed by the CEN Standard Method EN 15662 and GC–MS. Correia-Sá et al. [60] also reduced the sample amount from 10 g to 5 g because no volume of supernatant could be taken, but they added only 3 mL of H_2O to hydrate the sample and 7 mL of ACN as the extraction solvent, plus 4 g of $MgSO_4$, 1 g of NaCl, 1 g of sodium citrate tribasic dihydrate, and 0.5 g of sodium citrate dibasic sesquihydrate. Chen et al. [45], using a 5gram sample amount, obtained recoveries in the range of 72–108%, but the limits of quantification (LOQs) were relatively high, between 80 and 400 μg/kg for the simultaneous determination of 25 multiclass pesticides followed by HPLC–MS/MS. By contrast, Yu et al. [31], applying a 5-gram amount, reached low LOQs, within the range of 0.1–5 μg/kg, for 58 multiclass pesticides by using the AOAC buffer method combined with GC–MS/MS. For smaller sample amounts, Rouvière [20] obtained worse LOQs in the range 6.9–2118 μg/kg using 2 g of sample by the EN citrate buffer method and GC–MS.

3.2. Modifications of Water Addition during Extraction

The QuEChERS method was originally developed for matrices with a high water content (above 80%) [5]. Later, it was applied to dry matrices, such as cereal samples, in which a sample rehydration step was implemented by shaking before extraction [61–63]. Because soil is a matrix with a low moisture content, the addition of water has also been considered in most pesticide extractions from soil samples. This additional step makes it possible to promote a moisturizing process. In addition, it alters the formation of H-bonds between the functional groups of non-ionic polar pesticides and those containing oxygen and hydroxyl of humic substances to achieve maximum extraction yield and accurate results [64]. However, although the QuEChERS approach recommends that the amount of water added should be the same as the mass of the sample, different ratios of soil to water have been studied with different volumes of water. Yang et al. [43] studied different amounts of water (2, 4, 6, 8, 10, and 15 mL) added to 5 g of soil sample. The results showed that 10 mL of water provided a cleaner extract and an increase in the signal-to-noise ratio (15.0 mL did not improve the results) for the six chloroacetamide herbicides analyzed by GC–MS/MS. Łozowicka et al. [36] tested cold-water dosages of 5, 7.5, and 10 mL with 5 g of soil sample. The use of cold water prevents the degradation of heat-sensitive pesticides that occurs when anhydrous $MgSO_4$ is added during extraction. When 10 mL of water were added, better recoveries were obtained for about 40% of the 216 multiclass pesticides compared to 7.5 mL. In the case of 5 mL of water, no supernatant was obtained. Correia-Sá [60] found that the best recoveries for all the tested pesticides were obtained with the hydration step with a ratio of 5 g to 3 mL (recoveries ranged from 77 to 130% versus 20 to 46% without H_2O addition). By contrast, Acosta-Dacal et al. [50] added water to aliquots of an air-dried soil sample to reach 10, 20, 30, 40, and 50% moisture. As the percentage of moisture increased, the authors did not observe significant differences in the recovery values of the pesticides determined by UHPLC/MS-MS. Instead, the recoveries were worse for many of the pesticides analyzed in GC–MS/MS with the increase in moisture, which was related to the reduction in the matrix load in the sample and, therefore, the sensitivity. These apparently contradictory results confirm the importance of optimizing the hydration step for the successful extraction of pesticides from soils.

3.3. Modifications of the Extraction-Solvent Type

As is well known, the selection of an appropriate extraction solvent plays a decisive role in achieving the maximum recovery of pesticides. Several solvents, such as ethyl acetate (EtOAc) [16], MeOH [46], dichloromethane (DCM) [20], or different mixtures [15,35] have been used for multiresidue pesticide analysis in soil samples by the QuEChERS method. However, EtOAc poorly extracts the most highly polar pesticides, MeOH coextracts large amounts of interfering substances from the matrix, and DCM is a highly toxic organochlorine solvent. Instead, ACN is the default extraction solvent used in this method because it efficiently isolates a wide range of polar and nonpolar pesticides while minimizing the amount of coextracted undesirable lipophilic compounds; hence, it provides higher selectivity for pesticide analyses [5]. In this context, Chen et al. [45] compared MeOH, DCM, and ACN as extraction solvents for 25 herbicides, obtaining poor recoveries between 54–108% and 37–110% for the MeOH and DCM, respectively, but acceptable and consistent recoveries in the range of 71–113% when the ACN was used. Similarly, Guan et al. [41] found that ACN gave higher extraction efficiencies than acetone, EtOAc, acetone/hexane, and acetone/DCM for the determination of diniconazole, fipronil, flutriafol, hexaconazole, picoxystrobin, tebuconazole, and triadimenol by UHPLC–MS/MS. ACN was also selected by Chai et al. [25] for the extraction of ten OCPs, eight OPPs, and six PYPs, obtaining satisfactory recoveries in the range of 80–120%, 82–118%, and 87–112%, respectively, with RSD values lower than 11% in all cases. Other ACN-based QuEChERS methods have also been successfully validated for the simultaneous extraction of 216 [36], 225 [49], and 218 [50] pesticides belonging to very diverse chemical families. In addition, ACN is less toxic than DCM, which was one of the most widely used solvents for many years [13], making

QuEChERS more environmentally friendly. It should also not be forgotten that ACN can be easily separated from water by adding salt and subsequent centrifugation, which allows the more efficient removal of residual water compared to other solvents [5], and it is highly compatible with GC and HPLC/UHPLC analysis. Thus, the implementation of additional evaporation and reconstitution steps is not necessary. As disadvantages, ACN has a large solvent-expansion volume for GC analysis, and it is expensive. However, ACN is still the most commonly employed extraction solvent in the QuEChERS method for pesticide-residue analysis in soils using relatively small volumes, usually between 5 and 15 mL, with a sample-to-solvent ratio of 1 g per mL [17,22,23,44] or 0.5 g per mL [24,27,38,43,52]. The optimization step has also included yield experiments with acidified ACN. On one hand, HAc has been added, normally at 1%, to form the HAc/NaOAC buffer, which is the basis of the AOAC version, to prevent the degradation of alkali-sensitive pesticides, but it has also been included without the subsequent addition of NaOAc [18,21,25,26,30,34]. On the other hand, formic acid (FA) has also been added to stabilize pesticides that tend to degrade under basic conditions [36,37,49–51], even in higher proportions. Xu et al. [40] studied the recoveries of fluopicolide, cyazofamid, and their metabolites with various concentrations of FA (0%, 2%, 2.5%, and 3%), while Acosta-Dacal et al. [50] compared extractions using ACN containing HAc (1%), FA (0.5%, 1% and 2.5%), and no added acid for the analysis of 218 multiclass pesticides. In both works, the addition of FA at 2.5% was the best choice. Combinations of ACN with other solvents, such as EtOAc [15] and DCM [35], have rarely been applied for very volatile pesticides.

3.4. Modifications of the Salting-Out Effect

As stated at the beginning of this section, the three main versions of the QuEChERS method, each with its characteristic salts, have been widely applied to extract pesticides from soil samples. However, other combinations of the same salts, or even different salts, have been assayed to promote ACN/water-phase separation during extraction. In this sense, the combination of anhydrous $MgSO_4$ and NaCl in 1:1 [39] or 2:1 [32,34,52,53,57] ratios (*w/w*) have also been used as alternatives to the original ratio, and both salts have even been successfully used alone. As an example, García Pinto et al. [16] performed a series of experiments with different combinations of anhydrous $MgSO_4$ with and without NaCl for the extraction of chloroform, 1,2-dichlorobenzene, and hexachlorobenzene (HCB). The results showed that there were no significant differences between them, and only anhydrous $MgSO_4$ was used in the final method. Nevertheless, it is well known that the use of $MgSO_4$ alone can lead to the presence of higher co-extractives [5]. For its part, NaCl alone has been directly used in other works, even without any previous study or optimization, due to its ability to improve the recoveries of polar compounds [23,35,40,43,45]. In the work carried out by Salama et al. [65], the authors used a central composite design to optimize the humidity (4, 5, and 6 mL of water), shaking time (3, 5, and 7 min), and amount of NaCl (1, 1.5 and 2 g) for the extraction of 30 multiclass pesticides. Although the humidity and shaking time had the most significant effects on the selected responses, the amount of NaCl had no significant effect. The most favorable extraction performance was obtained using 6 mL of water, a 7-minite shaking time, and 1 g NaCl. In the case of the citrate and acetate buffers, the salts were mostly added in the same 2:1 and 3:1 ratios (*w/w*), respectively, as in the official methods, but the combination of anhydrous $MgSO_4$ and NaOAc has also been added in a 4:1 ratio (*w/w*) [31]. In this last case, this combination was compared with that of the EN and original versions and gave better recoveries in the range of 72–121% versus 67–123% (the recoveries were slightly lower for several pesticides) and lower than 70% for approximately 30% of the 58 pesticides studied, respectively. It is also important to mention the work of Feride et al. [66], which tested the extraction efficiency of different salts ($MgSO_4$, NaCl, K_2CO_3, Na_2SO_4, and NaOAc) for the simultaneous extraction of 42 multiclass pesticides and 23 multiclass industrial chemicals. The higher extraction efficiency was obtained using ACN containing 1% HAc and a combination of $MgSO_4$, NaCl, and NaOAc (4:1:1, *w/w*). Much less commonly, some authors have applied the QuEChERS method without including extraction salts [41,42,56]. Evidently, these authors did not add water to the soil sample as a hydration step.

4. The Clean-Up Step

Soil is an extremely complex matrix that typically requires a clean-up step prior to injection into the chromatographic system to remove undesired coextracted substances and minimize the matrix effect. These substances can act as interferences and negatively affect the reproducibility and sensitivity of the pesticide quantification, as well as increasing the need for equipment maintenance [67,68]. Therefore, in addition to the recovery assessment, the impact on the instrumental performance must also be considered. In this sense, the higher the organic matter content of the soil, the greater the attention that should be paid to the format and sorbent formulations of the clean-up step of the QuEChERS method.

4.1. The d-SPE Approach

In the first publication of the QuEChERS method, Anastassiades et al. [5] introduced the concept of dSPE as a powerful clean-up procedure to adsorb interferences by adding a small quantity of sorbents into an extract, while the target pesticides remained in the liquid phase. The supernatant was then separated by centrifugation. Therefore, dSPE does not require the use of columns and frits, vacuum manifolds, preconditioning steps, the collection and evaporation of solvent fractions, etc. Consequently, dSPE is a shorter, simpler and more environmentally friendly procedure than conventional SPE. As in the most popular versions, anhydrous $MgSO_4$ and PSA have been effectively used during pesticide analysis in soil samples [14,17,19,22,53,54,56]. For example, Słowik-Borowiec et al. [54] recently used 180 mg of $MgSO_4$ and 30 mg of PSA per mL of extract for the determination of 94 multiclass pesticides and 13 polycyclic aromatic hydrocarbons in soil samples. The sample extraction was carried out using a modified EN QuEChERS version, and the final instrumental analysis was completed by GC–MS/MS, achieving satisfactory recoveries from 70% to 117% and RSD values in the range of 0.6–15.4%. However, this should not be the default choice because the use of PSA can lead to the hydrolysis of alkali-sensitive pesticides due to its basicity [69]. Consequently, the type and amount of cleaning sorbent have been the factors that have received the most attention for these applications. In this regard, a considerable number of studies have been published on the combination of $MgSO_4$ and/or PSA with other common sorbents, such as octadecylsilane (C_{18}) [23,24,27,28,31,34,44,51,55] and/or graphitized carbon black (GCB) [41,43,57]. On one hand, since C_{18} is a reversed-phase sorbent which has been particularly effective at removing nonpolar interferences from fatty extracts [4], it would be very useful to clean extracts from soils with high organic matter. In fact, C_{18} has been used with $MgSO_4$ only [28,34,51] and even alone [35,42,45]. In a representative study, Yu et al. [31] evaluated the addition of (1) 900 mg of $MgSO_4$ + 150 mg of PSA + 150 mg of C_{18}, (2) 900 mg of $MgSO_4$ + 150 mg of PSA, and (3) 900 mg of $MgSO_4$ + 150 mg of C_{18} per 6 mL of supernatant in terms of the matrix effect and recoveries for the determination of 58 multiclass pesticides using the AOAC extraction version and GC–MS/MS. According to Figure 2, although most of the analytes exhibited matrix-enhancement effects, the $MgSO_4$ + PSA + C_{18} combination gave lower matrix effects than the other two combinations. The recoveries were in the range between 70% and 120% for all three sets for most of the pesticides. On the other hand, GCB is a planar molecule that has been added to remove pigments (chlorophyll and carotenoids), but its addition should be carefully evaluated because it has a strong affinity for pesticides with planar structures, such as HCB, and can cause low recoveries [43,70]. For example, Chen et al. [45] demonstrated that more than half of the 25 herbicides they analyzed exhibited notable recovery loss when 50 mg of PSA (23–80%) or GCB (23–74%,) per mL of extract were used. Instead, the addition of 50 mg of C_{18} per mL of extract achieved satisfactory recoveries in the range of 72–108%, but the corresponding combinations were not evaluated. By contrast, Yang et al. [43] validated the purification effect of the same sorbents alone and in different ratios. The results showed that higher recoveries, in the range of 87–108%, were obtained for most of the six herbicides by combining PSA/GCB/C_{18} with 75–80% using C_{18}, 86–96% using GCB, and 90–103% using PSA. The results in terms of the matrix effect were consistent, with values of −11% to 5%, 18% to −25%, −20% to −25%, and −20% to 13%, respectively.

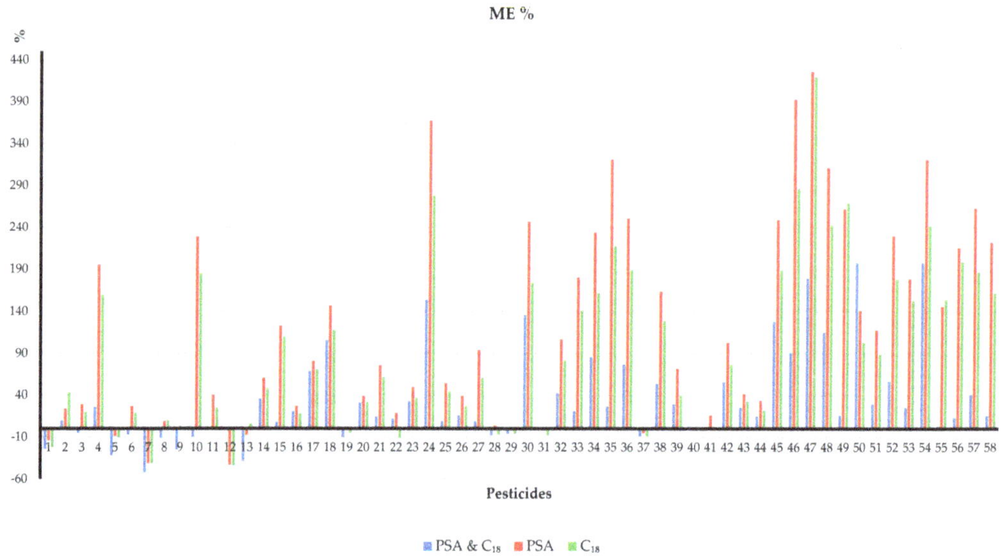

Figure 2. Matrix effects of the comparisons between different combinations of clean-up sorbents in soil samples. When matrix-effect (%) values are 0%, there is no matrix effect. Matrix-effect (%) values between 20% and 20% are mild. Matrix-effect (%) values between −50% and −20% or 20% and 50% are medium. Matrix-effect (%) values below 50% or above 50% are strong. Reprinted from [31], with permission from The Royal Society of Chemistry.

Other, less commonly used sorbents for the removal of interfering substances from soil samples are alumina [30], chitosan [26], nanosheets of graphitic carbon nitride (GCN) [32], and Florisil [37,38]. In a related comparative study, Łozowicka et al. [37] evaluated eight clean-up sorbents, namely PSA, GCB, C_{18}, alumina, chitosan, Florisil, diatomaceous earth, VERDE, and ChloroFiltr, for the determination of spirotetramat and its four metabolites (β-enol, β -keto, β -mono, and β -glu) in terms of the matrix effect and recoveries. The results showed that the Florisil (200 mg; 6 mL extract) provided the lowest matrix effect and recoveries between 76 and 94%, with RSD < 12%. Analogously, Dong et al. [38] demonstrated that Florisil gave better results in terms of extraction efficiency for the determination of metaldehyde and niclosamide ethanolamine than PSA, GCB, and multi-walled carbon nanotubes (MWCNTs). Nevertheless, Oliveira-Arias et al. [26] found that chitosan or diatomaceous earth achieved better results in terms of extraction efficiency and matrix effect compared to PSA, chitin (50 mg each together with 150 mg of $MgSO_4$ per 2 mL of extract), and no clean-up step for the determination of 17 pesticides from rice-paddy soil by HPLC–MS/MS. Furthermore, Guan et al. [41] compared Florisil (100 mg) with PSA (100 mg), C_{18} (100 mg), GCB (100 mg), PSA + C_{18} (100 mg, 1:1, *w/w*), and PSA + C_{18} + GCB (150 mg, 1:1:1, *w/w/w*) for the analysis of seven pesticides; slightly higher recoveries were obtained when the mixture of PSA, C_{18}, and GCB was used. Subsequently, the amounts of these sorbents were optimized, and the best proportion was a mixture of 50 mg of PSA, 50 mg of C_{18}, and 200 mg of GCB, together with 300 mg of $MgSO_4$. All the amounts above correspond to a volume of 5 mL of acetonitrile layer. One of the most complete comparative studies regarding the use of different sorbents in the d-SPE step in soil samples was developed by Kaczyński [30]. This study evaluated the purification effect of 14 combinations ((1) 25 mg

PSA and 2.5 GCB; (2) 25 mg PSA and 25 mg C_{18}; (3) 25 mg PSA + 7.5 mg GCB + 25 mg C_{18}; (4) 25 mg PSA; (5) 75 mg Z-Sep; (6) 50 mg Z-Sep+; (7) 20 mg Z-Sep; 50 mg C_{18}; (8) 200 mg Florisil; (9) 200 mg silica gel; (10) 200 mg C_{18}; (11) 200 mg C_8; (12) 200 mg alumina neutral; (13) 200 mg alumina acidic; and (14) 200 mg alumina basic per 2 mL of extract without $MgSO_4$ in all cases) for the determination of 26 acid herbicides with UHPLC–MS/MS. As a novelty, the authors tested the use of Z-Sep and Z-Sep+ sorbents based on zirconium dioxide, which have been used for commodities containing high amounts of fat [71,72]. However, as can be seen in Figure 3, the use of acidic alumina gave recoveries in an acceptable 70–120% range for all the pesticides (Figure 3a), and the matrix effects were either not significant or mild for the highest number of pesticides (Figure 3b). Acosta-Dacal et al. [50] also tested, for the first time in soil, another new sorbent, called Enhanced Matrix Removal-Lipid (EMR-lipid), specifically designed for high-fat matrices. However, the recoveries of 218 multiclass pesticides determined in an agricultural soil sample from the Canary Islands (clay loam soil) by UHPLC–MS/MS and GC–MS/MS were not improved. In fact, none of the other sorbents evaluated (PSA, C_{18}, and GCB) improved; therefore, a one-step QuEChERS-based method without clean-up was selected.

4.2. Other Clean-Up Approaches

Conventional SPE is one of the most commonly used alternatives to d-SPE for different applications despite its operational shortcomings, including the packaging of higher amounts of sorbents in order to obtain good clean-up effects. In the case of soil samples, a SPE method (1000 mg Florisil; 6 mL) was compared with a d-SPE method (150 $MgSO_4$, 50 mg PSA and 50 mg C_{18}; 1 mL) by Di et al. [29]. The recoveries of all the 10 OCPs analyzed were in the range of 95–115%, with RSD values lower than 5% for the Florisil–SPE cartridge, but lower for the d-SPE approach (31–87%, RSD < 10%). Ma et al. [47] also found slightly better recoveries using an SPE column packed with Florisil (94–99%) compared to a SPE column filled with a mix of $MgSO_4$, PSA, and GCB (83–100%) for the analysis of six pesticides. In work developed by Sun et al. [33], the authors compared two different SPE cartridges (HLB and C_{18}) to quantify benzobicyclon in soil and sediment samples, and the HLB cartridge showed a slightly better purification effect than the C_{18} cartridge.

In addition to the above-mentioned commercial sorbents, magnetic nanoparticles (MNPs) have been synthetized in the laboratory to selectively remove interference from soil samples. MNPs are also directly introduced in the extract and, once appropriately dispersed, they can easily be separated from it using an external magnet without additional centrifugation. Next, the analytes are eluted with an appropriate solvent. In this relatively novel approach, named magnetic d-SPE, bare magnetite (Fe_3O_4) is the most widely used MNP for a number of applications in pesticide-residue analysis, but its selectivity is relatively poor. In the study by Hubetska et al. [48], Fe_3O_4@Triton was compared with C_{18}, GCB, and Fe_3O_4 for the determination of 16 OCPs in avocado and strawberry samples. The nonionic surfactant, Triton X-100, was used as a precursor for the synthesis of the functionalized MNPs because it contains several functional complexes that can selectively bind to pesticides. The use of Fe_3O_4@Triton gave higher clean-up efficiency and recoveries than the addition of C_{18} and GCB in the d-SPE format and Fe_3O_4 in the magnetic d-SPE format. The use of Fe_3O_4@Triton was subsequently validated in soil samples, achieving good recoveries between 65 and 103%.

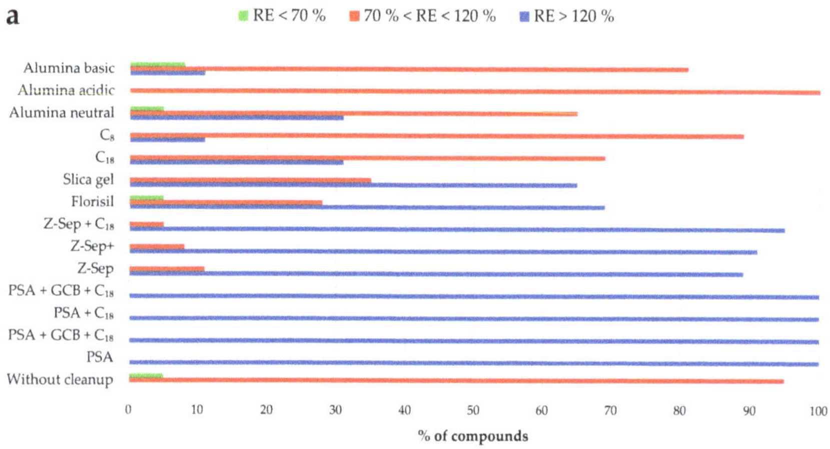

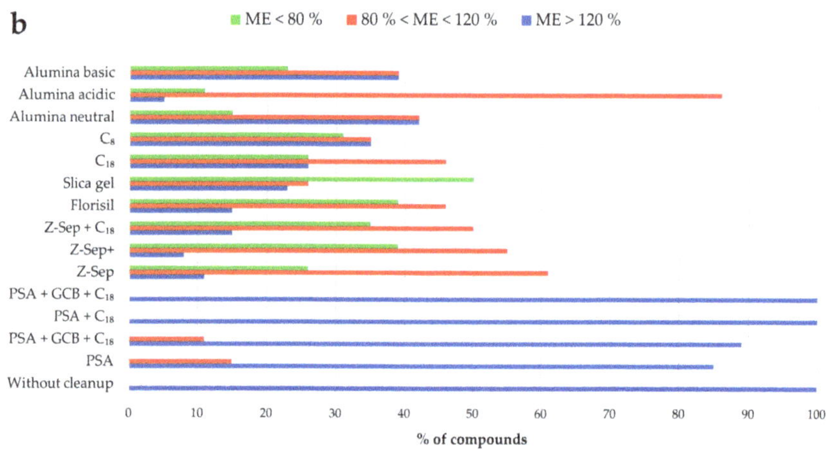

Figure 3. (**a**) Recoveries and (**b**) matrix effects of acid herbicides from various d-SPE sorbents. When matrix-effect (%) values are near to 100%, there is no matrix effect. Matrix-effect (%) values between 80% and 120% are mild. Matrix-effect (%) values below 80% or above 120% are strong. Reprinted from [30], with permission from Elsevier.

Disposable pipette extraction (DPX) is a practical SPE method that uses disposable pipette tips, in which the sorbent is contained. The sample extract is then aspirated and thoroughly mixed in a dynamic dispersive manner to achieve rapid equilibration. Consequently, undesirable compounds are concentrated on the sorbent and a clean extract is dispensed directly, without centrifugation. In a key study, Fernández et al. [24] compared, for the first time, the DPX and d-SPE procedures in soil samples using a composition of $MgSO_4$, PSA, and C_{18} in both cases. The results demonstrated that there was no significant difference for the two clean-up procedures in terms of recoveries for 36 multiclass pesticides in two types of soil (agricultural and organic) by GC–MS/MS. Another study reported the use of a glass Pasteur pipette packed with 200 mg of PSA for OPPs and 200 mg of silica gel for OCPs and PYPs in mineral and peat soils. For d-SPE using 25 mg PSA, the recoveries were in the range of 20–81%. By contrast, acceptable recoveries between 80 and 120% were obtained for all the pesticides in both soils using the DPX alternative [25]. Considering that the DPX procedure provides faster extraction times and is easy to perform, it is an alternative that should be considered for future applications. However, DPX provides poor filtration due to its screen mesh and does not transfer volumes ideally.

4.3. No-Clean-Up Approaches

As seen in the work published by Acosta-Dacal et al. [50], the clean-up step can be omitted to make the QuEChERS method less expensive, simpler, and faster, without compromising the analytical performance. This is not the case for most of the soil samples reported in the literature, but some other cases have been described in soils with relatively simple matrix compositions, mostly combined with MS/MS detection or selective detectors. Rouvière et al. [20] purified an extract from peat samples after extraction with ACN or DCM by d-SPE on PSA, but the recoveries of the 34 OCPs analyzed by GC–MS showed that the clean-up step was not necessary. Similarly, Caldas et al. [18] studied the influence of PSA and C_{18} during the analysis of clomazone, fipronil, tebuconazole, propiconazole, and azoxystrobin by HPLC-MS/MS, but these d-SPE sorbents did not have a significant influence on the recovery of the pesticides. Łozowicka [36] compared the EN QuEChERS version with and without the d-SPE step followed by GC–MS/MS and GC–μECD/NPD to determine 216 pesticides. Different combinations of PSA, C_{18}, and GCB together with $MgSO_4$ were tested, but the use of these sorbents did not have a significant influence on the recoveries or the matrix effect. Subsequently, the QuEChERS procedure without the d-SPE step was successfully validated and applied to the analysis of 263 soil samples. It should be noted that the authors placed the sample extracts in the freezer at −60 °C for 30 min right after the extraction, which is a clean-up process. In fact, this is the simplest method for fat removal from extracts [73,74]. However, it is clearly time-consuming and complicates the procedure.

5. Comparison of the QuEChERS Method with Other Extraction Methods

The analytical performance of the QuEChERS method for the analysis of pesticide residues in soils has been compared with other extraction methods, such as accelerated solvent extraction (ASE) [20,29], MAE [29], PLE [14,21,27,51,55], solid–liquid extraction (SLE) [21,44], Soxhlet extraction [44], and USE [14,21,29,51]. Although ASE, MAE, and USE were developed as more practical, faster, and more environmentally friendly procedures than the Soxhlet method, the QuEChERS method has since become the first choice of analytical chemists due to its high-throughput performance and easy modification according to the analytical needs of specific combinations of analytes and matrices. Đurović-Pejčev et al. [44], for instance, reported that the QuEChERS method provided higher extraction efficiency than traditional SLE and Soxhlet for most of the twelve pesticides belonging to the eight chemical groups analyzed in soil samples by GC–MS. Concretely, the recoveries applying the QuEChERS method were in the range of 54–103%, while the recoveries using SLE and Soxhlet were 40–91% and 12–92%, respectively. In turn, Rouvière [20] compared a previously optimized QuEChERS version using DCM as the extraction solvent with an ASE procedure for the analysis of 34 OCPs in soil by GC–MS. The average recovery varied between 60% and 100% when using QuEChERS, which, according to the authors, it proved to be simpler and faster. The ASE was a more tedious procedure and provided worse recoveries for most of the pesticides, ranging from 42% to 85%. In the study developed by García-Valverde [51], a modified QuEChERS version, when compared with an ultrasonic cylindrical probe and PLE combined with UHPLC–MS/MS (see Figure 4), proved to be more efficient (with higher recoveries with up to 12 samples by run) in the determination of 30 organic contaminants of emerging concern, including 13 multiclass pesticides, in agricultural soils, with LOQ < 0.1 ng/g in most cases. Homazava et al. [27] compared the performance, extraction efficiency, and matrix effect of a modified QuEChERS method with PLE followed by UHPLC–MS/MS for the analysis of 25 pesticides. QuEChERS was shown to be less time-consuming and demonstrated a higher sample throughput; recoveries between 79% and 113%, with RSDs of 1.0–12.2%, were obtained. By contrast, PLE extraction only reached recoveries between

65% and 122%, with RSDs of 1.7–23.4%. Moreover, the QuEChERS extracts were clearer and lower matrix effects were obtained (−54.5–7.0% versus −71.7–113.4%). Di et al. [29], for instance, compared the extraction efficiency of QuEChERS, MAE, ASE, and USE procedures combined with GC–ECD and GC–MS/MS for the determination of 10 OCPs. The QuEChERS and MAE procedures were found to achieve recoveries in the ranges of 78–124% (except for o,p′-DDD, with 57%) and 95–115%, respectively, while ASE and USE provided lower recoveries (47–118% and 44–128%, respectively). The authors highlighted the use of purging with nitrogen, in the case of ASE, and the application of ultrasounds, in the case of USE, as possible reasons, particularly for volatile pesticides. Despite the good results of the QuEChERS method, it yielded slightly higher RSD values compared to MAE. Hence, MAE was selected for the further analysis of real soil samples. This shows that, although the QuEChERS method covers a broader scope of pesticides in diverse sample types, providing higher recoveries and better analytical performance than traditional extraction procedures in most cases, for specific applications, there may be more appropriate methodologies. However, the QuEChERS method will continue to be one of the best options for the analysis of pesticide residues in soil.

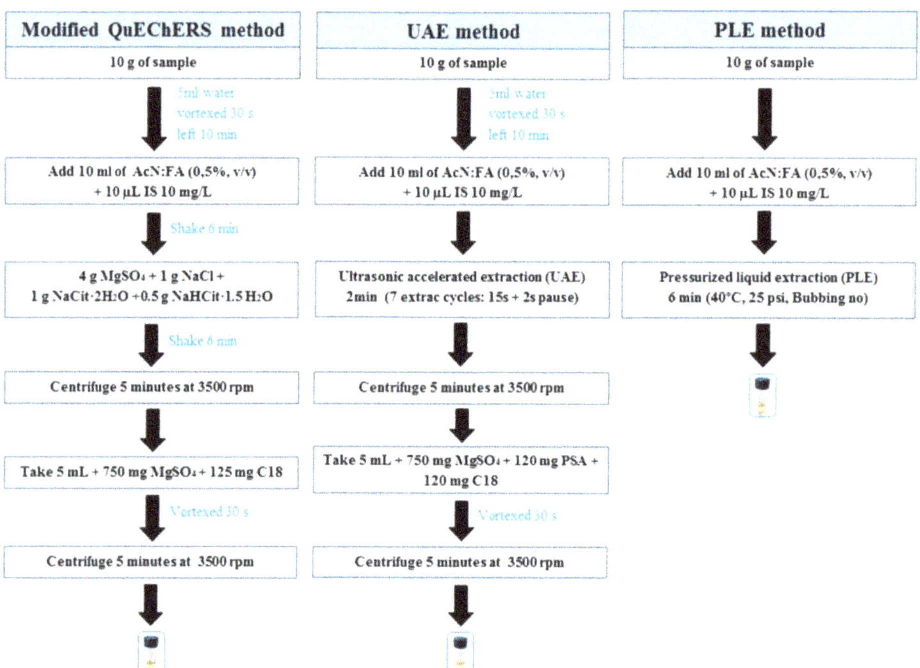

Figure 4. Diagram of the three methods used for the agricultural soil sample extraction. Reprinted from [51] with permission from Elsevier.

6. Conclusions and Future Trends

The three primary QuEChERS versions have been successfully applied for the simultaneously analysis of multiclass pesticides residues in soil samples due to their short operation time, simplicity, and low cost. The QuEChERS method is also aligned with green chemistry because it decreases the need for toxic solvents and reagents and generates much less waste. In addition, it has been easily adapted to a wide variety of pesticide/soil combinations to yield higher and more robust recovery rates. In this sense, ACN is the principal extraction solvent choice, even when compared to other organic solvents commonly used in this field, for the extraction of soil samples of just 5 or 10 g previously hydrated with similar amounts of water. Furthermore, ACN has been modified by adding HAc or FA as a preventive measure in many cases. Partitioning has been mostly achieved using the characteristic salts of the three main versions. In this sense, the use of a mild citrate buffer is another of the most commonly used default measures. Regarding the clean-up step, $MgSO_4$ and PSA are the most commonly used sorbents, whereas C_{18} and GCB have been used for soils with high content of organic matter and pigments, respectively. The use of traditional sorbents, such as Florisil in d-SPE and SPE formats, as well as nanotechnology-based sorbents, such as MNPs, has also been shown to be effective for cleaning purposes. Finally, DPX is a faster alternative to d-SPE and SPE that has hardly been used for soil samples, but for soils with low organic load, the clean-up step might not be necessary.

The QuEChERS method was originally applied to fruits and vegetables and, since then, most of the significant advances have been developed for these matrices before any others. This is the case of MWCNTs, Z-Sep, Z-Sep+, and EMR-lipid, as well as approaches such as magnetic d-SPE and DPX. Therefore, we should consider innovations for the analysis of pesticides and other organic analytes in fruits and vegetables. As an example, the first author of this review and the father of the QuEChERS method developed and validated a new version that uses ammonium formate instead of $MgSO_4$, NaCl, NaOAc, or citrate salts to induce phase separation and extraction [75]. Ammonium salts are more volatile, which prevents their deposition as solids in the GC inlet and in the MS ion source, which in turn increases equipment performance and minimizes the need for maintenance and liner replacement. Moreover, ammonium ions can enhance the formation of ammonium adducts instead of undesirable sodium adducts. As in the two official versions of the QuEChERS method, the addition of formic acid achieves suitable buffering. In fact, the performance of the ammonium formate version is similar to that of the QuEChERS AOAC Official Method 2007.01. Therefore, this is an alternative that could be adopted to improve the compatibility between the extraction of pesticides from soils using the QuEChERS method and MS detection. As a better alternative to DPX, filter-vial d-SPE was developed soon afterwards to quickly and conveniently clean and filter extracts in autosampler vials [76]. This approach eliminates centrifugation by combining d-SPE with in-vial filtration. QuEChERS automation is another trend that has gained strength in recent years, making QuEChERS an even faster approach to the analysis of large numbers of samples. Lehotay et al. [77] applied an automated mini-cartridge SPE cleanup combined with low-pressure (LP)GC–MS/MS to yield high-throughput capabilities and to reduce pesticide degradation in long instrumental sequences. Miniaturization is another feature that could be enhanced in the QuEChERS method for the analysis of pesticides in soil samples. Furthermore, the use of non-toxic extraction solvents, such as ionic liquids and deep eutectic solvents, would significantly reduce its waste disposal and costs. A much more recent method is the so-called QuEChERSER, which is an efficient and robust evolution that covers a wider polarity range than the QuEChERS method [78]. QuEChERSER relies on automation and miniaturization simultaneously, employing 1–5-gram samples extracted with 5 mL/g 4:1 (v/v) ACN-water solution and 1 g per g sample of 4/1 (w/w) $MgSO_4$/NaCl, followed by clean-up using automated instrument-top sample preparation (ITSP or μ-SPE) with 45 mg 20:12:12:1 of $MgSO_4$-PSA-C_{18}-CarbonX per 300 μL extract for GC (with no additional extraction salts or clean-up step for LC). The QuEChERSER mega-method has already been successfully validated for the analysis of pesticides in fruits and vegetables [79], other pesticides, veterinary drugs, environmental contaminants in bovine muscle [80], catfish muscle [81], tilapia [82], and pesticides in hemp and hemp products [83]. In summary, this is a field in constant evolution that deserves the continuous exploration of new greener, broader-coverage, faster, and cheaper approaches.

Author Contributions: Conceptualization, M.Á.G.-C.; methodology, M.Á.G.-C.; validation, M.Á.G.-C.; formal analysis, M.Á.G.-C. and D.A.V.-M.; investigation, M.Á.G.-C., D.A.V.-M. and D.A.R.-H.; data curation, M.Á.G.-C. and D.A.V.-M.; writing—original draft preparation, M.Á.G.-C.; writing—review and editing, M.Á.G.-C. and D.A.R.-H.; supervision, M.Á.G.-C. and D.A.V.-M.; project administration,

M.Á.G.-C. and D.A.V.-M.; funding acquisition, D.A.V.-M. All authors have read and agreed to the published version of the manuscript.

Funding: This research was funded by Ministerio de Ciencia, Tecnología e Innovación de Colombia, grant number 1223-777-57906-contract 619-2018. The APC was funded by Ministerio de Ciencia, Tecnología e Innovación de Colombia.

Institutional Review Board Statement: Not applicable.

Informed Consent Statement: Not applicable.

Data Availability Statement: Not applicable.

Acknowledgments: The authors would like to thank the Universidad EAN for their invaluable support.

Conflicts of Interest: The authors declare no conflict of interest.

References

1. Meijer, S.N.; Ockenden, W.A.; Sweetman, A.; Breivik, K.; Grimalt, J.O.; Jones, K.C. Global distribution and budget of PCBs and HCB in background surface soils: Implications for sources and environmental processes. *Environ. Sci. Technol.* **2003**, *37*, 667–672. [CrossRef] [PubMed]
2. Silva, V.; Mol, H.G.J.; Zomer, P.; Tienstra, M.; Ritsema, C.J.; Geissen, V. Pesticide residues in European agricultural soils—A hidden reality unfolded. *Sci. Total Environ.* **2019**, *653*, 1532–1545. [CrossRef] [PubMed]
3. Von Lindern, I.; Spalinger, S.; Stifelman, M.L.; Stanek, L.W.; Bartrem, C. Estimating children's soil/dust ingestion rates through retrospective analyses of blood lead biomonitoring from the Bunker Hill superfund site in Idaho. *Environ. Health Perspect.* **2016**, *124*, 1462–1470. [CrossRef]
4. Varela-Martínez, D.A.; González-Sálamo, J.; González-Curbelo, M.Á.; Hernández-Borges, J. Quick, Easy, Cheap, Effective, Rugged, and Safe (QuEChERS) Extraction. In *Gas Chromatography (Handbooks in Separation Science)*; Poole, C., Ed.; Elsevier: Amsterdam, The Netherlands, 2020; pp. 399–437.
5. Anastassiades, M.; Lehotay, S.J.; Štajnbaher, D.; Schenck, F.J. Fast and easy Mmultiresidue method employing acetonitrile extraction/partitioning and 'dispersive solid-phase extraction' for the determination of pesticide residues in produce. *J. AOAC Int.* **2003**, *86*, 412–431. [CrossRef]
6. Lehotay, S.J.; De Kok, A.; Hiemstra, M.; Van Bodegraven, P. Validation of a fast and easy method for the determination of residues from 229 pesticides in fruits and vegetables using gas and liquid chromatography and mass spectrometric detection. *J. AOAC Int.* **2005**, *88*, 595–614. [CrossRef] [PubMed]
7. Lehotay, S.J.; O'Neil, M.; Tully, J.; García, A.V.; Contreras, M.; Mol, H.; Heinke, V.; Anspach, T.; Lach, G.; Fussell, R.; et al. Determination of pesticide residues in foods by acetonitrile extraction and partitioning with magnesium sulfate: Collaborative study. *J. AOAC Int.* **2007**, *90*, 485–520. [CrossRef] [PubMed]
8. *Standard Method EN 15662*; Food of Plant Origin–Determination of Pesticide Residues using GC–MS and/or LC–MS/MS following Acetonitrile Extraction/Partitioning and Clean-Up by Dispersive SPE-QuEChERS Method. European Committee for Standardization (CEN): Brussels, Belgium, 2008.
9. Rajput, S.; Sharma, R.; Kumari, A.; Kaur, R.; Sharma, G.; Arora, S.; Kaur, R. Pesticide residues in various environmental and biological matrices: Distribution, extraction, and analytical procedures. *Environ. Dev. Sustain.* **2021**, *24*, 6032–6052. [CrossRef]
10. Mohamed, A.H.; Noorhisham, N.A.; Yahaya, N.; Mohamad, S.; Kamaruzzaman, S.; Osman, H.; Aboul-Enein, H. Sampling and sample preparation techniques for the analysis of organophosphorus pesticides in soil matrices. *Crit. Rev. Anal. Chem.* **2021**, 1–22.
11. Pszczolinska, K.; Michel, M. The QuEChERS approach for the determination of pesticide residues in soil samples: An overview. *J. AOAC Int.* **2016**, *99*, 1403–1414. [CrossRef]
12. Bruzzoniti, M.C.; Checchini, L.; De Carlo, R.M.; Orlandini, S.; Rivoira, L.; Del Bubba, M. QuEChERS sample preparation for the determination of pesticides and other organic residues in environmental matrices: A critical review. *Anal. Bioanal. Chem.* **2014**, *406*, 4089–4116. [CrossRef]
13. Tadeo, J.L.; Pérez, R.A.; Albero, B.; García-Valcárcel, A.I.; Sánchez-Brunete, C. Review of sample preparation techniques for the analysis of pesticide residues in soil. *J. AOAC Int.* **2012**, *95*, 1258–1271. [CrossRef] [PubMed]
14. Lesueur, C.; Gartner, M.; Mentler, A.; Fuerhacker, M. Comparison of four extraction methods for the analysis of 24 pesticides in soil samples with gas chromatography-mass spectrometry and liquid chromatography-ion trap-mass spectrometry. *Talanta* **2008**, *75*, 284–293. [CrossRef] [PubMed]
15. Rashid, A.; Nawaz, S.; Barker, H.; Ahmad, I.; Ashraf, M. Development of a simple extraction and clean-up procedure for determination of organochlorine pesticides in soil using gas chromatography-tandem mass spectrometry. *J. Chromatogr. A* **2010**, *1217*, 2933–2939. [CrossRef] [PubMed]
16. Pinto, C.G.; Laespada, M.E.F.; Martín, S.H.; Ferreira, A.M.C.; Pavón, J.L.P.; Cordero, B.M. Simplified QuEChERS approach for the extraction of chlorinated compounds from soil samples. *Talanta* **2010**, *81*, 385–391. [CrossRef]

17. Wang, J.; Zhao, L.; Li, X.; Jiang, Y.; Li, N.; Qin, Z.; Pan, C. Residue dynamic of pyrimorph on tomatoes, cucumbers and soil under greenhouse trails. *Bull. Environ. Contam. Toxicol.* **2011**, *86*, 326–330. [CrossRef] [PubMed]
18. Caldas, S.S.; Bolzan, C.M.; Cerqueira, M.B.; Tomasini, D.; Furlong, E.B.; Fagundes, C.; Primel, E.G. Evaluation of a modified QuEChERS extraction of multiple classes of pesticides from a rice paddy soil by LC-APCI-MS/MS. *J. Agric. Food Chem.* **2011**, *59*, 11918–11926. [CrossRef]
19. Temur, C.; Tiryaki, O.; Uzun, O.; Basaran, M. Adaptation and validation of QuEChERS method for the analysis of trifluralin in wind-eroded soil. *J. Environ. Sci. Health Part B* **2012**, *47*, 842–850. [CrossRef]
20. Rouvire, F.; Bulete, A.; Cren-Olivé, C.; Arnaudguilhem, C. Multiresidue analysis of aromatic organochlorines in soil by gas chromatography-mass spectrometry and QuEChERS extraction based on water/dichloromethane partitioning. Comparison with accelerated solvent extraction. *Talanta* **2012**, *93*, 336–344. [CrossRef]
21. Prestes, O.D.; Padilla-Sánchez, J.A.; Romero-González, R.; Grio, S.L.; Frenich, A.G.; Martínez-Vidal, J.L. Comparison of several extraction procedures for the determination of biopesticides in soil samples by ultrahigh pressure LC-MS/MS. *J. Sep. Sci.* **2012**, *35*, 861–868. [CrossRef]
22. Wang, L.; Zhao, P.; Zhang, F.; Du, F.; Pan, C. Diafenthiuron residue and decline in pakchoi and soil under field application. *Ecotoxicol. Environ. Saf.* **2012**, *79*, 75–79. [CrossRef]
23. Guan, W.; Zhang, H. Determination and study on residue and dissipation of benazolin-ethyl and quizalofop-p-ethyl in rape and soil. *Int. J. Environ. Anal. Chem.* **2013**, *93*, 679–691. [CrossRef]
24. Fernandes, V.C.; Domingues, V.F.; Mateus, N.; Delerue-Matos, C. Multiresidue pesticides analysis in soils using modified QuEChERS with disposable pipette extraction and dispersive solid-phase extraction. *J. Sep. Sci.* **2013**, *36*, 376–382. [CrossRef] [PubMed]
25. Chai, L.K.; Elie, F.; Jinang, C. Determination of 24 pesticides residues in mineral and peat soils by modified QuEChERS method and gas chromatography. *Int. J. Environ. Anal. Chem.* **2014**, *94*, 519–530. [CrossRef]
26. De Oliveira Arias, J.L.; Rombaldi, C.; Caldas, S.S.; Primel, E.G. Alternative sorbents for the dispersive solid-phase extraction step in quick, easy, cheap, effective, rugged and safe method for extraction of pesticides from rice paddy soils with determination by liquid chromatography tandem mass spectrometry. *J. Chromatogr. A* **2014**, *1360*, 66–75. [CrossRef] [PubMed]
27. Homazava, N.; Gachet Aquillon, C.; Vermeirssen, E.; Werner, I. Simultaneous multi-residue pesticide analysis in soil samples with ultra-high-performance liquid chromatography–tandem mass spectrometry using QuEChERS and pressurised liquid extraction methods. *Int. J. Environ. Anal. Chem.* **2014**, *94*, 1085–1099. [CrossRef]
28. De Carlo, R.M.; Rivoira, L.; Ciofi, L.; Ancillotti, C.; Checchini, L.; Del Bubba, M.; Bruzzoniti, M.C. Evaluation of different QuEChERS procedures for the recovery of selected drugs and herbicides from soil using LC coupled with UV and pulsed amperometry for their detection. *Anal. Bioanal. Chem.* **2015**, *407*, 1217–1229. [CrossRef]
29. Di, S.; Shi, S.; Xu, P.; Diao, J.; Zhou, Z. Comparison of different extraction methods for analysis of 10 organochlorine pesticides: Application of MAE-SPE method in soil from Beijing. *Bull. Environ. Contam. Toxicol.* **2015**, *95*, 67–72. [CrossRef]
30. Kaczyński, P.; Łozowicka, B.; Jankowska, M.; Hrynko, I. Rapid determination of acid herbicides in soil by liquid chromatography with tandem mass spectrometric detection based on dispersive solid phase extraction. *Talanta* **2016**, *152*, 127–136. [CrossRef]
31. Yu, Y.; Liu, X.; He, Z.; Wang, L.; Luo, M.; Peng, Y.; Zhou, Q. Development of a multi-residue method for 58 pesticides in soil using QuEChERS and gas chromatography-tandem mass spectrometry. *Anal. Methods* **2016**, *8*, 2463–2470. [CrossRef]
32. Pang, N.; Wang, T.; Cui, Y.; Hu, J. New dispersive solid phase extraction sorbent of graphitic carbon nitride for field evaluation and dissipation kinetics of pesticides in wheat ecosystem by liquid chromatography tandem mass spectrometry. *Int. J. Environ. Anal. Chem.* **2016**, *96*, 1156–1169. [CrossRef]
33. Sun, Q.; Wang, Y.; Tian, C.; Gui, W.; Guo, Y.; Zhu, G. Residue analysis of benzobicyclon in soil and sediment samples by ultra high-performance liquid chromatography tandem mass spectrometry. *J. AOAC Int.* **2016**, *99*, 1628–1635. [CrossRef] [PubMed]
34. Pang, N.; Wang, T.; Hu, J. Method validation and dissipation kinetics of four herbicides in maize and soil using QuEChERS sample preparation and liquid chromatography tandem mass spectrometry. *Food Chem.* **2016**, *190*, 793–800. [CrossRef] [PubMed]
35. Abdel-Ghany, M.F.; Hussein, L.A.; El Azab, N.F. Multiresidue analysis of five neonicotinoid insecticides and their primary metabolite in cucumbers and soil using high performance liquid chromatography with diode-array detection. *J. AOAC Int.* **2017**, *100*, 176–188. [CrossRef] [PubMed]
36. Łozowicka, B.; Rutkowska, E.; Jankowska, M. Influence of QuEChERS modifications on recovery and matrix effect during the multi-residue pesticide analysis in soil by GC/MS/MS and GC/ECD/NPD. *Environ. Sci. Pollut. Res.* **2017**, *24*, 7124–7138. [CrossRef] [PubMed]
37. Łozowicka, B.; Mojsak, P.; Kaczyński, P.; Konecki, R.; Borusiewicz, A. The fate of spirotetramat and dissipation metabolites in Apiaceae and Brassicaceae leaf-root and soil system under greenhouse conditions estimated by modified QuEChERS/LC–MS/MS. *Sci. Total Environ.* **2017**, *603*, 178–184. [CrossRef] [PubMed]
38. Dong, B.; Shao, X.; Lin, H.; Hu, J. Dissipation, residues and risk assessment of metaldehyde and niclosamide ethanolamine in pakchoi after field application. *Food Chem.* **2017**, *229*, 604–609. [CrossRef]
39. Gong, J.; Yang, G.; Yu, W.; Huang, M.; Li, F.; Jin, M.; Hu, D.; Zhang, K. The development and validation of a liquid chromatography–tandem mass spectrometry procedure for the determination of dioctyldiethylenetriamine acetate residues in soil, green and cured tobacco leaves using a modified QuEChERS approach. *Chromatographia* **2018**, *81*, 1035–1041.
40. Xu, T.; Feng, X.; Pan, L.; Jing, J.; Zhang, H. Residue and risk assessment of fluopicolide and cyazofamid in grapes and soil using LC-MS/MS and modified QuEChERS. *RSC Adv.* **2018**, *8*, 35485–35495. [CrossRef]

41. Guan, J.; Li, D.; Guo, Y.; Shi, S.; Yan, F. Simultaneous determination of seven pesticide residues in soil samples using ultrasound-assisted dispersive solid-phase extraction combined with UHPLC-MS/MS. *Sep. Sci. Plus* **2018**, *1*, 296–305. [CrossRef]
42. Song, S.; Chen, Z.; Wei, J.; Lei, Y.; Deng, C.; Tan, H.; Li, X. Determination of polyoxin b in cucumber and soil using liquid chromatography tandem mass spectrometry coupled with a modified quechers method. *Acta Chromatogr.* **2019**, *31*, 157–163. [CrossRef]
43. Yang, J.; Li, Q.; Wang, L.; Shao, J.; Mei, W.; Wang, L. Development and application of a dispersive solid-phase extraction method for the simultaneous determination of chloroacetamide herbicide residues in soil by gas chromatography-tandem mass spectrometry (GC-MS/MS). *Int. J. Environ. Anal. Chem.* **2019**, *99*, 282–296. [CrossRef]
44. Đurović-Pejčev, R.D.; Bursić, V.P.; Zeremski, T.M. Comparison of QueChers with traditional sample preparation methods in the determination of multiclass pesticides in soil. *J. AOAC Int.* **2019**, *102*, 46–51. [CrossRef]
45. Chen, Y.; Yu, Y.; Liu, X.; Yang, Y.; Lu, P.; Hu, D. Ddevelopment and validation of a liquid chromatography–tandem mass spectrometry method for multiresidue determination of 25 herbicides in soil and tobacco. *Chromatographia* **2020**, *83*, 229–239. [CrossRef]
46. Fang, N.; Lu, Z.; Zhang, Z.; Hou, Z.; Liang, S.; Wang, B.; Wang, S.; Lu, Z. Dissipation and residues of dimethyl disulfide in tomatoes and soil under greenhouse and open field conditions. *J. Environ. Sci. Health Part B* **2020**, *55*, 566–573. [CrossRef] [PubMed]
47. Ma, T.; Li, S.; Li, Y.; Li, X.; Luo, Y. Simultaneous determination and exposure assessment of six common pesticides in greenhouses through modified QuEChERS and gas chromatography–mass spectrometry. *Stoch. Environ. Res. Risk Assess.* **2020**, *34*, 1967–1982. [CrossRef]
48. Hubetska, T.S.; Kobylinska, N.G.; Menendez, J.R.G. Application of hydrophobic magnetic nanoparticles as cleanup adsorbents for pesticide residue analysis in fruit, vegetable, and various soil samples. *J. Agric. Food Chem.* **2020**, *68*, 13550–13561. [CrossRef]
49. Acosta-Dacal, A.; Rial-Berriel, C.; Díaz-Díaz, R.; Bernal-Suárez, M.M.; Zumbado, M.; Henríquez-Hernández, L.A.; Macías-Montes, A.; Luzardo, O.P. Extension of an extraction method for the determination of 305 organic compounds in clay-loam soil to soils of different characteristics. *MethodsX* **2021**, *8*, 101476. [CrossRef]
50. Acosta-Dacal, A.; Rial-Berriel, C.; Díaz-Díaz, R.; Bernal-Suárez, M.; Luzardo, O.P. Optimization and validation of a QuEChERS-based method for the simultaneous environmental monitoring of 218 pesticide residues in clay loam soil. *Sci. Total Environ.* **2021**, *753*, 142015. [CrossRef]
51. García Valverde, M.; Martínez Bueno, M.J.; Gómez-Ramos, M.M.; Aguilera, A.; Gil García, M.D.; Fernández-Alba, A.R. Determination study of contaminants of emerging concern at trace levels in agricultural soil. A pilot study. *Sci. Total Environ.* **2021**, *782*, 146759. [CrossRef]
52. Luo, J.; Bian, C.; Rao, L.; Zhou, W.; Li, Y.; Li, B. Determination of the residue behavior of isocycloseram in Brassica oleracea and soil using the QuEChERS method coupled with HPLC. *Food Chem.* **2021**, *367*, 130734. [CrossRef]
53. Do Amaral, B.; Peralta-Zamora, P.; Nagata, N. Simultaneous multi-residue pesticide analysis in southern Brazilian soil based on chemometric tools and QuEChERS-LC-DAD/FLD method. *Environ. Sci. Pollut. Res.* **2022**, *29*, 39102–39115. [CrossRef] [PubMed]
54. Słowik-Borowiec, M.; Szpyrka, E.; Książek-Trela, P.; Podbielska, M. Simultaneous determination of multi-class pesticide residues and PAHs in plant material and soil samples using the optimized QuEChERS method and tandem mass Spectrometry Analysis. *Molecules* **2022**, *27*, 2140. [CrossRef]
55. Lafay, F.; Daniele, G.; Fieu, M.; Pelosi, C.; Fritsch, C.; Vulliet, E. Ultrasound-assisted QuEChERS-based extraction using EDTA for determination of currently-used pesticides at trace levels in soil. *Environ. Sci. Pollut. Res.* **2022**, 1–15. [CrossRef] [PubMed]
56. Han, L.; Wu, Q.; Wu, X. Dissipation and Residues of Pyraclostrobin in Rosa roxburghii and Soil under Filed Conditions. *Foods* **2022**, *11*, 669. [CrossRef]
57. Fu, D.; Zhang, Y.; Zhang, S.; Wang, M.; Liang, X.; Zhang, Y.; Huang, S.; Zhang, C. Validation and application of a QuEChERS-based method for estimation of the half-lives of cyromazine and acetamiprid in cowpeas and soil by LC-ESI-MS/MS. *Int. J. Environ. Anal. Chem.* **2022**, *102*, 650–666. [CrossRef]
58. Lehotay, S.J.; Maštovská, K.; Lightfield, A.R. Use of buffering and other means to improve results of problematic pesticides in a fast and easy method for residue analysis of fruits and vegetables. *J. AOAC Int.* **2005**, *88*, 615–629. [CrossRef] [PubMed]
59. González-Curbelo, M.Á.; Varela-Martínez, D.A.; Socas-Rodríguez, B.; Hernández-Borges, J. Recent applications of nanomaterials in capillary electrophoresis. *Electrophoresis* **2017**, *38*, 2431–2446. [CrossRef]
60. Correia-Sá, L.; Fernandes, V.C.; Carvalho, M.; Calhau, C.; Domingues, V.M.F.; Delerue-Matos, C. Optimization of QuEChERS method for the analysis of organochlorine pesticides in soils with diverse organic matter. *J. Sep. Sci.* **2012**, *35*, 1521–15307. [CrossRef]
61. González-Curbelo, M.Á.; Dionis-Delgado, S.; Asensio-Ramos, M.; Hernández-Borges, J. Pesticide analysis in toasted barley and chickpea flours. *J. Sep. Sci.* **2019**, *35*, 299–307. [CrossRef]
62. González-Curbelo, M.Á.; Socas-Rodríguez, B.; Herrero, M.; Herrera-Herrera, A.V.; Hernández-Borges, J. Dissipation kinetics of organophosphorus pesticides in milled toasted maize and wheat flour (gofio) during storage. *Food Chem.* **2019**, *229*, 854–859. [CrossRef]
63. González-Curbelo, M.Á.; Hernández-Borges, J.; Borges-Miquel, T.M.; Rodríguez-Delgado, M. Determination of pesticides and their metabolites in processed cereal samples. *Food Addit. Contam. Part A* **2012**, *29*, 104–116. [CrossRef] [PubMed]
64. Gevao, B.; Semple, K.T.; Jones, K.C. Bound pesticide residues in soils: A review. *Environ. Pollut.* **2000**, *108*, 3–14. [CrossRef]
65. Salama, G.; El Gindy, A.; Abdel Hameed, E.A. The use of experimental design for optimisation of QuEChERS extraction of commonly used pesticides in Egyptian soil and drainage water and their determination by GC/MS. *Int. J. Environ. Anal. Chem.* **2020**, 1–12. [CrossRef]

66. Sefiloglu, F.O.; Tezel, A.; Balcıoğlu, I.A. Validation of an Analytical Workflow for the Analysis of Pesticide and Emerging Organic Contaminant Residues in Paddy Soil and Rice. *J. Agric. Food Chem.* **2021**, *69*, 3298–3306. [CrossRef]
67. Kwon, H.; Lehotay, S.J.; Geis-Asteggiante, L. Variability of matrix effects in liquid and gas chromatography-mass spectrometry analysis of pesticide residues after QuEChERS sample preparation of different food crops. *J. Chromatogr. A* **2012**, *1270*, 235–245. [CrossRef]
68. Tsuchiyama, T.; Katsuhara, M.; Sugiura, J.; Nakajima, M.; Yamamoto, A. Combined use of a modifier gas generator, analyte protectants and multiple internal standards for effective and robust compensation of matrix effects in gas chromatographic analysis of pesticides. *J. Chromatogr. A* **2019**, *1589*, 122–133. [CrossRef]
69. Słowik-Borowiec, M.; Szpyrka, E. Multiresidue Analysis of Pesticides in Wine and Grape Using Gas Chromatography with Microelectron Capture and Nitrogen–Phosphorus Detection. *Food Anal. Methods* **2018**, *11*, 3516–3530. [CrossRef]
70. Cabrera, L.D.C.; Caldas, S.S.; Prestes, O.D.; Primel, E.G.; Zanella, R. Evaluation of alternative sorbents for dispersive solid-phase extraction clean-up in the QuEChERS method for the determination of pesticide residues in rice by liquid chromatography with tandem mass spectrometry. *J. Sep. Sci.* **2016**, *39*, 1945–1954. [CrossRef]
71. Rajski, R.; Lozano, A.; Uclés, A.; Ferrer, C.; Fernández-Alba, A.R. Determination of pesticide residues in high oil vegetal commodities by using various multi-residue methods and clean-ups followed by liquid chromatography tandem mass spectrometry. *J. Chromatogr. A* **2013**, *1304*, 109–120. [CrossRef]
72. Sapozhnikova, Y.; Lehotay, S.J. Multi-class, multi-residue analysis of pesticides, polychlorinated biphenyls, polycyclic aromatic hydrocarbons, polybrominated diphenyl ethers and novel flame retardants in fish using fast, low-pressure gas chromatography-tandem mass spectrometry. *Anal. Chim. Acta* **2013**, *758*, 80–92. [CrossRef]
73. Nguyen, T.D.; Lee, M.H.; Lee, G.H. Rapid determination of 95 pesticides in soybean oil using liquid-liquid extraction followed by centrifugation, freezing and dispersive solid phase extraction as cleanup steps and gas chromatography with mass spectrometric detection. *Microchem. J.* **2010**, *95*, 113–119. [CrossRef]
74. Koesukwiwat, U.; Lehotay, S.J.; Mastovska, K.; Dorweiler, K.J.; Leepipatpiboon, N. Extension of the QuEChERS method for pesticide residues in cereals to flaxseeds, peanuts, and doughs. *J. Agric. Food Chem.* **2010**, *58*, 5950–5958. [CrossRef]
75. González-Curbelo, M.Á.; Lehotay, S.J.; Hernández-Borges, J.; Rodríguez-Delgado, M.Á. Use of ammonium formate in QuEChERS for high-throughput analysis of pesticides in food by fast, low-pressure gas chromatography and liquid chromatography tandem mass spectrometry. *J. Chromatogr. A* **2014**, *1358*, 75–84. [CrossRef]
76. Han, L.; Sapozhnikova, Y.; Lehotay, S.J. Streamlined sample cleanup using combined dispersive solid-phase extraction and in-vial filtration for analysis of pesticides and environmental pollutants in shrimp. *Anal. Chim. Acta* **2014**, *827*, 40–46. [CrossRef]
77. Lehotay, S.J.; Han, L.; Sapozhnikova, Y. Automated mini-column solid-phase extraction cleanup for high-throughput analysis of chemical contaminants in foods by low-pressure gas chromatography-tandem mass spectrometry. *Chromatographia* **2016**, *79*, 1113–1130. [CrossRef]
78. Lehotay, S.J. The QuEChERSER Mega Method. *LCGC N. Am.* **2022**, *40*, 13–19.
79. Lehotay, S.J.; Michlig, N.; Lightfield, A.R. Assessment of test portion sizes after sample comminution with liquid nitrogen in an improved high-throughput method for analysis of pesticide residues in fruits and vegetables. *J. Agric. Food Chem.* **2020**, *68*, 1468–1479. [CrossRef] [PubMed]
80. Monteiro, S.H.; Lehotay, S.J.; Sapozhnikova, Y.; Ninga, E.; Lightfield, A.R. High-throughput mega-method for the analysis of pesticides, veterinary drugs, and environmental contaminants by ultra-high-performance liquid chromatography—tandem mass spectrometry and robotic mini-solid-phase extraction cleanup + low-pressure gas chromatography—tandem mass spectrometry, Part 1: Beef. *J. Agric. Food Chem.* **2021**, *69*, 1159–1168. [PubMed]
81. Ninga, E.; Sapozhnikova, Y.; Lehotay, S.J.; Lightfield, A.R.; Monteiro, S.H. High-throughput mega-method for the analysis of pesticides, veterinary drugs, and environmental contaminants by ultra-high-performance liquid chromatography–tandem mass spectrometry and robotic mini-solid-phase extraction cleanup + low-pressure gas chromatography–tandem mass spectrometry, Part 2: Catfish. *J. Agric. Food Chem.* **2021**, *69*, 1169–1174.
82. Monteiro, S.H.; Lehotay, S.J.; Sapozhnikova, Y.; Ninga, E.; Moura Andrade, G.C.R.; Lightfield, A.R. Validation of the QuEChERSER mega-method for the analysis of pesticides, veterinary drugs, and environmental contaminants in tilapia (*Oreochromis Niloticus*). *Food Addit. Contam. Part A Chem. Anal. Control Expo. Risk Assess.* **2022**, *39*, 699–709. [CrossRef]
83. Michlig, N.; Lehotay, S.J.; Lightfield, A.R.; Beldoménico, H.; Repetti, M.R. Validation of a high-throughput method for analysis of pesticide residues in hemp and hemp products. *J. Chromatogr. A* **2021**, *1645*, 462097. [CrossRef] [PubMed]

molecules

Article

Dissipation, Processing Factors and Dietary Risk Assessment for Flupyradifurone Residues in Ginseng

Nan Fang [1,2], Changpeng Zhang [2], Zhongbin Lu [1], Zhou Lu [3], Zhongbei Zhang [4], Bo Wang [1], Zhiguang Hou [1,*] and Xueping Zhao [2,*]

1 College of Plant Protection, Jilin Agricultural University, Changchun 130118, China
2 State Key Laboratory for Managing Biotic and Chemical Threats to the Quality and Safety of Agro-Products, Ministry of Agriculture and Rural Affairs Key Laboratory for Pesticide Residue Detection, Institute of Agro-Products Safety and Nutrition, Zhejiang Academy of Agricultural Sciences, Hangzhou 310021, China
3 Laboratory of Quality & Safety Risk Assessment for Ginseng and Antler Products, Jilin Agricultural University, Changchun 130118, China
4 Agricultural Product Quality Inspection and Monitoring Center, Baishan Municipal Bureau of Agriculture and Rural Affairs, Baishan 134399, China
* Correspondence: zhiguangh@jlau.edu.cn (Z.H.); zhaoxueping@tom.com (X.Z.); Tel.: +86-0431-84510942 (Z.H.)

Abstract: The massive use of pesticides has brought great risks to food and environmental safety. It is necessary to develop reliable analytical methods and evaluate risks through monitoring studies. Here, a method was used for the simultaneous determination of flupyradifurone (FPF) and its two metabolites in fresh ginseng, dried ginseng, ginseng plants, and soil. The method exhibited good accuracy (recoveries of 72.8–97.5%) and precision (relative standard deviations of 1.1–8.5%). The field experiments demonstrated that FPF had half-lives of 4.5–7.9 d and 10.0–16.9 d in ginseng plants and soil, respectively. The concentrations of total terminal residues in soil, ginseng plants, dried ginseng, and ginseng were less than 0.516, 2.623, 2.363, and 0.641 mg/kg, respectively. Based on these results, the soil environmental risk assessment shows that the environmental risk of FPF to soil organisms is acceptable. The processing factors for FPF residues in ginseng were 3.82–4.59, indicating that the concentration of residues increased in ginseng after drying. A dietary risk assessment showed that the risk of FPF residues from long-term and short-term dietary exposures to global consumers were 0.1–0.4% and 12.07–13.16%, respectively, indicating that the application of FPF to ginseng at the recommended dose does not pose a significant risk to consumers.

Keywords: flupyradifurone; ginseng; analytical method; mass spectrometry; pesticide residue analysis

Citation: Fang, N.; Zhang, C.; Lu, Z.; Lu, Z.; Zhang, Z.; Wang, B.; Hou, Z.; Zhao, X. Dissipation, Processing Factors and Dietary Risk Assessment for Flupyradifurone Residues in Ginseng. *Molecules* **2022**, 27, 5473. https://doi.org/10.3390/molecules27175473

Academic Editor: Miguel Ángel González-Curbelo

Received: 31 July 2022
Accepted: 23 August 2022
Published: 26 August 2022

1. Introduction

Ginseng (*Panax ginseng* C. A. Mey.) is one of the most commonly used ginseng botanicals in the world, mainly consumed in fresh and processed form. Because of the special cultivation environment (loose and fertile brown forest soil with a deep humus layer and high-water content), insect pests and fungal diseases are the biggest problems that affect ginseng cultivation [1]. Pesticide application is essential to ensure high yield and quality of ginseng but also causes environmental and food safety problems. Therefore, the residue analysis and risk assessment of pesticides in ginseng and its products are very important.

Flupyradifurone (FPF, Figure S1) is the first representative of the novel butenolide class of insecticides developed by Bayer [2]. It is effective on the pests resistant to neonicotinoid insecticides and has less adverse effects on honeybee colonies [3–5]. FPF has been applied to many agricultural and horticultural crops such as apples, cotton, rice, tomatoes, potatoes, and berries (strawberries, blackberries, and raspberries), and has been registered in the US, EU, and Australia [6,7]. The maximum residue limits (MRLs) of FPF for crops in these countries are 0.01–3 mg/kg. China has set MRLs for FPF in some crops in 2021 [8].

Difluoroacetic acid (DFA) and 6-chloronicotinic acid (6-CNA) were the main metabolites of FPF [9]. DFA was high leachability, very mobile, moderate aquatic ecotoxicology, and moderate mammals acute toxicity, and 6-CNA was moderately mobile, moderate aquatic ecotoxicology, and low mammals acute toxicity [10]. Therefore, the residue definition of FPF was the sum of FPF, DFA, and 6-CNA, and expressed as FPF [9], and the determination of DFA and 6-CNA were important.

Currently, there are few studies on the dissipation, terminal residues, processing factors (PFs), and dietary risk assessment of FPF in food and agricultural products [11]. Most of the related data for FPF comes from the Bayer and Joint FAO/WHO Meeting on Pesticide Residues (JMPR). Li et al. [12] developed a method for the determination of FPF and two other metabolites in fruits, vegetables, and grains. However, relevant data on FPF in ginseng and its products (dried ginseng) have not yet been reported. Therefore, the aims of this study were to (a) determine the residue levels of FPF and its metabolites in ginseng (fleshy taproot) and processed commodities (dried ginseng); (b) evaluate the dissipation of FPF and its metabolites in ginseng plants (the part of the stem and leaves on the ground) and soil, and (c) assess the dietary risk to consumers.

2. Results and Discussion

2.1. Optimization of Sample Preparation

It was reported that 6-CNA and DFA were difficult to extract in water using the QuEChERS method [13] because of their octanol–water partition coefficients (Log P of the 6-CNA and DFA was 0.98 and −0.11, respectively) [10]. We also found that the extraction efficiencies of DFA and 6-CNA improved significantly upon the addition of formic acid, and the recoveries did not increase when the concentration of formic acid reached 2%. Thus, a mixture of acetonitrile and water containing 2% formic acid was used for extraction, and the water content (0–50%) in the mixture was studied. The results demonstrate that as the content of water in the extraction solution increased, the FPF recovery decreased and was less than 70% when the content of water was 50% (Figure 1). However, the recoveries of DFA and 6-CNA increased significantly. Satisfactory recoveries of the three compounds were obtained when the water:acetonitrile ratio in the extraction solution was 1:4 (v/v).

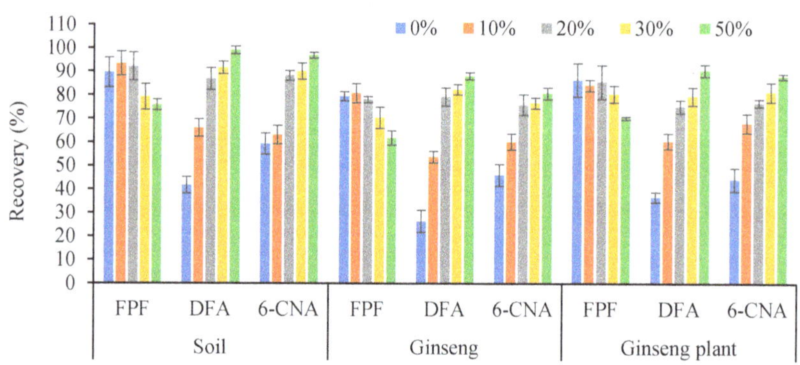

Figure 1. Recovery of flupyradifurone, difluoroacetic acid, and 6-chloronicotinic acid in soil, fresh ginseng, and ginseng plants for the method using different proportions of water (0–50%) in the extraction solution.

Primary secondary amine (PSA) has been proven to have a strong adsorption effect on 6-CNA and DFA, and the purification effect of Sorbents octadecyl silica (C_{18}) and graphitised carbon black (GCB) in several common sorbents was enhanced [12,13]. It is probably due to the presence of two amino groups in PSA, which leads to a strong adsorption to polar compounds. Therefore, C_{18} was used for the purification of soil and ginseng (the recoveries were 73.4–96.7%), and a combination of C_{18} and GCB was used for the purification of ginseng plants (the recoveries were 72.8–94.8%) in this study (Table S3).

2.2. Validation Results of Analytical Method

The mean recoveries of FPF, 6-CNA, and DFA from each sample spiked at all levels were 72.8–97.5%, with intraday and interday relative standard deviations (RSDs) of 1.1–5.7% and 3.3–8.5%, respectively (Table S3). The limit of quantitation (LOQs) for FPF and 6-CNA as per the developed method were 0.01 mg/kg, and that for DFA was 0.05 mg/kg in each matrix. Good linearity (R^2 = 0.9991–0.9999) was observed from all matrix-matched calibration curves (Table S4).

2.3. Dissipation of FPF and Its Metabolites in Ginseng Plants, Soil, and Ginseng

The validated method was successfully applied to the determination of analytes in soil and ginseng plant samples in the dissipation experiment. The results demonstrate that the dissipation of FPF in soil and fresh ginseng plants followed a first-order kinetic model (Figure 2 and Table 1).

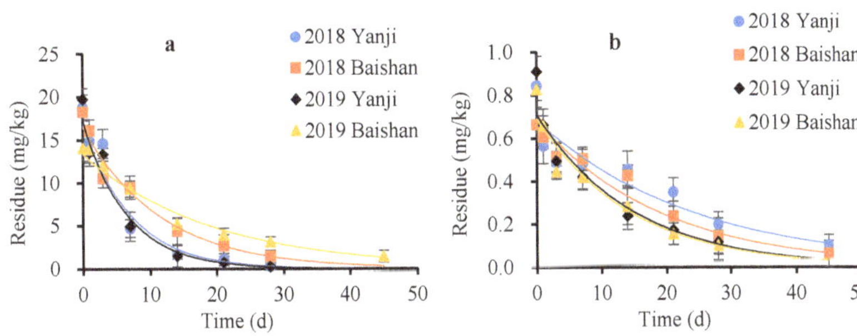

Figure 2. Degradation kinetic curve of flupyradifurone in ginseng plants and soil: (**a**) ginseng plants; (**b**) soil.

Table 1. Dissipation kinetics of flupyradifurone in soil and ginseng plants.

Year	Location	Matrix	Regression Equation	Coefficient (R^2)	Half-Life (d)
2018	Baishan	Soil	$C = 0.6854\ e^{-0.0524t}$	0.9777	13.2
		Ginseng plants	$C = 16.5064\ e^{-0.0874t}$	0.9884	7.9
	Yanji	Soil	$C = 0.6833\ e^{-0.0409t}$	0.9382	16.9
		Ginseng plants	$C = 17.0917\ e^{-0.1428t}$	0.9759	4.9
2019	Baishan	Soil	$C = 0.6918\ e^{-0.0693t}$	0.9722	10.0
		Ginseng plants	$C = 13.3104\ e^{-0.1522t}$	0.9816	4.5
	Yanji	Soil	$C = 0.7105\ e^{-0.0678t}$	0.9647	10.2
		Ginseng plants	$C = 17.2706\ e^{-0.1555t}$	0.9889	4.5

The initial deposits of FPF in ginseng plants were 14.03–19.27 mg/kg. The residue of FPF in ginseng plants decreased by 77.82–98.78% to 0.24–3.11 mg/kg on day 28. The calculated half-lives of FPF in ginseng plants were 4.5–7.9 d, indicating that FPF is an easily degradable pesticide in ginseng plants ($t_{1/2}$ < 30 d), and the different climate had no obvious effect on the degradation of FPF in ginseng plants. The metabolic behaviour of FPF in ginseng plants may involve cleavage of the –CN group to form DFA and 6-CNA. The production and dissipation of DFA and 6-CNA in ginseng plants are illustrated by the curves in Figure 3. The results indicated that after FPF is degraded in ginseng plants, the main residue is DFA, and the residue of 6-CNA is small. FPF and its two metabolites were metabolised rapidly by ginseng plants. The concentration of FPF metabolites in the ginseng plants was determined by the degradation rate of FPF and its metabolites.

The climate difference in different years and regions was the main factor, which mainly affected the degradation rate of FPF and its metabolites by affecting the growth of ginseng (light, temperature, and rainfall). This may be the main reason for the difference in the concentration change of DFA and 6-CNA at Baishan and Yanji in 2018 and 2019.

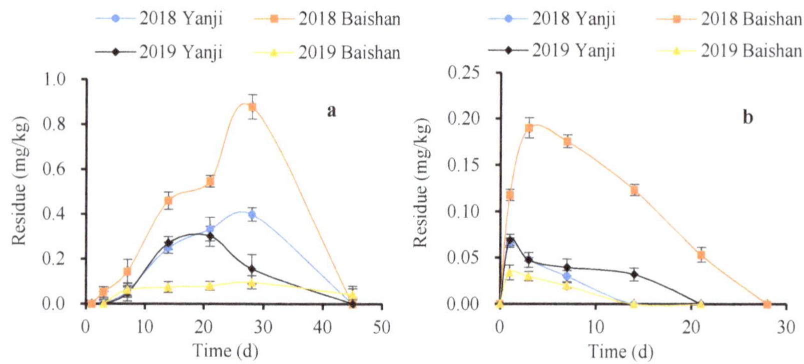

Figure 3. Change of concentration of difluoroacetic acid and 6-chloronicotinic acid in ginseng plants: (**a**) difluoroacetic acid; (**b**) 6-chloronicotinic acid.

The concentration of initial deposits of FPF in soil was 0.66–0.91 mg/kg, which dropped to below the LOQ (0.01 mg/kg) after 45 d. The calculated half-lives of FPF in soil were 10.0–16.9 d. Many studies show that microorganisms and organic matter were significant contributors to pesticide degradation in soil [14–16]. The soil used to grow ginseng is artificially mixed according to a certain formula, which is typically rich in minerals and organic matter, and treated before planting ginseng. Therefore, the small quantity of microorganisms and the adsorption of organic matter and minerals may explain why FPF has a longer half-life in soil than in ginseng plants. In addition, ginseng plants contain various enzymes [17–20], and enzyme-catalysed detoxification by ginseng plants might play a dominant role in the rapid degradation of FPF. DFA and 6-CNA were not detected in the soil, probably because of the lower initial FPF deposits in the soil, which cause the concentrations of DFA and 6-CNA in the soil to be lower than the LOQ.

FPF in fresh ginseng did not decompose according to a first-order kinetic relationship. In the dissipation experiment, FPF was sprayed on the surface of the leaves without direct contact with fresh ginseng. Therefore, the residue of FPF in fresh ginseng was accumulated by transport from the leaves and absorption from the soil. The process of transport and absorption is complex and easily affected by the natural environment in the field, resulting in the random change of the FPF residue in fresh ginseng over time.

2.4. Terminal Residues of FPF, DFA and 6-CNA in Ginseng Plants, Soil, and Ginseng

The residue definition (for estimation of dietary intake in plant commodities) of FPF was the sum of FPF, DFA, and 6-CNA, and expressed as FPF [11]. Therefore, the total residues were calculated based on molecular weight (Table S5).

The terminal FPF, DFA, and 6-CNA residues in ginseng plants were detected because ginseng plants may be used for the extraction of saponins (the main medicinal component of ginseng). The concentrations of terminal FPF and 6-CNA residues in ginseng plants were less than the LOQ, and those of the terminal DFA residues were 0.081–0.601 mg/kg (21 d) and 0.096–0.863 mg/kg (28 d). The concentrations of total residues in ginseng plants were 0.272–1.835 mg/kg (21 d) and 0.317–2.623 mg/kg (28 d), which is expressed as FPF.

The concentration of terminal FPF residues in soil was 0.01–0.155 mg/kg (21 d) and 0.01–0.347 mg/kg (28 d), and those of the terminal DFA and 6-CNA residues were lower than the LOQ. The concentration of total residues in the soil was 0.179–0.516 mg/kg (21 d) and 0.179–0.324 mg/kg (28 d), which are expressed as parent equivalents. These data can be used for environmental risk assessments of soil organisms based on the RQ. The RQ was calculated from the predicted environmental concentration (PEC, mg/kg) of the soil and the predicted no-effect concentration (PNEC, mg/kg): RQ = PEC/PNEC. The PNEC was calculated from the toxicity endpoint obtained from ecotoxicological research and the corresponding uncertainty factor (UF): PNEC = endpoint/UF. According to the principle of the maximum risk, the high residue determined in this study was used instead of the PEC. The LC_{50} of FPF for earthworms (acute 14 d) was selected as the toxicity endpoint, and the UF for LC_{50} was 10 [21]. The RQ was <1 (0.019), indicating that the environmental risk of FPF to soil organisms is acceptable.

The concentrations of terminal FPF, DFA, and 6-CNA residues in fresh ginseng were 0.118–0.436, 0.056–0.110, and 0.013–0.019 mg/kg (21 d); 0.022–0.4, 0.042–0.165, and 0.014–0.018 mg/kg (28 d), respectively. The concentrations of total terminal residues (parent equivalents) were 0.296–0.525 mg/kg (21 d) and 0.228–0.641 mg/kg (28 d). These results were used for the calculation of PFs and dietary risk assessment.

2.5. *Effect of Processing on Residue Levels in Fresh Ginseng*

The PFs were determined from fresh and dried ginseng at intervals of 21 and 28 d; Table 2 shows the PFs of FPF after fresh ginseng was dried. Based on the results, all PFs can be considered comprehensively because the variation is small, and the median value can be used as the best estimate of the PF [22].

Table 2. Effect of processing on flupyradifurone residues in ginseng.

Year	Location	Processed Fractions	Total Residues (mg kg^{-1}) Pre-Harvest Interval (d) 21	Total Residues (mg kg^{-1}) Pre-Harvest Interval (d) 28	PFs Pre-Harvest interval (d) 21	PFs Pre-Harvest interval (d) 28	Best Estimate
2018	Baishan	raw	0.296 ± 0.025	0.228 ± 0.033	/	/	/
		dried	1.277 ± 0.019	0.813 ± 0.028	4.31	3.57	3.94
	Yanji	raw	0.414 ± 0.017	0.641 ± 0.042	/	/	/
		dried	1.634 ± 0.109	2.363 ± 0.027	3.95	3.69	3.82
2019	Baishan	raw	0.461 ± 0.024	0.406 ± 0.019	/	/	/
		dried	2.269 ± 0.092	1.727 ± 0.105	4.92	4.25	4.59
	Yanji	raw	0.525 ± 0.017	0.452 ± 0.033	/	/	/
		dried	2.298 ± 0.024	1.705 ± 0.107	4.38	3.77	4.07

In the fresh ginseng processing study, drying increased the concentration of the residues to 0.813–2.363 mg/kg at two locations in 2018 and 2019, with PFs ranging from 3.82 to 4.59 (median). The data were in accordance with those reported by Kim et al. [23], who found that the PFs of difenoconazole in ginseng for drying were 2.00–5.16. Alister et al. [24] reported that more stable pesticides (high hydrolysis DT_{50}) were the least reduced during the drying step. FPF is a stable insecticide used for hydrolysis and at high temperature (degradation point is 270 °C) [10]. Therefore, the FPF residues in dried ginseng increased because of water evaporation.

2.6. *Dietary Rrisk Assessment of FPF in Dried Ginseng*

Dried ginseng is typically used in food, health products, and medicine and has gradually become a staple in many countries such as China, Japan, and Korea. With increasing concern from the public over pesticide residues in dried ginseng, various countries such as the USA, EU, Korea, Japan, and China have set MRLs for many pesticides in dried ginseng to protect consumer health. However, the relevant data for FPF are unavailable so far. Therefore, to provide the necessary information for establishing regulations, the dietary risk assessment of FPF in dried ginseng was performed in this study.

The acceptable daily intake (ADI) for FPF established by JMPR was 0–0.08 mg/kg bw [11]. The total national estimated daily intake (NEDI) of FPF were calculated using the

STMRs and MRLs (Table 3). The STMRs were obtained from terminal residues experiments in this paper, and the selection of reference MRLs (of the relevant registered crops in China) adhering to the following priority order: China, Codex Alimentarius Commission (CAC), US, Australia, Korea, EU, and Japan [25]. The average body weight of Chinese adults was calculated to be 63 kg [26]. Therefore, the total NEDI (2.0045 mg) was 39.77% of the maximum ADI (5.04 mg) for FPF. The acute reference dose (ARfD) of FPF established by JMPR was 0.2 mg/kg bw [11]. The national estimated short-term intake (NESTI) values of FPF were calculated for dried ginseng using the high residue (estimated in this study) and large portion consumed (obtain from an IESTI calculator was 0.6 g/kg bw/day, available at: https://zwfw.nhc.gov.cn/kzx/tzgg/tzggqb/, accessed on 22 July 2020) (Table 3). The NESTI (1.4478 mg) was 11.49% of the maximum ARfD (12.6 mg) for FPF. The results showed that based on the information provided by this study, the chronic and acute dietary risk of FPF in dried ginseng is acceptable and the long-term and short-term dietary exposures to FPF residues is not a public health risk for typical Chinese consumers.

Table 3. The long-term and short-term dietary intake risk assessment of flupyradifurone based on the Chinese dietary pattern.

Food Category	FI (kg day^{-1}) [a]	Commodity	MRLs [b] (mg kg^{-1})	STMR [b] (mg kg^{-1})	HR [b] (mg kg^{-1})	Source of Reference Limit
Rice cereals and rice products	0.2399	Rice	3			USA
Wheat cereals and wheat products	0.1385	Maize	0.01			CAC [d]
Other cereal grains	0.0233	Cereal grains	3			CAC
Potatoes	0.0495	Potato	0.05			CAC
Dried beans and their products	0.016	Beans (dry)	0.4			CAC
Dark-colored vegetables	0.0915	Tomatoes	3			China
Light-colored vegetables	0.1837	Lettuce	4			CAC
Pickles	0.0103					
Fruits	0.0457	Oranges	1			China
Nuts	0.0039	Pecan	0.01			CAC
Livestock and poultries	0.0795	Poultry	0.8			CAC
Milk and milk products	0.0263	Milk	0.7			CAC
Egg and egg products	0.0236	Egg	0.7			CAC
Fish and fish products	0.0301					
Oilseeds and oil	0.0327	Cotton seed	0.8			CAC
Animal origin oil and fat	0.0087	Poultry fat	1			CAC
Sugars and starch	0.0044					
Salt	0.012					
Soy sauce	0.009	Ginseng		1.667 1.801	2.413 2.394	PHI [c] of 21 days PHI of 28 days
Total FI (kg day^{-1}) [a]	1.0286					
Total NEDI (mg)	2.0045					
NESTI [e] (mg)	1.4478					
ADI (mg/kg bw)	0.08					
ARfD (mg/kg bw)	0.2					
Body weiht (kg bw)	63					
%ADI (%)	39.77%					
%ARfD (%)	11.49%					

[a] The consumption values of ginseng and other crops referred to the recommended dietary food intake (FI) of an adult (63 kg) per day for its corresponding food classification (data from the dietary guideline published by Health Ministry of the People's Republic of China). [b] The supervised trials median residue (STMR) in ginseng and the maximum residue limits (MRLs) in other crops were used to calculate the national estimated daily intake (NEDI). The high residue (HR) in ginseng was used to calculate the national estimated short-term intake (NESTI). [c] PHI: Pre-harvest interval. [d] CAC: Codex Alimentarius Commission. [e] The large portion consumed of ginseng for calculating the NESTI was 0.6 g/kg bw/day, available at: https://zwfw.nhc.gov.cn/kzx/tzgg/tzggqb/, accessed on 22 July 2020.

In addition, the international estimated daily intake (IEDI) and international estimated short-term intake (IESTI) were calculated by the IEDI and IESTI calculator [27] to estimate the dietary exposures of FPF residues to global consumers. The dietary intake data of ginseng in the GEMS/Food regional consumption data (available at: https://extranet.who.int/gemsfood/, accessed on 29 October 2020) were not reported. The Announcement No. 17 (National Health Commission, China, available at: https://zwfw.nhc.gov.cn/kzx/tzgg/tzggqb/, accessed on 11 November 2021) reported that the maximum daily dietary intake did not exceed 3 g/kg bw. Therefore, according to the principle of the maximum dietary risk, this data was used for calculating the IEDI and IESTI. The calculated total IEDIs for the 17 GEMS/Food cluster diets were 0.1–0.4% of the maximum ADI (Table 4). The calculated IESTIs for the 17 GEMS/Food cluster diets of ginseng were 12.07–13.16% of ARfD. The results indicated that the long-term and short-term intake of residues of FPF resulting from its proposed uses is unlikely to present a public health concern for global consumers.

Table 4. The long-term and short-term dietary intake risk assessment of flupyradifurone based on the Chinese dietary pattern.

Codex Code	Commodity Description	STMR [a] (mg/kg)	G01		G02		G03		G04		G05		G06	
			Diet	Intake	Diet	Intake	Diet	Intake	Diet	Intake	Diet	Intake	Diet	Intake
VR 0604	Ginseng, raw	1.801	3.00	5.40	3.00	5.40	3.00	5.40	3.00	5.40	3.00	5.40	3.00	5.40
FS 0013	Subgroup of Cherries, raw	0.555	0.92	0.51	9.15	5.08	0.01	0.01	0.61	0.34	0.06	0.03	6.64	3.69
FS 0014	Subgroup of Plums, raw (including dried plums)	0.23	2.67	0.61	8.77	2.02	0.07	0.02	3.03	0.70	0.70	0.16	4.34	1.00
DF 0014	Plums, dried (prunes)	1.15	0.09	0.10	0.06	0.07	0.01	0.01	0.18	0.21	0.04	0.05	0.06	0.07
FS 2001	Subgroup of peaches, raw (including dried apricots)	0.39	8.01	3.12	5.87	2.29	0.18	0.07	8.19	3.19	1.64	0.64	22.46	8.76
Total intake (ug/person)				9.8		14.9		5.5		9.8		6.3		18.9
Bodyweight per region (kg bw)				60		60		60		60		60		60
ADI (ug/person)				4800		4800		4800		4800		4800		4800
%ADI				0.2%		0.3%		0.1%		0.2%		0.1%		0.4%

Codex Code	Commodity Description	STMR [a] (mg/kg)	G07		G08		G09		G10		G11		G12	
			Diet	Intake	Diet	Intake	Diet	Intake	Diet	Intake	Diet	Intake	Diet	Intake
VR 0604	Ginseng, raw	1.801	3.00	5.40	3.00	5.40	3.00	5.40	3.00	5.40	3.00	5.40	3.00	5.40
FS 0013	Subgroup of Cherries, raw	0.555	1.40	0.78	4.21	2.34	0.04	0.02	2.93	1.63	1.50	0.83	NC	-
FS 0014	Subgroup of Plums, raw (including dried plums)	0.23	5.55	1.28	4.37	1.01	6.08	1.40	3.66	0.84	3.93	0.90	0.46	0.11
DF 0014	Plums, dried (prunes)	1.15	0.61	0.70	0.35	0.40	0.05	0.06	0.35	0.40	0.49	0.56	0.13	0.15
FS 2001	Subgroup of peaches, raw (including dried apricots)	0.39	13.03	5.08	16.29	6.35	8.29	3.23	12.95	5.05	5.35	2.09	0.04	0.02
Total intake (ug/person)				13.2		15.5		10.1		13.3		9.8		5.7
Bodyweight per region (kg bw)				60		60		55		60		60		60
ADI (ug/person)				4800		4800		4400		4800		4800		4800
%ADI				0.3%		0.3%		0.2%		0.3%		0.2%		0.1%

Table 4. *Cont.*

Codex Code	Commodity Description	STMR[a] (mg/kg)	G13		G14		G15		G16		G17	
			Diet	Intake	Diet	Intake	Diet	Intake	Diet	Intake	Diet	Intake
VR 0604	Ginseng, raw	1.801	3.00	5.40	3.00	5.40	3.00	5.40	3.00	5.40	3.00	5.40
FS 0013	Subgroup of Cherries, raw	0.555	0.01	0.01	0.01	0.01	5.96	3.31	0.01	0.01	NC	-
FS 0014	Subgroup of Plums, raw (including dried plums)	0.23	0.07	0.02	0.02	0.00	16.65	3.83	0.01	0.00	NC	-
DF 0014	Plums, dried (prunes)	1.15	0.01	0.01	0.01	0.01	0.37	0.43	0.01	0.01	NC	-
FS 2001	Subgroup of peaches, raw (including dried apricots)	0.39	0.02	0.01	0.01	0.00	10.76	4.20	0.01	0.00	NC	-
Total intake (ug/person)				5.4		5.4		17.2		5.4		5.4
Bodyweight per region (kg bw)				60		60		60		60		60
ADI (ug/person)				4800		4800		4800		4800		4800
%ADI				0.1%		0.1%		0.4%		0.1%		0.1%

[a] The supervised trials median residue (STMR) in ginseng were obtained from the terminal residues experiments in this paper and in other crops were obtained from the JMPR report 2017 (https://www.fao.org/agriculture/crops/thematic-sitemap/theme/pests/lpe/lpe-f/en/, accessed on 15 July 2021).

3. Materials and Methods

3.1. Chemicals and Reagents

The FPF standard (99.5%) and the 17% FPF soluble concentrate were provided by Bayer (Leverkusen, Germany). The 6-CNA (99.2%) and DFA (98.0%) standards were obtained from Chem Service (West Chester, PA, USA). Chromatographic-grade methanol and acetonitrile were purchased from Thermo Fisher Scientific (Waltham, MA, USA). Formic acid was purchased from Sigma-Aldrich (St. Louis, MO, USA). C_{18} and GCB were purchased from Agela Technologies (Tianjin, China). Analytical-grade sodium chloride and anhydrous magnesium sulfate were purchased from Sinopharm Chemical Reagent (Beijing, China). Stock solutions (1000 mg/L) of FPF, 6-CNA, and DFA were prepared in methanol and stored at 4 ± 3 °C (replaced after three months).

3.2. Field Experiments

Open field trials on ginseng were carried out from 2018 to 2019 in a mountainous region in Baishan (42°38 N, 126°79 E) and Yanji (42°98 N, 129°49 E) in Jilin Province. The field trials were designed in accordance with the NY/T 788-2018 Guidelines [28]. The sites consisted of treatment plots and control plots of sufficient size to obtain representative samples for each sampling interval (50 m^2), and each treatment comprised three replicate plots.

For the terminal residue experiments, the 17% FPF soluble concentrate was applied twice at a dosage of 102 g active ingredients per hectare (g a.i./hm^2) foliar spray. The recommended application interval was 7 d. At least 2.0 kg of soil (at depths of 0–10 cm), 500 g of harvested fresh ginseng samples and 500 g of ginseng plants were randomly collected from 12 points in the test plots at 21 and 28 d after the last application.

For the dissipation experiments, the 17% FPF soluble concentrate was applied once at a dosage of 102 g a.i./hm^2 foliar spray and soil (no ginseng was planted). At least 2.0 kg of soil (at depths of 0–10 cm), 500 g fresh ginseng, and 500 g ginseng plants were randomly collected from 12 points in the test plots at 0, 1, 3, 7, 14, 21, 28, and 45 d after application.

All samples were placed in sealed sample bags and labelled. The samples were stored at −18 °C before analysis. The storage stability report of JMPR showed that FPF, 6-CNA, and DFA were stable for at least 52 months in high-water, high-acid, high-oil, high-protein, and high-starch-content matrices (representative of plants) when stored in the frozen form at approximately −18 °C [11].

3.3. Processing of Fresh Ginseng

Each sample from the terminal residue experiments was divided into two parts for direct analysis and processing procedures. Drying is the most common and simple processing method for fresh ginseng, and the air-drying method is superior to far-infrared and freeze-drying methods [29]. To study the FPF residue in ginseng after processing, fresh ginseng was washed with tap water and dried in a forced air-drying oven (GZX-9070MBE, Shanghai, China) at 50 °C for 10 h. Dried ginseng was cooled at room temperature, sealed, and stored at −18 °C.

3.4. Sample Preparation

Processed sample: FPF, DFA and 6-CNA in dried ginseng were analysed by the method developed in our previous study [13].

Raw agricultural commodity (RAC) sample: The fresh ginseng samples were shredded with an electric grinder (FP3010, Braun, Germany), and the ginseng plant samples were crushed using dry ice and an electric grinder before extraction. The prepared fresh ginseng, ginseng plant, and soil samples (10.0 g) were extracted twice with 10 mL of acetonitrile:water (4:1, *v*/*v*) containing 2% formic acid, followed by dilution with water to 25 mL. The diluted extract (1 mL) was purified by the dispersed solid phase extraction method (50 mg of C_{18} for soil and fresh ginseng, 50 mg of C_{18} and 50 mg of GCB for ginseng plant). After centrifugation, the supernatant of the purified solution was filtered using a 0.22 μm syringe filter and analysed by HPLC-MS/MS.

3.5. Instrumental

HPLC-MS/MS was performed using an Agilent 1260-6470 triple quadrupole mass spectrometer (Santa Clara, CA, USA) equipped with an Agilent C_{18} column (3.0 mm × 100 mm, 1.8 μm, ZORBAX RRHD Eclipse Plus). The injection volume was 5 μL. The temperature at both ends of the column was maintained at 30 °C. The mobile phase was a mixture of 0.1% formic acid aqueous solution (phase A) and acetonitrile (phase B). The flow rate was 0.3 mL/min, and gradient elution was carried out as Table S1. The total elution time was 15 min, and FPF and its two metabolites were separated within 10 min.

MS was performed using an electrospray ionisation (ESI) source. FPF was ionised in the positive ion mode, and DFA and 6-CNA were ionised in the negative ion mode. The parameters for the ESI source and for the determination of FPF, DFA, and 6-CNA are listed in Tables S1 and S2, respectively.

3.6. Analytical Method Validation

The accuracy (recovery), precision (intraday and interday repeatability), matrix effect, and sensitivity (LOQ) of the method were verified by recovery experiments according to SANTE/11813/2017 [30]. The external standard method was used for the quantification of FPF, 6-CNA, and DFA. Linearity was evaluated by solvent and matrix-matched standard calibration curves (0.001, 0.005, 0.01, 0.05, 0.1, 0.5, and 1 mg/L for FPF and 6-CNA; 0.01, 0.02, 0.05, 0.1, 0.2, 0.5, and 1 mg/L for DFA). The accuracy and precision of the method were determined by fortification experiments that involved spiking blank samples at several levels of FPF (0.01, 0.05, 0.5, and 20 mg/kg), 6-CNA (0.01, 0.05, and 0.5 mg/kg), and DFA (0.05, 0.1, and 1 mg/kg). Add 0.1 mL of the working solution mixture to the blank sample to bring the FPF, DFA and 6-CNA in the blank sample to the respective spiked levels, and then the sample is treated according to the procedure in Section 3.4. Each treatment was performed five times. Precision was expressed as the intraday and interday RSD. The LOQ is defined as the lowest spiked level of the validation, meeting the method performance acceptability criteria. The matrix effect is a common problem that hinders quantitative HPLC-MS/MS analysis (Niessen et al., 2010). At present, the most common methods to compensate for matrix effect include isotope labelling, echo peak technique, extraction solution dilution, and matrix-matching calibration. Matrix-matching calibration was used to obtain more representative results in this study because of its accuracy and convenience, and the matrix effect was calculated using the Equation (1):

$$\text{Matrix effect (\%)} = (S_{matrix}/S_{solvent} - 1) \times 100 \quad (1)$$

where S_{matrix} is the slope of the matrix-matched calibration curve, and $S_{solvent}$ is the slope of the solvent calibration curve.

3.7. Calculation

The degradation kinetics of FPF can be described by a first-order reaction (Equation (2)). When $C_t = 1/2C_0$, the formula for half-life ($t_{1/2}$, Equation (3)) can be obtained by taking the logarithm on both sides of Equation (2):

$$C_t = C_0 \times e^{\wedge(-kt)} \quad (2)$$

$$t_{1/2} = \ln 2/k \quad (3)$$

where C_0 is the initial pesticide residue concentration (mg/kg), C_t is the concentration of pesticide residue (mg/kg) at time t (d), and k is the dissipation rate constant.

The PFs were calculated using Equation (4) [31]:

$$\text{PF} = \text{residues (mg/kg) in processed product/residues (mg/kg) in RAC} \quad (4)$$

where RAC is the raw agricultural commodity.

The diet risk assessment is an estimate of the potential residue intake by consumers, including the estimate of both the long-term and short-term dietary exposures. The IEDI and international estimated short-term intake (IESTI) for FPF were calculated for the 17 GEMS/Food cluster diets using the supervised trials median residues (STMRs) and high residues obtained from this paper calculated by an IEDI calculator [27]. The total NEDI and NESTI of FPF were calculated using the Chinese Diet Risk Assessment Model [26]. The risk of pesticide exposure to consumer is acceptable when the estimated dietary intake (per kilogram of body weight) of pesticide residues is less than the ADI or the ARfD [31].

$$\text{NEDI} = \sum(\text{STMR} \times \text{FI}) \quad (5)$$

$$\text{NESTI} = \text{HR} \times \text{LP} \quad (6)$$

where FI and LP are the average daily food intake per person (kg/day) and large portion consumed (kg/day).

4. Conclusions

In this study, the analytical method, dissipation, terminal residues, processing factor, and dietary risk assessment for FPF and its two metabolites in ginseng plants, soil, fresh ginseng, and its processed products were studied. The method was validated, and satisfactory linearity, repeatability, intermediate precision, and accuracy were obtained. The recoveries were 72.8–97.5%. The method precision was high in terms of repeatability and intermediate precision, with RSD values of 1.1–8.5%. The results of field experiments on dissipation and terminal residues indicated that FPF is an easily degradable pesticide, and it dissipated faster in the ginseng plant ($t_{1/2}$ = 4.5–7.9 d) than soil ($t_{1/2}$ = 10.0–16.9 d). According to the terminal residue study, in which the PFs of FPF in ginseng were studied, the FPF residues in dried ginseng were increased (PF = 3.82–4.59). In addition, chronic and acute dietary risk assessments for FPF in dried ginseng were conducted. The calculated NEDI (2.0045 mg) and NESTI (1.4478 mg) for Chinese consumers were 39.77% of the maximum ADI and 11.49% of ARfD, respectively. The calculated IEDI and IESTI were 0.1–0.4% of the maximum ADI and 12.07–13.16% of ARfD. This study shows that when the recommended dose of FPF was applied to ginseng field, the environmental risk of FPF to soil organisms is acceptable and the harvested fresh ginseng and its products (dried ginseng) would not pose a significant potential risk to global consumers.

Supplementary Materials: The following supporting information can be downloaded at: https://www.mdpi.com/article/10.3390/molecules27175473/s1, Figure S1: Chemical structure of flupyradifurone; Table S1: The MRM and gradient elution conditions for analysis of flupyradifurone, DFA and 6-CNA; Table S2: The parameters for the ESI source in HPLC-MS/MS; Table S3: Mean recoveries and RSD for target compounds from different matrices at three spiked levels; Table S4: Calibration information of FPF, 6-CNA, or DFA in different matrices; Table S5: Terminal residues of FPF, 6-CNA and DFA in soil, ginseng, and ginseng plants ($n = 3$).

Author Contributions: Conceptualization, N.F. and Z.H.; methodology, N.F.; validation, Z.L. (Zhou Lu), Z.Z. and B.W.; resources, Z.L. (Zhongbin Lu) and X.Z.; writing—original draft preparation, N.F.; writing—review and editing, C.Z.; project administration, Z.H.; funding acquisition, Z.L. (Zhongbin Lu). All authors have read and agreed to the published version of the manuscript.

Funding: This research was funded by the Agricultural Industry Standard of the Chinese Ministry of Agriculture (2018) (grant number: 181721301092371097) and the Special Fund for Ginseng Industry Development of Jilin Province (2015) (grant number: 201501).

Institutional Review Board Statement: Not applicable.

Informed Consent Statement: Not applicable.

Data Availability Statement: Not applicable.

Conflicts of Interest: The authors declare no conflict of interest.

Sample Availability: Samples of the compounds are not available from the authors.

References

1. Yanan, L.; Dongyue, Z.; Min, L.; Chenyun, H.; Yong, L.; Wanlong, D. Investigation of Pests and Diseases Occurrence and Pesticides Application in Main Producing Areas of Panax ginseng. *Chin. Agric. Sci. Bull.* **2014**, *30*, 294–298.
2. Nauen, R.; Jeschke, P.; Velten, R.; Beck, M.E.; Ebbinghaus-Kintscher, U.; Thielert, W.; Wolfel, K.; Haas, M.; Kunz, K.; Raupach, G. Flupyradifurone: A brief profile of a new butenolide insecticide. *Pest Manag. Sci.* **2015**, *71*, 850–862. [CrossRef] [PubMed]
3. Tan, K.; Wang, C.; Dong, S.; Li, X.; Nieh, J.C. The pesticide flupyradifurone impairs olfactory learning in Asian honey bees (Apis cerana) exposed as larvae or as adults. *Sci. Rep.* **2017**, *7*, 17772. [CrossRef] [PubMed]
4. Hesselbach, H.; Scheiner, R.J.S.R. Effects of the novel pesticide flupyradifurone (Sivanto) on honeybee taste and cognition. *Sci. Rep.* **2018**, *8*, 4954. [CrossRef]
5. Hesselbach, H.; Scheiner, R. The novel pesticide flupyradifurone (Sivanto) affects honeybee motor abilities. *Ecotoxicology* **2019**, *28*, 354–366. [CrossRef]
6. EFSA (European Food Safety Authority). Conclusion on the peer review of the pesticide risk assessment of the active substance flupyradifurone. *EFSA J.* **2016**, *13*, 4020.
7. EFSA (European Food Safety Authority). Setting of new maximum residue levels for flupyradifurone in strawberries, blackberries and raspberries. *EFSA J.* **2016**, *14*, 4423.
8. ICAMA (Institute Control of Agrochemicals, Ministry of Agriculture, P.R. China). National Food Safety Standard-Maximum Residue Limits for Pesticides in Food. Available online: http://202.127.42.84/tbt-sps/mrlsdb/mrlsdb.do (accessed on 13 April 2021).
9. JMPR (Joint FAO/WHO Meeting on Pesticide Residues). Flupyradifurone (Report). Available online: http://www.fao.org/fileadmin/templates/agphome/documents/Pests_Pesticides/JMPR/Report2017/5.18_FLUPYRADIFURONE__285_.pdf (accessed on 8 September 2019).
10. IUPAC (International Union of Pure and Applied Chemistry). Pesticides Properties Data Base. Available online: http://sitem.herts.ac.uk/aeru/ppdb/en/index.htm (accessed on 13 October 2020).
11. Dal Bello, F.; Medana, C.; Guarino, B.; Dioni, A.; Fabbri, D.; Calza, P. Investigation of sulfoxaflor, flupyradifurone and their transformation products in plant-based food matrices. *Food Control* **2022**, *132*, 108537. [CrossRef]
12. Li, Y.; Liu, X.; Wu, X.; Dong, F.; Xu, J.; Zheng, Y. Simultaneous determination of flupyradifurone and its two metabolites in fruits, vegetables, and grains by a modified quick, easy, cheap, effective, rugged, and safe method using ultra high performance liquid chromatography with tandem mass spectrometry. *J. Sep. Sci.* **2016**, *39*, 1090–1098. [CrossRef]

13. Fang, N.; Lu, Z.; Zhang, Z.; Hou, Z.; Liang, S.; Wang, B.; Lu, Z. Determination of the Novel Insecticide Flupyradifurone and Its Two Metabolites in Traditional Chinese Herbal Medicines Using Modified QuEChERS and High-Performance Liquid Chromatography-Tandem Mass Spectrometry. *Int. J. Anal. Chem.* **2020**, *2020*, 8812797. [CrossRef]
14. Zhang, P.; Ren, C.; Sun, H.; Min, L. Sorption, desorption and degradation of neonicotinoids in four agricultural soils and their effects on soil microorganisms. *Sci. Total Environ.* **2018**, *615*, 59–69. [CrossRef] [PubMed]
15. Fang, N.; Lu, Z.; Zhang, Z.; Hou, Z.; Liang, S.; Wang, B.; Wang, S.; Lu, Z. Dissipation and residues of dimethyl disulfide in tomatoes and soil under greenhouse and open field conditions. *J. Environ. Sci. Health. Part. B Pestic. Food Contam. Agric. Wastes* **2020**, *55*, 566–573. [CrossRef] [PubMed]
16. Wang, S.; Jin, F.; Cao, X.; Shao, Y.; Wang, J.; She, Y.; Qi, Y.; Zhang, C.; Li, H.; Jin, M.; et al. Residue behaviors and risk assessment of flonicamid and its metabolites in the cabbage field ecosystem. *Ecotoxicol. Environ. Saf.* **2018**, *161*, 420–429. [CrossRef] [PubMed]
17. Chang, K.H.; Jee, H.S.; Lee, N.K.; Park, S.H.; Lee, N.W.; Paik, H.D. Optimization of the enzymatic production of 20(S)-ginsenoside Rg(3) from white ginseng extract using response surface methodology. *New Biotechnol.* **2009**, *26*, 181–186. [CrossRef] [PubMed]
18. Ali, M.B.; Hahn, E.J.; Paek, K.Y. CO_2-induced total phenolics in suspension cultures of Panax ginseng C. A. Mayer roots: Role of antioxidants and enzymes. *Plant Physiol. Biochem. PPB* **2005**, *43*, 449–457. [CrossRef] [PubMed]
19. Lee, L.S.; Stephenson, K.K.; Fahey, J.W.; Parsons, T.L.; Lietman, P.S.; Andrade, A.S.; Lei, X.; Yun, H.; Soon, G.H.; Shen, P.; et al. Induction of chemoprotective phase 2 enzymes by ginseng and its components. *Planta Med.* **2009**, *75*, 1129–1133. [CrossRef]
20. Guo, S.; Ce, S.; Mei, H.; Limin, Y. The Effect of the Fermented Liquid and Mycelium Extract of FX-139 on the Growth and Defense Enzymes of Ginseng callus. *Chin. Biotechnol. Bull.* **2015**, *31*, 128–133. [CrossRef]
21. ICAMA (Institute Control of Agrochemicals, Ministry of Agriculture, P.R. China). Guidance on Environment Risk Assessment for Pesticide Registration. Available online: http://www.chinapesticide.org.cn/hjfxpg/9561.jhtml (accessed on 13 October 2020).
22. ICAMA (Institute Control of Agrochemicals, Ministry of Agriculture, P.R. China). Manual on Processing Factor of Pesticide Residue. Available online: https://www.icama.cn/portal/homepage/index.do/ (accessed on 13 October 2020).
23. Kim, J.G.; Park, H.R.; Yang, K.W.; Kim, S.S.; Kwon, C.H.; Jeong, Y.H.; Hur, J.H. Processing and reducing factors of difenoconazole during ginseng processing. *Korean J. Food Sci. Technol.* **2011**, *43*, 263–270. [CrossRef]
24. Alister, C.; Araya, M.; Becerra, K.; Volosky, C.; Saavedra, J.; Kogan, M. Industrial prune processing and its effect on pesticide residue concentrations. *Food Chem.* **2018**, *268*, 264–270. [CrossRef]
25. ICAMA (Institute Control of Agrochemicals, Ministry of Agriculture, P.R. China). China Pesticide Information Network. Available online: http://www.icama.org.cn/hysj/index.jhtml (accessed on 9 July 2021).
26. ICAMA (Institute Control of Agrochemicals, Ministry of Agriculture, P.R. China). Diet Risk Assessment Model. Available online: http://sitem.herts.ac.uk/aeru/iupac/search_alt.htm (accessed on 19 December 2020).
27. WHO (World Health Organization). Template for the Evaluation of Chronic Exposure (IEDI). Available online: https://www.who.int/foodsafety/areas_work/chemical-risks/gems-food/en/ (accessed on 25 October 2020).
28. MARA (Ministry of Agriculture and Rural Affairs of the People's Republic of China). Guideline for the Testing of Pesticide Residues in Crops (NY/T 788-2018). Available online: http://down.foodmate.net/standard/sort/5/54761.html (accessed on 25 October 2020).
29. Kim, S.J.; Murthy, H.N.; Hahn, E.J.; Lee, H.L.; Paek, K.Y. Effect of processing methods on the concentrations of bioactive components of ginseng (*Panax ginseng* C.A. Meyer) adventitious roots. *LWT Food Sci. Technol.* **2008**, *41*, 959–964. [CrossRef]
30. EU (Eupropean Commission). Guidance Document on Analytical Quality Control and Method Validation Procedures for Pesticide Residues and Analysis in Food and Feed (SANTE/11813/2017). Available online: https://ec.europa.eu/food/sites/food/files/plant/docs/pesticides_mrl_guidelines_wrkdoc_2017-11813.pdf (accessed on 9 March 2019).
31. Li, Y.; Xu, J.; Zhao, X.; He, H.; Zhang, C.; Zhang, Z. The dissipation behavior, household processing factor and risk assessment for cyenopyrafen residues in strawberry and mandarin fruits. *Food Chem.* **2021**, *359*, 129925. [CrossRef] [PubMed]

molecules

MDPI

Article

Simultaneous Preconcentration of Triazole Fungicide Residues Using In-Situ Coacervative Extraction Based on a Double-Solvent Supramolecular System Prior to High Performance Liquid Chromatographic Analysis

Rachaya Buppasang [1], Jaruwan Palasak [1], Rawikan Kachangoon [1], Kraingkrai Ponhong [1,2], Norio Teshima [3], Rodjana Burakham [4], Supalax Srijaranai [4] and Jitlada Vichapong [1,2,*]

1 Creative Chemistry and Innovation Research Unit, Department of Chemistry and Center of Excellence for Innovation in Chemistry, Faculty of Science, Mahasarakham University, Maha Sarakham 44150, Thailand
2 Multidisciplinary Research Unit of Pure and Applied Chemistry (MRUPAC), Department of Chemistry and Center of Excellent for Innovation in Chemistry, Faculty of Science, Mahasarakham University, Maha Sarakham 44150, Thailand
3 Department of Applied Chemistry, Aichi Institute of Technology, 1247 Yachigusa, Yakusa-cho, Toyota 470-0392, Japan
4 Materials Chemistry Research Center, Department of Chemistry and Center of Excellence for Innovation in Chemistry, Faculty of Science, Khon Kaen University, Khon Kaen 40002, Thailand
* Correspondence: jitlada.v@msu.ac.th; Tel./Fax: +66-4375-4246

Abstract: An in situ coacervative extraction (IS-CAE) based on a double-solvent supramolecular system coupled to liquid–liquid microextraction is investigated for extraction and enrichment of triazole fungicides. The formation of a double-solvent supramolecular system was generated by in situ formation and used as an extraction solvent for the coacervative extraction method. No disperser solvent was required. This new double-solvent supramolecular system has a higher extraction ability than any of its components alone. The different factors that could affect the extraction capability were studied and optimized, including the type of double extractant and its volume, salt addition, vortex time, and centrifugation time. Under optimum extraction conditions, this method provides high enrichment factors (EFs) of 73–318 with low limits of detection (LODs) of 0.3–1 μg L^{-1} and limits of quantitation (LOQs) of 1–3 μg L^{-1}. In addition, the proposed method was prosperously applied for the determination of triazole fungicides in water, fruit juice, and soy milk samples.

Keywords: in situ coacervative extraction; double-solvent supramolecular system; triazole fungicides; extraction; HPLC

Citation: Buppasang, R.; Palasak, J.; Kachangoon, R.; Ponhong, K.; Teshima, N.; Burakham, R.; Srijaranai, S.; Vichapong, J. Simultaneous Preconcentration of Triazole Fungicide Residues Using In-Situ Coacervative Extraction Based on a Double-Solvent Supramolecular System Prior to High Performance Liquid Chromatographic Analysis. *Molecules* **2022**, *27*, 6273. https://doi.org/10.3390/molecules27196273

Academic Editor: Miguel Ángel González-Curbelo

Received: 20 August 2022
Accepted: 20 September 2022
Published: 23 September 2022

Publisher's Note: MDPI stays neutral with regard to jurisdictional claims in published maps and institutional affiliations.

1. Introduction

The selection of a suitable sample preparation method is important because it has a significant effect on the method's sensitivity, selectivity, accuracy, reproducibility, and reliability [1]. Due to complex interfering substances and the presence of analytes at an ultra-trace level in real samples, various sample pretreatment techniques are needed [2] for clean up and matrix removal and preconcentration of target analytes. Many sample preparation techniques require high consumption of hazardous organic solvents which often generate waste during the process and are time-consuming. To overcome these problems, miniaturized extraction techniques have been investigated. Nowadays, modern trends in sample preparation techniques are affected by the concepts of green and sustainable solvents, especially in liquid–liquid microextraction [3]. According to the requirements of the sustainable sample preparation process, green alternative solvents should have various characteristics, which include nontoxicity, low energy consumption, dissolution of a large spectrum of solutes, and fewer steps [4]. The requirement of environmentally friendly solvents is gradually improving. Consequently, it is important to design and to develop an environmentally friendly alternative solvent in sample preparation methods.

Recently, new classes of extraction solvents, namely supramolecular solvents (SUPRAS), have been investigated. They are nanostructured liquids that form automatically in colloidal suspensions of amphiphiles via the phenomena of self-assembly and coalescence [5]. Due to their unique properties, SUPRAS have been better substitutes for conventional organic solvents for sample preparation before chromatographic techniques [6]. Their physico-chemical properties, which make them very attractive as an alternative extraction solvent in microextraction techniques, include: (i) their ability to interact with analytes via several interactions such as ionic bonding, hydrogen bonding, π-cation, and hydrophobic interaction, leading to an improvement in extraction efficiency and (ii) tunability by altering either the type or concentration of amphiphiles [7]. In addition, SUPRAS are tunable solvents and the properties of the solvents can be easily changed by altering the group of the amphiphiles [8]. Moreover, they are environmentally friendly solvents produced from inexpensive amphiphiles, in which the coacervation occurs rapidly at room temperature [9], in which the pH, salt, and the solvent are also affecting the coacervation.

Triazole fungicides are a group of highly effective systemic fungicides that contain a hydroxyl group (ketone group), a substituted phenyl group, and a 1,2,4-triazole group in the main chain [10]. They have a wide fungicidal spectrum and good control effects on a variety of crop diseases. Owing to their antifungal properties, they are widely used for preventing and controlling diseases and are widely used in agriculture for control of various fungal diseases such as powdery mildew, gray mold, spotted deciduous disease, black star disease, brown spot disease, and rust disease, in agricultural products such as fruits, vegetables, legumes, and grain crops [11,12]. However, triazole fungicides have high stability and lipophilicity, long residual duration, and are not easily degraded, which leads to easy accumulation in human and environmental media [10]. In order to protect human health, the Codex Alimentarius Commission (CAC) has established standards/regulations for the maximum residue limits (MRLs) of triazole fungicides in different matrices. For example, the MRL of hexaconazole, triadimefon, and bitertanol is 0.01–0.02 mg kg^{-1}; the MRL of tebuconazole is 0.02–5.0 mg kg^{-1}; and the MRL of myclobutanil is 0.05–3.0 mg kg^{-1} [10]. Therefore, it is necessary to establish a fast and efficient method for analyzing triazole fungicides in agricultural products [13].

In this study, we developed an in situ coacervative extraction (IS-CAE) based on a double-solvent supramolecular system coupled to liquid–liquid microextraction for extraction and enrichment of triazole fungicides prior to high-performance liquid chromatographic analysis. The phase separation obtained after centrifugation was formed by mixing the double-solvent supramolecular system. No organic solvent or heating were required. The proposed coacervative extraction strategy is far greener and more sustainable than the currently employed coacervative extraction. The important parameters affecting the IS-CAE were optimized and the resulting method was also applied to water, fruit juice, and soy milk samples.

2. Results and Discussion

2.1. Optimization of In Situ Extraction (IS-CAE) Procedure

In order to obtain high extraction efficiency, different experimental factors that affect the efficiency of the in situ coacervative extraction (IS-CAE) procedure were investigated and optimized. The peak area of the studied triazoles was used for the evaluation based on the one variable-at-a-time method, and all experiments were performed in triplicate using standard solution at a concentration of 100 μg L^{-1} of each analyte.

The choice of a suitable double extraction solvent is important because this is a significant parameter in the proposed method. A double extraction solvent must have a melting point close to room temperature, high extraction efficiency, less toxicity, and low solubility in the aqueous phase [14]. Therefore, 1-dodecanol (melting point 24 °C), and 1-undecanol (melting point 24 °C) were selected as extraction solvents in this work. First, each of the solvents was studied as an extraction solvent, and the results were compared with their double mixture with a specific ratio (as shown in Figures 1–3). It was found that the extraction efficiency of triazoles using the double mixture resulted in a higher extraction efficiency than the single solvent. Therefore, a double mixture solvent (1-dodecanol and 1-undecanol) was used for further study. The formation of the double-solvent supramolecular system was generated by in situ formation. Therefore, the 1-undecanol and 1-dodecanol volumes were studied. In this work, the 1-undecanol volume was studied in the range of 25–200 μL (as shown in Figure 4). The results showed that a high extraction efficiency in terms of peak area was obtained with 50 μL of 1-undecanol. The volume of 1-dodecanol was investigated in the range of 25–200 μL (as shown in Figure 5). The results showed that with 25 μL of 1-dodecanol the phase did not occur. A high extraction efficiency in terms of peak area was obtained with 50 μL of 1-dodecanol. Therefore, the most suitable proportion of double extractant was selected to be 1:1 of 1-undecanol/1-dodecanol to achieve the best extraction efficiency in this method.

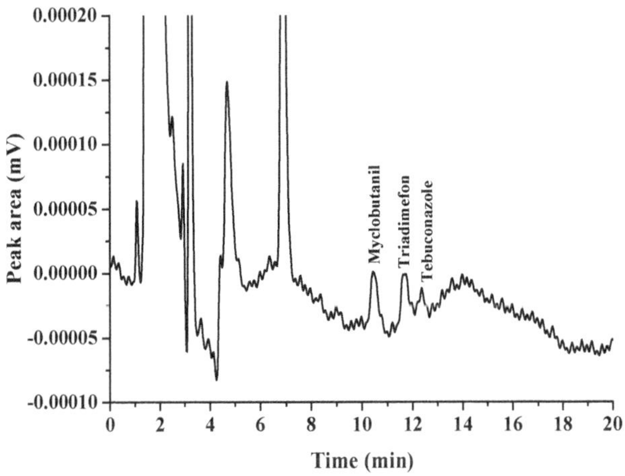

Figure 1. Extraction of triazole fungicides using 1-dodecanol as a solvent.

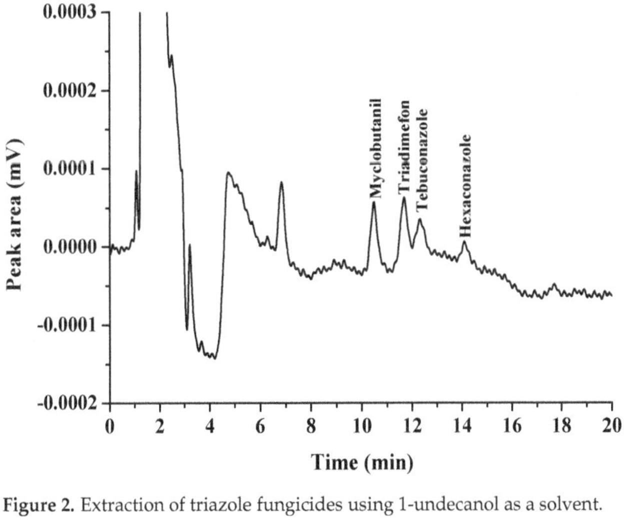

Figure 2. Extraction of triazole fungicides using 1-undecanol as a solvent.

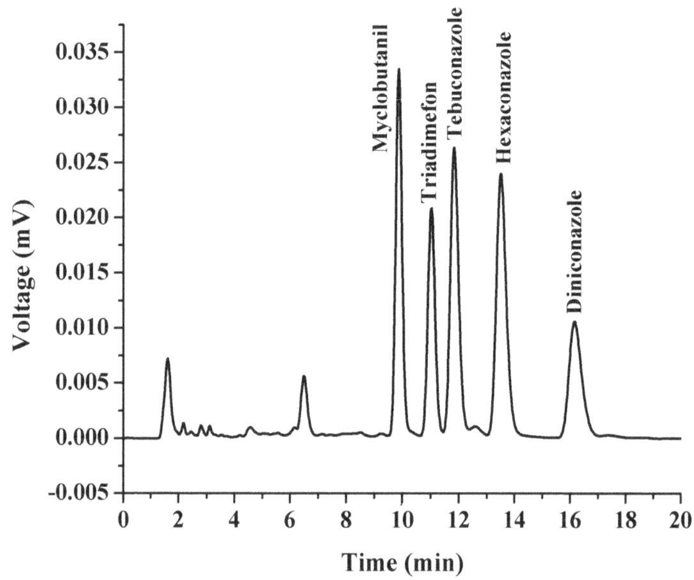

Figure 3. Extraction of triazole fungicides using mixture solvent (1-dodecanol and 1-undecanol).

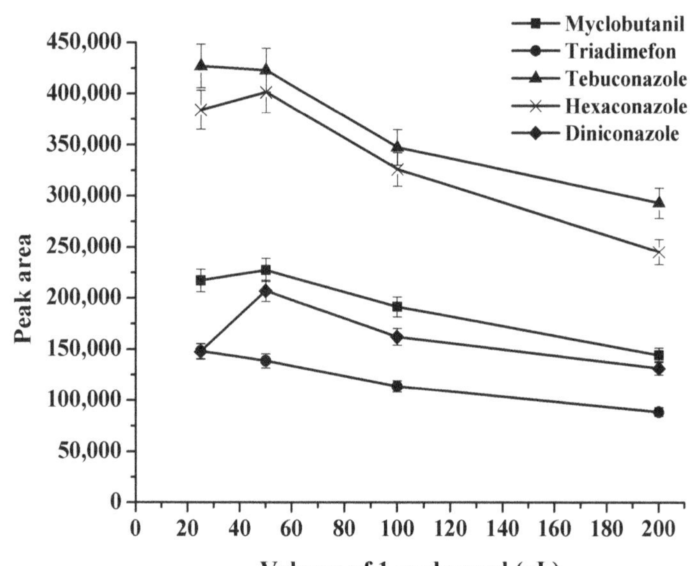

Figure 4. Effect of volume of 1-undecanol on extraction efficiency.

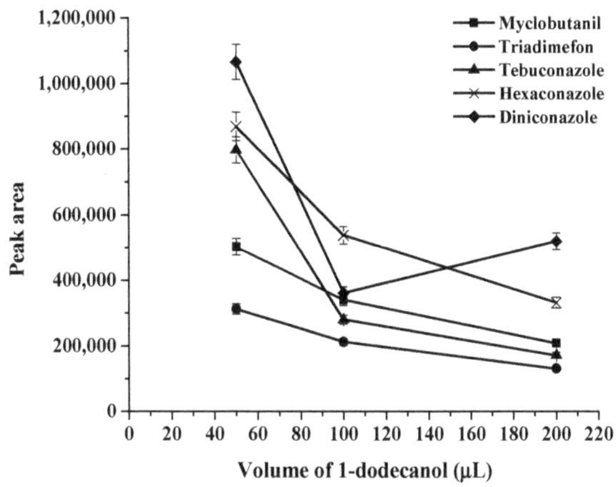

Figure 5. Effect of volume of 1-dodecanol on extraction efficiency.

To evaluate the salt effect on the efficiency of the in situ extraction procedure, various tests were carried out using different concentrations of salt in the range of 0–10% (*w/v*) NaCl (data not shown). The results indicated that by increasing NaCl from 0 to 5% (*w/v*), the peak area of triazoles remained nearly constant. At higher percentages, the analytical signal of the analytes decreased due to the dilution effect. Therefore, the experiments were carried out in the absence of any salt.

The vortex of the solution can accelerate the transfer of an analyte from an aqueous solution to the double-solvent supramolecular phase. An appropriate dispersion occurs in the presence of a strong vortex. Therefore, the vortex time was examined at 0, 15, 30, and 45 s. The results obtained (Figure 6) showed that the maximal analytical signals were observed at 30 s. Therefore, 30 s of vortex time was chosen for the next experiments.

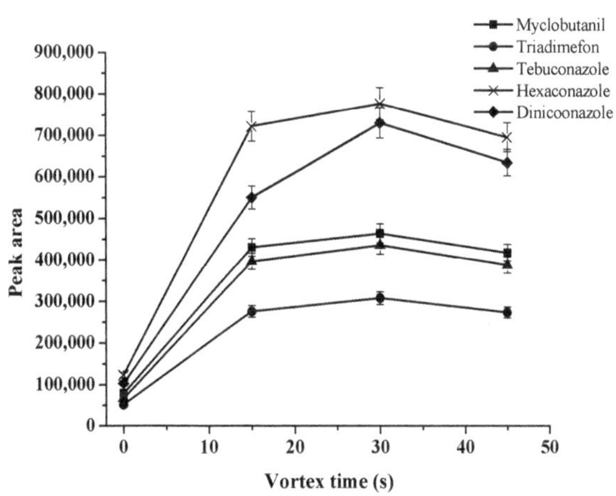

Figure 6. Effect of vortex time on extraction efficiency.

The centrifugation times were studied at 0, 5, 10, and 15 min at 2500 rpm. There were no significant differences in extraction efficiency found by increasing the centrifugation time from 5 to 15 min (as can be seen in Figure 7). Incomplete phase separation was obtained at 0 min (without centrifugation). In order to minimize the extraction time, therefore, 5 min was selected as the optimum centrifugation time.

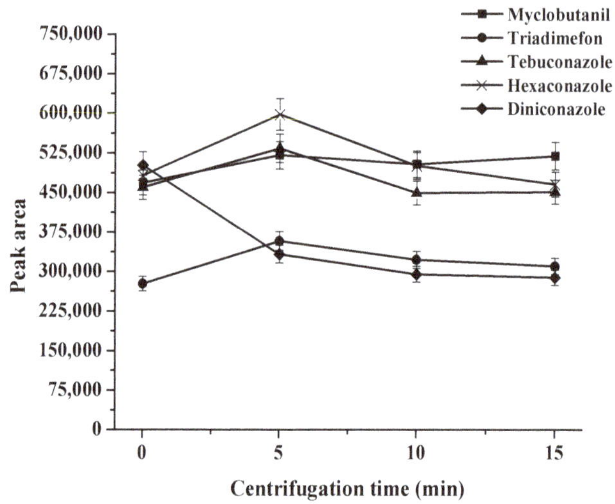

Figure 7. Effect of centrifugation time on extraction efficiency.

2.2. Analytical Performance of the Proposed Extraction Method

Linear ranges (LR), coefficient of determination (R^2), limit of detection (LOD), limit of quantification (LOQ), relative standard deviation (RSD) and enrichment factors (EFs) were calculated to validate the proposed method. All the data were obtained by conducting three replicates for each experimental test and the results are shown in Table 1. The calibration curve was constructed by plotting the peak area ratios against concentrations of triazoles. The linearity range was found to be from 0.3 to 1000.0 μg L^{-1}, with a high coefficient of determination (R^2 > 0.999), which showed an excellent level of linearity. The LODs and LOQs of the analytes were determined according to signal-to-noise ratios of 3 and 10, respectively. The results showed that the LODs ranged from 0.3 to 1.0 μg L^{-1}, while the LOQs were within 1–3 μg L^{-1}. The precision was studied by intra-day RSDs (n = 3) and inter-day RSDs (n = 3 × 3), which were lower than 4.84% and 4.95%, respectively. The EFs, were calculated using the ratio of the extracted analyte concentration in extraction phase to its initial concentration in aqueous sample solution, and were in the range of 73–318. The chromatograms of the triazoles obtained by direct HPLC and the proposed in situ coacervative extraction procedure are presented in Figures 8 and 9, respectively.

Table 1. Analytical performances of the present method.

Analyte	Linear Range (μg L^{-1})	R^2	LOD (μg L^{-1})	LOQ (μg L^{-1})	Intra-Day Precision (n = 3), RSD (%)		Inter-Day Precision (n = 3 × 3), RSD (%)		EF (C_{ex}/C_o)
					t_R	Peak Area	t_R	Peak Area	
Myclobutanil	3–1000	0.9999	1.0	3.0	1.89	2.50	1.96	3.42	74.82
Triadimefon	3–1000	0.9995	0.3	1.0	1.98	4.84	1.99	4.84	103.50
Tebuconazole	3–1000	0.9995	0.3	1.0	1.03	3.89	1.04	4.62	317.49
Hexaconazole	3–1000	0.9998	0.3	1.0	0.56	2.70	0.65	3.13	137.33
Diniconazole	3–1000	0.9995	1.0	3.0	0.61	3.52	0.75	4.95	73.81

R^2: coefficient of determination; LOD: limit of detection; LOQ: limit of quantification; RSD: relative standard deviation; EF: enrichment factor; t_R: retention time.

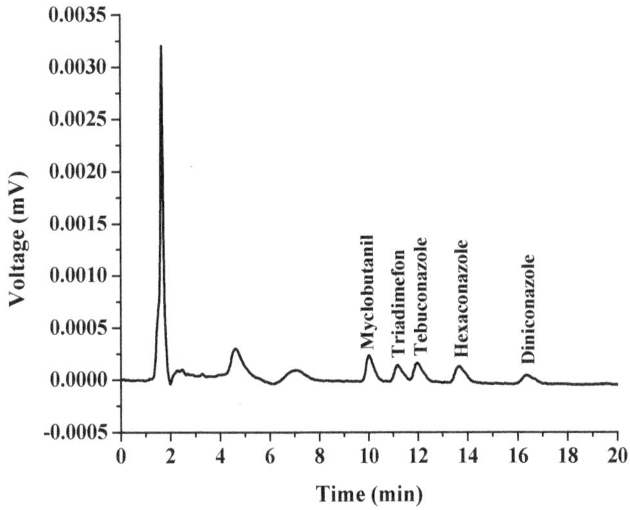

Figure 8. Chromatogram of standard triazole fungicides obtained by direct HPLC analysis. The concentration of each standard was 100 μg L^{-1}.

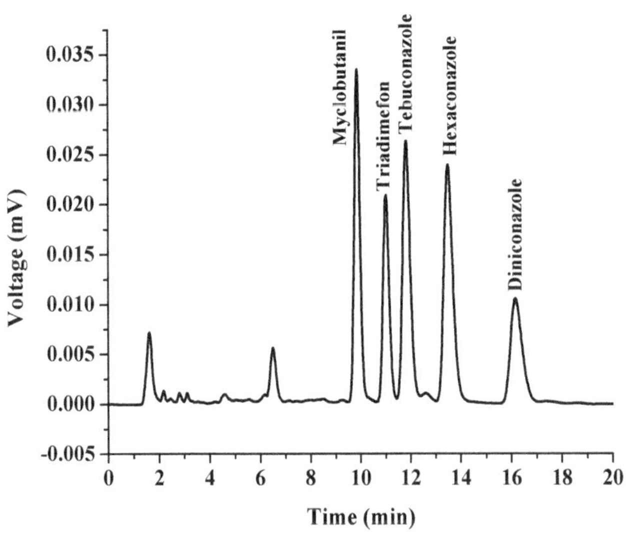

Figure 9. Chromatogram of standard triazole fungicides obtained with preconcentration using the proposed in situ coacervative extraction based on a double-solvent supramolecular system. The concentration of all standards was 100 μg L^{-1}. Conditions: Sample 10 mL, double SUPRA (50 μL of 1-dodecanol and 50 μL of 1-undecanol), vortex time 30 s, and centrifugation 2500 rpm for 5 min. Finally, collection of the top layer for HPLC analysis.

2.3. *Real Sample Analysis*

The applicability of the proposed in situ coacervative extraction (IS-CAE) coupled to the HPLC method was investigated to determine triazole fungicide residues in water, fruit juice, and soy milk samples. To investigate the matrix effect of real samples, a matrix-match calibration procedure was carried out. A set of matrix-matched calibration curves was prepared by extracting representative water, fruit juice, and soy milk samples spiked with 3.0–1000.0 μg L^{-1} of each target analyte. The studied triazole fungicides exhibit wide calibration capability and good linearity, with R^2 values greater than 0.99 for all studied samples.

The matrix effect (ME) was calculated by comparing the ratio of the slopes of the matrix-matched curve to that of the solvent (as shown in Equation (3)). Generally, An ME between 80–120% indicates no matrix effects, an ME between 50–80% or 120–150% refers to minor matrix effects, and an ME < 50% or >150% indicates major matrix effects [15,16]. As shown in Table 2, from no ME to a minor ME was observed for the water and fruit juice samples and major MEs were found in the soy milk samples.

Table 2. Matrix effect (ME, %).

Sample	Myclobutanil	Triadimefon	Tebuconazole	Hexaconazole	Diniconazole
Water I	78.83	77.95	71.31	82.18	82.89
Water II	87.78	71.37	79.63	84.38	86.63
Grape juice	75.00	75.00	83.33	75.00	100.00
Soy milk I	71.12	49.94	155.59	73.45	155.53
Soy milk II	72.25	48.83	152.14	77.72	145.55
Soy milk III	75.15	49.98	145.54	78.83	147.72

The accuracy and repeatability of the in situ extraction coupled to the HPLC method were evaluated by spiking the real samples with five triazole fungicides at concentration levels of 10, 30, and 50 μg L^{-1}. The results were shown in Table 3. Extraction recoveries in the range of 77–117% were obtained with RSDs in the range of 0.1–10.7%. Figure 10 illustrates the chromatograms of the blank and spiked (grape juice) samples. Based on these observations, it can be concluded that the proposed in situ extraction coupled to the HPLC method has excellent applicability for the selective extraction of five triazole fungicides in various samples.

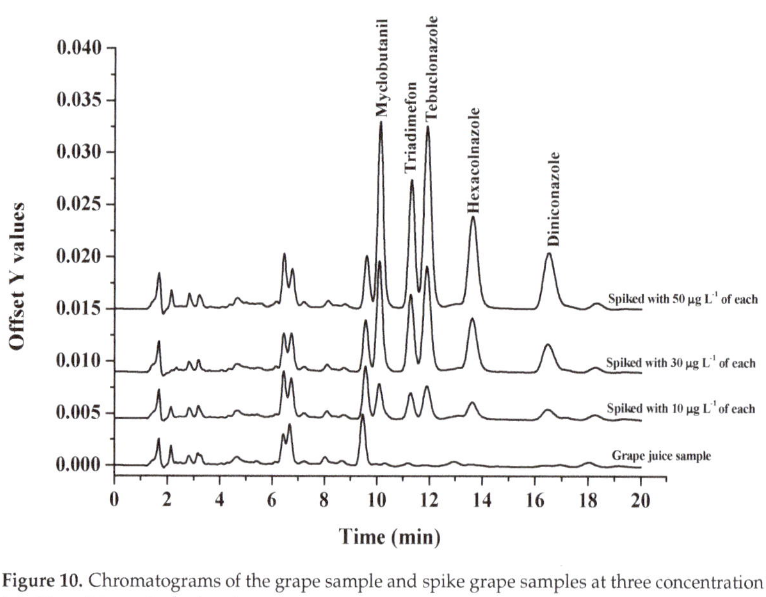

Figure 10. Chromatograms of the grape sample and spike grape samples at three concentration levels (10, 30, and 50 μg L^{-1} of each analyte).

Table 3. Comparisons of the proposed IS-CAE method with other methods for the determination of triazole fungicides.

Method	Analyte/ Sample	Linear Range	Limit of Detection (LOD)	%Recovery	Enrichment Factor (EF)	Reference
SVME	Triadimefon and triadimenol/beer samples	0.5–50 μg L^{-1} for triadimenol and 1.0–100 μg L^{-1} for triadimefon	0.24–0.99 μg L^{-1}	84–100	-	[17]
ATPS	Triazole fungicides/vegetable samples	0.100–30 μg mL^{-1}	0.03113–0.3525 μg mL^{-1}	71.57–107.8	-	[18]
SBSE	Triazole fungicides/grape and cabbage samples	0.1–500 μg L^{-1}	0.022–0.071 μg L^{-1}	80.7–111	49–57	[10]
VA-DLLME	Triazole fungicide, herbicide, pesticide and insecticide/fruit juice samples	149–500,000 ng L^{-1}	45–78 ng L^{-1}	55–89	1382–2246	[19]
CD-DLLME	Triazole and strobilurin fungicides/ water, juice, and vinegar samples	1–100 μg L^{-1}	0.3 μg L^{-1}	83.0–103.2	124	[20]
IS-CAE	Triazole fungicides	3–1000 μg L^{-1}	0.3–1.0 μg L^{-1}	77–117	73–318	This work

SVME-LC-MS/MS, Supramolecular solvent-based vortex-mixed microextraction coupled with liquid chromatography tandem mass spectrometer; ATPS-Online heart-cutting 2D-LC, aqueous two-phase system coupled with online heart-cutting two-dimensional liquid chromatography; SBSE- HPLC-DAD, stir bar sorption extraction combined with high-performance liquid chromatography-diode array detector; VA-DLLME, Vaporization assisted dispersive liquid-liquid microextraction coupled to gas chromatography-flame ionization detection; CD-DLLME-HPLC-DAD, cyclodextrin-based dispersive liquid–liquid microextraction coupled to high-performance liquid chromatography-diode array detector.

2.4. Comparison of the Proposed in Situ Coacervative Extraction (IS-CAE) Method with Other Previous Extraction Methods

To highlight the outstanding points of the developed method, some major characteristics were compared with those that have been obtained from other reported methods [10,17–20], as listed in Table 3. As compared with other methods, the established method has various advantages, such as the use of a green extraction solvent, a short extraction time (6 min), and avoidance of the use of a disperser solvent. Moreover, the proposed method exhibits a favorable linear range, low LOD, acceptable recovery, and high enrichment factor. Therefore, the proposed method is fast, simple, and environmentally friendly.

3. Experimental Methods

3.1. Chemicals and Reagents

All chemicals and reagents used in this work were of analytical grade. Five triazole fungicides (myclobutanil (MCBT), triadimefon (TDF), tebuconazole (TBZ), hexaconazole (HCZ), and diniconazole (DCZ)) from Dr. Ehrenstorfer GmbH (Augsburg, Germany)were used. Methanol (Merck, Darmstadt, Germany) was used to prepare the stock solution of each fungicide (1000 mg L^{-1}) and stored in refrigerator at 4 °C under light protection until analysis. HPLC-grade methanol and acetonitrile were obtained from Merck (Darmstadt, Germany). 1-Undecanol and 1-dodecanol were purchase from Sigma-Aldrich (Darmstadt, Germany). Deionized water with the resistivity of 18.2 MΩ.cm was obtained from a Type 1 Simplicity® ultrapure water system (Merck, Darmstadt, Germany). All solutions were filtered through a 0.45 μm nylon membrane filter before injected into the HPLC system.

3.2. Instrumentations

The chromatographic analysis of triazole fungicides was performed on a Waters 1525 Binary HPLC pump (Water, MA, USA) equipped with a diode array detector (DAD). The stationary-phase column was a Purospher® STAR RP-18 endcapped (4.6 × 150 mm^2, 5 μm) column (Merck, Darmstadt, Germany) with the column temperature maintained at ambient temperature. The mobile phase consisted of acetonitrile and water, and the separation was carried out under an isocratic elution of 50:50 (%v/v), and the flow rate was 1.0 mL min^{-1}, the injection volume was 20 μL, and the detection wavelength was set to 220 nm.

3.3. In-Situ Coacervative Extraction (IS-CAE) Procedure

The standard solution of triazoles (or sample solution) of 10.00 mL was mixed with 50 μL of 1-dodecanol and 50 μL of 1-undecanol in the centrifuge tube. Then, the solution was vortexed for 30 s. After that, the emulsion was centrifuged at 2500 rpm for 5 min to complete the phase separation. The reconstituted solution was collected before injecting into the HPLC system. A schematic diagram of the proposed microextraction procedure is shown in Figure 11.

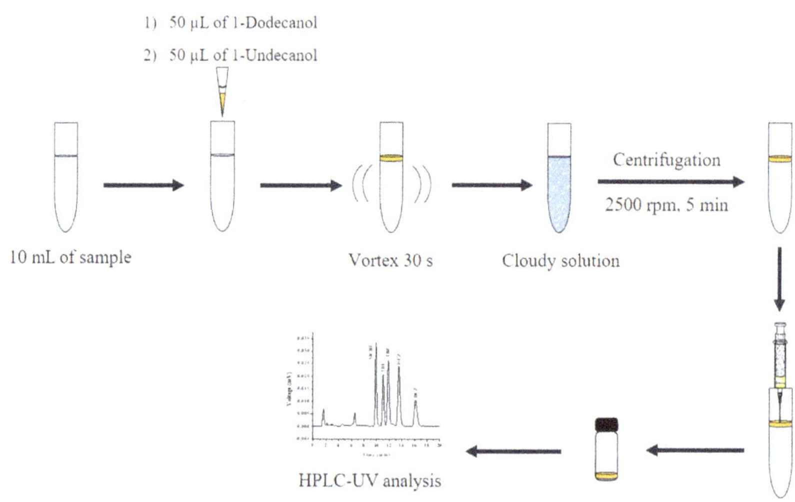

Figure 11. Schematic illustration of the proposed microextraction using in situ coacervative extraction based on a double-solvent supramolecular system for triazole fungicides and HPLC analysis. Conditions: Sample 10 mL, double SUPRA (50 μL of 1-dodecanol and 50 μL of 1-undecanol), vortex time 30 s, and centrifugation 2500 rpm for 5 min. Finally, collection of the top layer for HPLC analysis.

3.4. Sample Preparation

3.4.1. Water Samples

The water samples were collected from different areas located near rice fields in Maha Sarakham province, northeastern of Thailand, and were filtered through a 0.45 μm nylon membrane filter (Millipore, Burlington, MA, USA) before extraction using the proposed method.

3.4.2. Fruit Juice Samples

Commercial grape and apple juice samples, available in local supermarkets, were collected for analysis. Before analysis, a 30.0 mL aliquot of fruit juice was centrifuged at 3500 rpm for 15 min, and was filtered through a Whatman No. 42 filter paper. Then, the filtrate was filtered through a 0.45 μm nylon membrane filter before extraction using the proposed method.

3.4.3. Soy Milk Samples

Commercial soy milk samples were purchased from a local supermarket in Kantarawichai Distinct, Maha Sarakham Province, Northeast, Thailand. Proteins and fats in 1 mL samples were precipitated by shaking vigorously with acetonitrile and trifluoroacetic acid (5:1, *v*/*v*). then, the mixture was vortexed (1500 rpm, 3 min) and centrifuged at 4500 rpm for 10 min. The supernatant was extracted by using the coacervative extraction procedure (see Section 2.3). For the fortification of samples, standards of triazole were spiked into milk samples prior to protein and fat separation.

3.5. Calculation of Enrichment Factor (EF), Relative Recovery (RR), and Matrix Effect (ME)

The EF is the ratio between the concentration of analyte in the sediment phase (C_{sed}) and the initial concentration of analyte in the aqueous sample solution (C_0). To study the effect of experimental conditions on the extraction efficiency, the EFs were calculated according to the following equations:

$$EF = C_{sed}/C_0 \quad (1)$$

The %RR was defined as the %amount of analyte recovered from matrix (real samples) with reference to the extracted standard (standard spiked into the same matrix):

$$RR(\%) = \frac{C_{found} - C_{real}}{C_{added}} \times 100 \quad (2)$$

where C_{found} is the concentration of analyte after adding a known amount of working standard to real sample, C_{real} is the analyte concentration in real sample, and C_{added} represents the concentration of a known amount of working standard that was spiked into the real samples.

ME (%) is expressed as the ratio of the slopes obtained from calibration curves of each analyte spiked into the samples to the slopes obtained after extraction using the proposed method, according to the following equation:

$$ME(\%) = \frac{\text{slope of spiked real sample}}{\text{slope of standard solution}} \times 100 \quad (3)$$

4. Conclusions

In this study, an in situ coacervative extraction (IS-CAE) based on a double-solvent supramolecular system combined with HPLC was investigated for the analysis of triazole fungicides. The advantages of this method include a simple and inexpensive operational procedure, environmentally friendly, dispersive-solvent-free, and low organic solvent consumption. In this method, two long normal chain alcohols are in situ formed in the sample solution in which coacervative extraction was performed. This new supermolecule is used as an extractant system, which has a higher extraction power than any of its components alone. Therefore, IS-CAE fulfills the demand of green and sustainable analytical chemistry. In addition, this method was successfully applied to determine triazole fungicide residues in water, fruit juice, and soy milk samples, by providing satisfactory recoveries.

Author Contributions: R.K., J.V. participated in designing the study. R.B. (Rachaya Buppasang), J.P., R.K. and J.V. conducted the study. Data were collected and analyzed by J.V. The manuscript was written by J.V., K.P., N.T., R.B. (Rodjana Burakham), S.S. and J.V. All authors have read and agreed to the published version of the manuscript.

Funding: This research is supported by Mahasarakham University and Thailand Science Research and Innovation (TSRI).

Institutional Review Board Statement: Not applicable.

Informed Consent Statement: Not applicable.

Data Availability Statement: The data presented in this study are available on request from the corresponding author.

Acknowledgments: The authors gratefully acknowledge financial support for this research from Mahasarakham University, Thailand Science Research and Innovation (TSRI), and the Center of Excellence for Innovation in Chemistry (PERCH-CIC). We would like to thank the National Research Council of Thailand (NRCT; N41A640218). We also thank everyone for their assistance in the laboratory.

Conflicts of Interest: The authors declare no conflict of interest.

Sample Availability: Samples of the compounds are not available from the authors.

References

1. Hassan, M.; Uzcan, F.; Alshana, U.; Soylak, M. Switchable-hydrophilicity solvent liquid–liquid microextraction prior to magnetic nanoparticle-based dispersive solid-phase microextraction for spectrophotometric determination of erythrosine in food and other samples. *Food Chem.* **2021**, *348*, 129053. [CrossRef] [PubMed]
2. Pochivalov, A.; Davletbaeva, P.; Cherkashina, K.; Lezov, A.; Vakh, C.; Bulatov, A. Surfactant-mediated microextraction approach using switchable hydrophilicity solvent: HPLC-UV determination of Sudan dyes in solid food samples. *J. Mol. Liq.* **2018**, *271*, 807–814. [CrossRef]
3. Tang, W.; An, Y.; Row, K.H. Recoverable deep eutectic solvent-based aniline organic pollutant separation technology using choline salt as adsorbent. *J. Mol. Liq.* **2020**, *306*, 112910. [CrossRef]
4. Tang, W.; Row, K.H. Design and evaluation of polarity controlled and recyclable deep eutectic solvent based biphasic system for the polarity driven extraction and separation of compounds. *J. Clean. Prod.* **2020**, *268*, 122306. [CrossRef]
5. Cai, Z.-H.; Wang, J.-D.; Wang, L.-T.; Zhang, S.; Yan, X.-Y.; Wang, Y.-Q.; Zhao, P.-Q.; Fu, L.-N.; Zhao, C.-J.; Yang, Q.; et al. Green efficient octanoic acid based supramolecular solvents for extracting active ingredients from Zanthoxylum bungeanum Maxim. *Peels, J. Clean. Prod.* **2022**, *331*, 129731. [CrossRef]
6. Torres-Valenzuela, L.S.; Ballesteros-Gómez, A.; Rubio, S. Supramolecular solvent extraction of bioactives from coffee cherry pulp. *J. Food Eng.* **2020**, *278*, 109933. [CrossRef]
7. Seebunrueng, K.; Phosiri, P.; Apitanagotinon, R.; Srijaranai, S. A new environment-friendly supramolecular solvent-based liquid phase microextraction coupled to high performance liquid chromatography for simultaneous determination of six phenoxy acid herbicides in water and rice samples. *Microchem. J.* **2020**, *152*, 104418. [CrossRef]
8. Ballesteros-Gomez, A.; Sicilia, M.D.; Rubio, S. Supramolecular solvents in the extraction of organic compounds. A review. *Anal. Chim. Acta* **2010**, *677*, 108–130. [CrossRef] [PubMed]
9. Zohrabi, P.; Shamsipur, M.; Hashemi, M.; Hashemi, B. Liquid-phase micro-extraction of organophosphorus pesticides using supramolecular solvent as a carrier for ferrofluid. *Talanta* **2016**, *160*, 340–346. [CrossRef] [PubMed]
10. Wang, Y.; He, M.; Chen, B.; Hu, B. Hydroxyl-containing porous organic framework coated stir bar sorption extraction combined with high performance liquid chromatography-diode array detector for analysis of triazole fungicides in grape and cabbage samples. *J. Chromatogr. A* **2020**, *1633*, 461628. [CrossRef]
11. De Albuquerque, N.C.P.; Carrão, D.B.; Habenschus, M.D.; de Oliveira, A.R.M. Metabolism studies of chiral pesticides: A critical review. *J. Pharm. Biomed.* **2018**, *147*, 89–109. [CrossRef] [PubMed]
12. Dong, F.; Li, J.; Chankvetadze, B.; Cheng, Y.; Xu, J.; Liu, X.; Li, Y.; Chen, X.; Bertucci, C.; Tedesco, D.; et al. Chiral triazole fungicide difenoconazole: Absolute stereochemistry, stereoselective bioactivity, aquatic toxicity, and environmental behavior in vegetables and soil. *Environ. Sci. Technol.* **2013**, *47*, 3386–3394. [CrossRef] [PubMed]
13. Liu, G.; Li, L.; Huang, X.; Zheng, S.; Xu, D.; Xu, X.; Zhang, Y.; Lin, H. Determination of triazole pesticides in aqueous solution based on magnetic graphene oxide functionalized MOF-199 as solid phase extraction sorbents. *Microporous Mesoporous Mater.* **2018**, *270*, 258–264. [CrossRef]
14. Akramipour, R.; Fattahi, N.; Golpayegani, M.R. Sensitive determination of methotrexate in plasma of children with acute leukemia using double-solvent supramolecular system as a novel extractant for dispersive liquid-liquid microextraction. *J. Chromatogr. B* **2021**, *1171*, 122628. [CrossRef] [PubMed]

15. Surapong, N.; Santaladchaiyakit, Y.; Burakham, R. A water-compatible magnetic dual-template molecularly imprinted polymer fabricated from a ternary biobased deep eutectic solvent for the selective enrichment of organophosphorus in fruits and vegetables. *Food Chem.* **2022**, *384*, 132475. [CrossRef]
16. Boontongto, T.; Burakham, R. Eco-friendly fabrication of a magnetic dual-template molecularly imprinted polymer for the selective enrichment of organophosphorus pesticides for fruits and vegetables. *Anal. Chim. Acta* **2021**, *1186*, 339128. [CrossRef]
17. Zhao, W.; Zhao, J.; Zhao, H.; Cao, Y.; Liu, W. Supramolecular solvent-based vortex-mixed microextraction: Determination of chiral triazole fungicide in beer samples. *Chirality* **2018**, *30*, 302–309. [CrossRef] [PubMed]
18. Zeng, H.; Xie, X.; Huang, Y.; Chen, J.; Liu, Y.; Zhang, Y.; Mai, X.; Deng, J.; Fan, H.; Zhang, W. Enantioseparation and determination of triazole fungicides in vegetables and fruits by aqueous two-phase extraction coupled with online heart-cutting two-dimensional liquid chromatography. *Food Chem.* **2019**, *301*, 125265. [CrossRef] [PubMed]
19. Farajzadeh, M.A.; Kiavar, L.; Pezhhanfar, S. Development of a method based on dispersive liquid–liquid microextraction followed by partial vaporization of the extract for ultra–preconcentration of some pesticide residues in fruit juices. *J. Chromatogr. A* **2021**, *1653*, 462427. [CrossRef] [PubMed]
20. Jing, X.; Huang, X.; Zhang, Y.; Wang, M.; Xue, H.; Wang, X.; Jia, L. Cyclodextrin-based dispersive liquid–liquid microextraction for the determination of fungicides in water, juice, and vinegar samples via HPLC. *Food Chem.* **2022**, *367*, 130664. [CrossRef] [PubMed]

molecules

Article

Determination of Cyclaniliprole in Fruits and Vegetables Using Disposable Pipette Extraction Cleanup and Ultrahigh-Performance Liquid Chromatography-Tandem Mass Spectrometry

Zhou Lu [1,2,*], Weiqian Yue [2], Weiming Ren [1], Yanhong Wang [1] and Yueru Li [1]

[1] Center of Quality Standard and Testing Technology for Agro-Products, Jilin Agricultural University, Changchun 130118, China
[2] College of Plant Protection, Jilin Agricultural University, Changchun 130118, China
* Correspondence: joelu87@163.com

Abstract: Despite an outstanding agent for control of Lepidoptera, the diamide insecticide cyclaniliprole (CYCP) is a suspected carcinogen. In the present study, an analytical method was developed for the determination of CYCP in six fruits and vegetables (apple, grape, peach, bell pepper, lettuce, and tomato) using ultrahigh-performance liquid chromatography coupled with tandem mass spectrometry. Sample preparation was carried out by the acetonitrile-salting-out extraction followed by simple and fast cleanup of disposable pipette extraction tip containing styrene divinyl benzene and/or graphitized carbon black. Satisfactory linearity ($r > 0.99$) was obtained in the calibration range of 0.001–1 μg mL^{-1}. Matrix effects decreased from −9.9–−17.9% to −1.0–−7.6% after the cleanup. The recoveries of CYCP at three spike levels (0.01, 0.1, and 1 mg kg^{-1}) from different matrices were between 75.7% and 111.5%, with the intra-day ($n = 5$) and inter-day ($n = 15$) relative standard deviations lower than 12.1%. The limit of quantification was 0.01 mg kg^{-1}. The developed method provides a good reference for routine monitoring of CYCP in these fruits and vegetables.

Keywords: cyclaniliprole; diamide insecticide; residue analysis; DPX; UHPLC-MS/MS

Citation: Lu, Z.; Yue, W.; Ren, W.; Wang, Y.; Li, Y. Determination of Cyclaniliprole in Fruits and Vegetables Using Disposable Pipette Extraction Cleanup and Ultrahigh-Performance Liquid Chromatography-Tandem Mass Spectrometry. *Molecules* **2022**, *27*, 6464. https://doi.org/10.3390/molecules27196464

Academic Editor: Miguel Ángel González-Curbelo

Received: 1 September 2022
Accepted: 23 September 2022
Published: 30 September 2022

1. Introduction

Cyclaniliprole (CYCP; 2′,3-dibromo-4′-chloro-1-(3-chloro-2-pyridyl)-6′-{((1RS)-1-cyclopropylethyl)carbamoyl}pyrazole-5-carboxanilide) is a newly developed insecticide introduced by Ishihara Sangyo Kaisha, Ltd. (Osaka, Japan) [1]. Its chemical structure is shown in Figure 1a. CYCP belongs to anthranilic diamides, and members of this class act by binding to insect ryanodine receptors, leading to lethargy, paralysis, and death due to the unregulated loss of intracellular calcium stores [2]. CYCP effectively controls major agricultural Lepidoptera such as *Plutella xylostella*, *Mythimna separata*, and *Spodoptera litura* in a wide range of crops. Unlike other diamide compounds such as chlorantraniliprole and cyantraniliprole, the developer claimed that CYCP has a structural advantage that makes it more active against diamide-resistant insects [3]. As the resistance of pests to early diamides such as chlorantraniliprole has been continuously identified [4–6], CYCP could be an important complement to this insecticide class to overcome this issue.

The substantial contribution made by pesticides to agriculture is obvious. At least a third of our crop production will be wasted due to Damage by various pests and pathogens if there are no pesticides applied [7]. However, the unreasonable use of pesticides will exhibit adverse effects on food safety and further endanger human health. Although CYCP has low mammalian toxicity with the oral median lethal dose (LD_{50}) to rats > 2000 mg/kg b.w., it is suspected to be a carcinogen, as it induces C-cell adenoma in male rats according to the report from the European Food Safety Authority [8]. To avoid this risk, governments have established laws or regulations to specify the maximum residue limits (MRLs).

Residue analysis methods are essential means to ensure the pesticide residue level in foods is in compliance with MRLs.

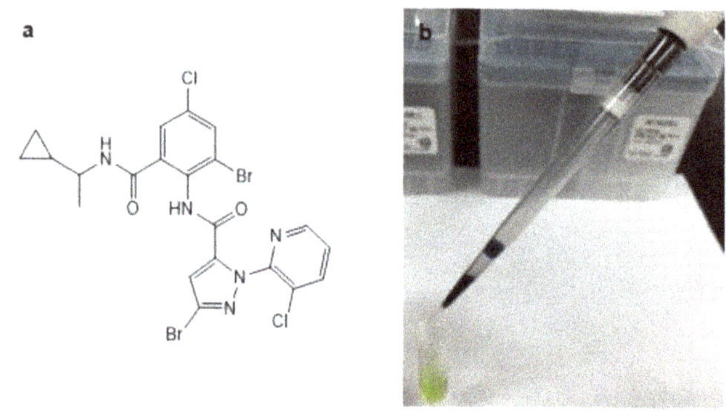

Figure 1. Molecular structure of CYCP (**a**) and a photograph of DPX apparatus (**b**).

Pretreatment is necessary for raw samples to transform into a certain status amenable to instrumental analysis. The Quick, Easy, Cheap, Rugged, and Safe (QuEChERS) method has been widely used in this process. Developed by Anastassiades et al. [9], QuEChERS extracts samples with acetonitrile (MeCN) and then partitions the aqueous/organic phases by adding $mgSO_4$ and NaCl. The MeCN extract is cleaned up by thoroughly mixing it with the adsorbent primary secondary amine to remove interfering co-extractives, which is called dispersive-solid phase extraction (d-SPE). Soon afterward, this methodology evolved into the buffering versions effectively dealing with some pH-sensitive pesticides [10,11], which finally led to two official methods, AOAC 2007.01 [12] and EN 15662 [13], depending on whether the acetate or the citrate buffering strategy was used. Nowadays, the QuEChERS is still an open source system as researchers continue to modify it to achieve higher efficiency in the extraction and cleanup of various types of samples [14–18].

Disposable pipette extraction (DPX) is a novel cleanup technique developed to incorporate d-SPE and solid-phase extraction (SPE) approaches. A photograph of the DPX apparatus is shown in Figure 1b. The efficiency of DPX cleanup is mainly attributed to the repeated d-SPE process that occurs when the sample extract is aspirated and dispensed through the pipette tip and fully contacts the freely moving adsorbents packed in it [19]. Furthermore, after mixing, adsorbents with higher weight will be quickly deposited at the end of the tip to form a "micro-SPE" layer so that when the liquid is dispensed, the samples undergo another cleanup step as they pass through the deposit adsorbent. The DPX technique also saves more time as the centrifugation required in the d-SPE process is not necessary here. So far, DPX has been successfully applied for the analysis of various organic contaminants in food and environmental samples [20–24].

The present study developed and validated an analytical method for the determination of a newly developed anthranilic diamide insecticide CYCP in fruits and vegetables using ultrahigh-performance liquid chromatography coupled with tandem mass spectrometry (UHPLC-MS/MS). Samples were extracted by MeCN-salting-out, and the extract was cleaned up using DPX tips packed with styrene divinyl benzene (SDVB) and/or graphitized carbon black (GCB). The effects of different adsorbents and instrumental conditions on the method performance were investigated. Real sample analysis was conducted as the application of the developed method.

2. Results and Discussion

2.1. MS/MS Optimization

MS/MS was run in both positive (ESI+) and negative (ESI−) modes to identify the appropriate CYCP precursor ion. As a result, the $[CYCP + H]^+$ of 599.9 Da was identified under ESI+ (Figure S1a), while [CYCP − H]$^-$ of 597.9 Da appeared under ESI—(Figure S1c). Afterwards, certain energy was provided for each precursor ion to fragment into product ions. As shown in the MS2 spectrum of ESI+ (Figure S1b), two product ions (283.8 Da and 514.8 Da) with reasonable intensities were observed. By comparison, only one product ion (256.1 Da) with an acceptable response was observed in that of ESI− (Figure S1d). Since MS analysis of residues usually requires the identification of an analyte by at least two characteristic product ions, ESI+ mode was used for the detection of CYCP in our method, and the product ion of 283.8 Da, which has a relatively high intensity, served as the quantifier.

A number of MS/MS parameters that could affect the signal strength of ion transitions including declustering potential (DP), collision energy (CE), and collision cell exit potential (CXP) were optimized by running a ramp over a certain range, and the value that yielded the highest response was adopted.

2.2. Chromatography Optimization

The addition of some volatile acids or salts in the LC mobile phase could affect either the peak shape or signal response of the analyte during an LC-MS analysis [25,26]. The shape of a chromatographic peak is important for its integration, leading to accurate quantification, and an increased response can improve the detection sensitivity. In this study, we tested five different aqueous-phase (solvent A) compositions (a. ultrapure water; b. 0.1% HCOOH; c. 0.2% HCOOH; d. 5 mM NH_4COOH; and e. 5 mM NH_4COOH + 0.1% HCOOH) to determine the one with the best performance. As shown in Figure 2, by comparison with ultrapure water, the addition of HCOOH suppressed the signal, while the LC-MS provided both good peak shape and the highest response for CYCP when 5 mM NH_4COOH was used as the additive in the LC aqueous phase. Therefore, the use of composition d was adopted by the developed method.

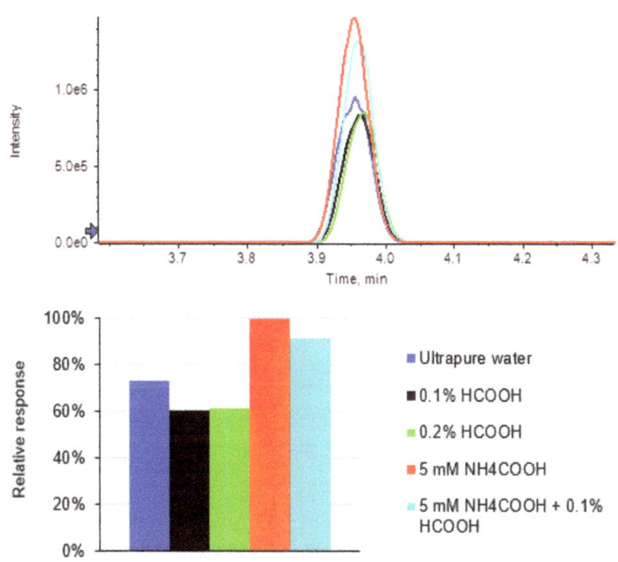

Figure 2. Effects of different LC aqueous phases on peak shape and response of CYCP.

2.3. Sample Preparation Optimization

The present study compared the efficiency of the citrate (Na_3Citrate/Na_2HCitr) buffering and non-buffering extraction method and the results are shown in Figure 3a. CYCP recoveries from six matrices extracted by both approaches were in the satisfactory range of 93.6–102.5%, while no significant difference in recoveries from each individual matrix was observed. This indicated that the stability of CYCP could be maintained in matrices with different pH values. Therefore, from an economical point of view, non-buffering extraction method was adopted by the developed method.

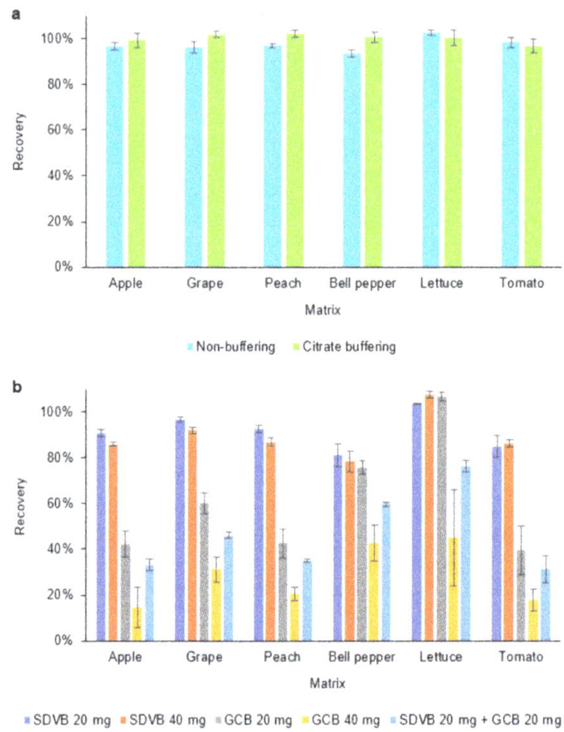

Figure 3. Effects of different extraction (**a**) and cleanup (**b**) strategies on recoveries of CYCP from fruit and vegetable samples.

SDVB is good at removing non-polar and weakly polar co-extractives in agricultural products such as lipids, waxes, and steroids [27], while GCB is commonly used for adsorbing pigments [28]. In this study, the cleanup performance of DPX tips containing five sets of adsorbents (a. 20 mg of SDVB; b. 40 mg of SDVB; c. 20 mg of GCB; d. 40 mg of GCB; and e. 20 mg of SDVB + 20 mg of GCB) was investigated. Anastassiades and Lehotay proposed an empirical rule according to a number of previous studies on d-SPE cleanup: every milliliter of sample extract combined with 50 mg of adsorbent provides satisfactory recoveries with a wide analyte scope [29]. As we have decided that 800 μL of extract is adequate for both in-tip mixing and instrumental analysis, the appropriate amount of adsorbent according to this rule would be 40 mg. Due to the limited space in the tip, the performance of less adsorbent use (20 mg) was also investigated.

The results (Figure 3b) showed that the use of SDVB of 20 and 40 mg yielded similar CYCP recoveries within a satisfactory range (78.3–107.4%) for all matrices. However, the use of GCB substantially reduced the recoveries of CYCP to 14.5–60.1% for apple, grape, peach, and tomato. This influence was attenuated for bell pepper and lettuce as the recoveries increased to 42.6–106.6%. Notably, when 20 mg of GCB was employed, acceptable recoveries (75.7% for bell pepper and 106.6% for lettuce) were obtained for these two green vegetables. We could extrapolate two possible reasons accounting for this: first, as apple, grape, peach, and tomato contain more organic acids than lettuce and bell pepper, the lower pH values of their extracts may facilitate the adsorption of CYCP on GCB; second, when the d-SPE process occurred between GCB and the matrices of lettuce and bell pepper, the carbon material preferentially adsorbed certain pigments such as chlorophyll, leading to higher CYCP recoveries.

Since a number of adsorbent sets provided good and similar recoveries for one specific matrix, we further compared their co-extractive removal ability with respect to ME reduction. As shown in Figure 4, 40 mg of SDVB reduced ME to the minimum for apple, grape, peach, and tomato, while 20 mg of GCB led to the lowest ME for bell pepper and lettuce. As a result, DPX tips containing 40 mg of SDVB were used for the cleanup of apple, grape, peach, and tomato extracts, and tips containing 20 mg of GCB were employed for the purification of bell pepper and lettuce extracts.

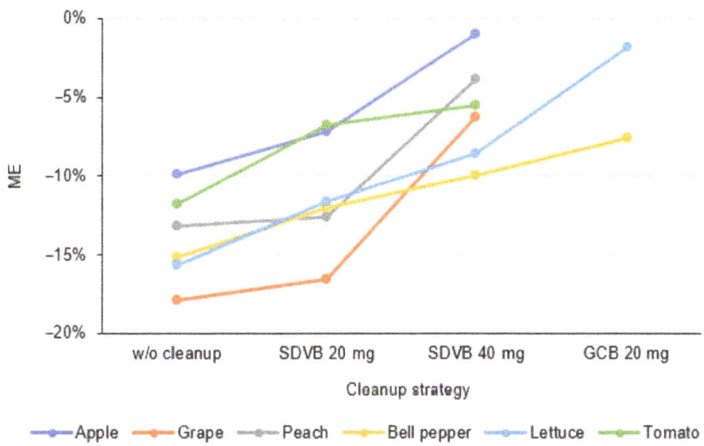

Figure 4. Effects of different cleanup strategies on the ME of fruit and vegetable samples.

2.4. Method Validation

Typical MRM chromatograms of spiked and blank samples are shown in Figure 5. No interference from the blank matrices appeared at the retention time of CYCP.

Information on calibration, ME, LOQs, and regulated MRLs of CYCP in different matrices is listed in Table 1. Good linearity ($r > 0.99$) was achieved for each matrix-matched calibration curve in the range of 0.001–1 μg mL^{-1}. MEs with positive and negative values correspond to signal enhancement and suppression, respectively. In this study, all matrices exhibited suppression to CYCP response, and ME decreased from −9.9–−17.9% to −1–−7.6% after DPX cleanup.

Table 1. Information on calibration, ME, LOQs, and MRLs of CYCP in different matrices.

Matrix	Calibration Equation	r	ME (%)	LOQ (mg kg^{-1})	MRL (mg kg^{-1}) (EU/US)
Acetonitrile	y = 6472x − 148	0.9987	-	0.01	-
Apple	y = 6407x + 384	0.9980	−1.0	0.01	0.01/0.3
Grape	y = 6062x + 335	0.9947	−6.3	0.01	0.01/0.8
Peach	y = 6218x + 923	0.9934	−3.9	0.01	0.01/1
Bell pepper	y = 5980x + 347	0.9972	−7.6	0.01	0.01/0.2
Lettuce	y = 6356x + 965	0.9921	−1.8	0.01	0.01/15
Tomato	y = 6118x + 291	0.9974	−5.5	0.01	0.01/0.2

Mean recoveries of CYCP from the six fruits and vegetables that were spiked at 0.01, 0.1, and 1 mg kg^{-1} were in the range of 75.7–111.5%, and the corresponding intra-day

(n = 5) and inter-day (n = 15) relative standard deviations (RSDs) ranged from 0.4% to 12.1%. The results conformed to the method performance acceptability criteria (70% ≤ recovery ≤ 120%; RSD ≤ 20%) required by SANTE [30]. Detailed Data are shown in Table 2. The LOQs was 0.01 mg kg^{-1} in all matrices according to SANTE guidelines [30], which were lower than relevant MRLs regulated by USA and identical to those regulated by EU.

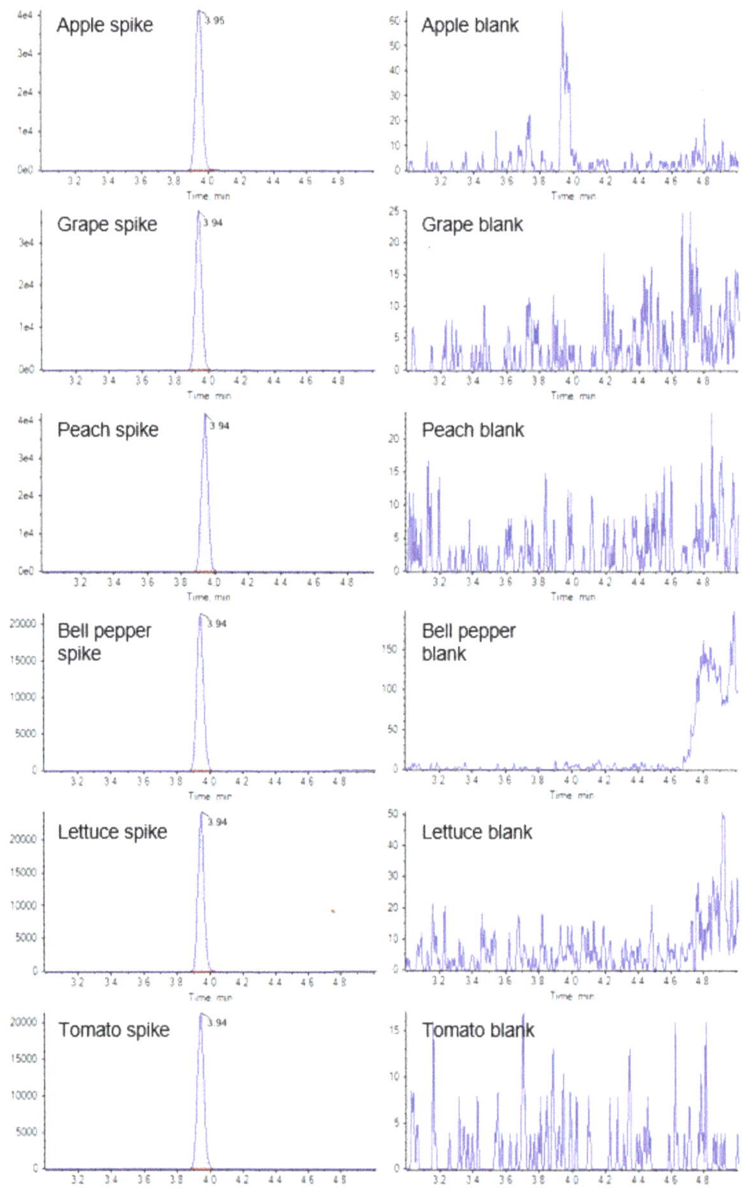

Figure 5. Typical MRM chromatograms of spiked (0.01 mg kg^{-1}) (left-hand column) and blank samples (right-hand column).

Table 2. Recoveries, RSD_a (intra-day, n = 5), and RSD_r (inter-day, n = 15) of CYCP from different fruit and vegetable samples analyzed by the developed method.

Matrix	Spike Level (mg kg^{-1})	Day 1		Day 2		Day 3		RSD_r (%)
		Rec. (%)	RSD_a (%)	Rec. (%)	RSD_a (%)	Rec. (%)	RSD_a (%)	
Apple	0.01	95.2	5.1	98.2	4.0	97.3	7.9	5.6
	0.1	85.8	0.9	102.6	1.7	102.1	1.3	7.5
	1	93.6	1.8	97.9	2.6	97.6	3.7	3.4
Grape	0.01	91.2	2.7	98.0	4.9	98.6	4.2	5.2
	0.1	91.9	1.3	107.5	0.8	105.8	0.4	6.4
	1	90.5	2.4	98.1	2.2	98.3	1.2	4.3
Peach	0.01	90.9	3.0	100.4	1.7	102.1	1.6	5.6
	0.1	86.7	1.9	110.8	4.5	102.7	2.6	9.9
	1	93.1	2.2	98.4	1.5	99.1	1.1	3.2
Bell pepper	0.01	86.2	2.0	102.2	0.8	101.1	1.9	7.9
	0.1	75.7	2.8	102.9	1.4	101.0	1.4	12.1
	1	88.2	1.6	102.3	1.1	103.8	0.5	7.5
Lettuce	0.01	88.4	2.3	108.4	1.0	107.2	2.1	9.5
	0.1	106.6	2.0	111.5	1.4	110.0	1.5	2.3
	1	88.7	2.3	105.2	0.9	106.3	0.9	8.4
Tomato	0.01	83.8	3.3	95.9	2.1	95.9	4.0	7.1
	0.1	86.2	1.7	100.4	1.5	100.0	1.7	6.5
	1	90.4	1.8	94.5	1.5	94.9	2.0	2.8

CYCP was not detected (> LOQ) in all collected real samples (10 for each matrix, 60 in total). As CYCP has not been registered for use in China [31], this result is thought to be rational.

3. Materials and Methods

3.1. Chemicals and Reagents

The analytical standard of CYCP (purity 99.0%) was provided by Shenyang Research Institute of Chemical Industry (Shenyang, China). Anhydrous $mgSO_4$, NaCl, sodium citrate tribasic dehydrate ($Na_3Citrate$), and sodium citrate dibasic sesquihydrate (Na_2HCitr) were purchased from Agilent Technologies (Palo Alto, CA, USA). LCMS-grade MeCN and formic acid (HCOOH) were obtained from DiKMA Technologies (Beijing, China). Ammonium formate (NH_4COOH; purity $\geq$ 99.995%) was purchased from Sigma-Aldrich (St. Louis, MO, USA). DPX tips (1250 μL) containing different amounts of SDVB and/or GCB were bought from DPX Technologies (Columbia, SC, USA). A manual pipette (100–1000 μL) used for DPX cleanup was obtained from INTEGRA Biosciences (Zizers, Switzerland). The water used in this study was ultrapure (18 MΩ cm) and was prepared by a LAB-UV-40 water purification system manufactured by Lab-Partner Technology Development (Changchun, China).

A 0.01 g sample of CYCP standard was dissolved in 10 mL of MeCN to prepare the stock solution at 1000 μg mL^{-1}. The solution was stored in an amber vial in a refrigerator at 4 °C, and its stability in three months was guaranteed by UHPLC-MS/MS monitoring. Working solutions with lower concentrations were freshly prepared from the stock solution before their use.

3.2. Sample Preparation

Samples of apple, grape, peach, bell pepper, lettuce, and tomato were bought from a local agro-product market and pureed in a food processor (Braun, Kronberg, Germany). For sample extraction, 10.0 g (±0.1 g) of a certain processed fruit or vegetable sample was taken into a centrifuge tube (50 mL) and then added to 10.0 mL of MeCN. The tube was capped and vigorously hand-shaken for 1 min. Afterward, 1 g of NaCl and 4 g of anhydrous $mgSO_4$ were added. The tube was shaken again for 30 s and then centrifuged at

5000 rpm for 5 min. Finally, 800 μL of the supernatant was transferred to a centrifuge tube (2 mL) waiting for DPX cleanup.

The manual pipette equipped with the DPX tips containing 40 mg of SDVB was used for cleanup of extracts of apple, grape, peach, and tomato, while the tips containing 20 mg of GCB were employed for purifying those of bell pepper and lettuce. Aspirating volume of the pipette was set to be 1000 μL, and the extract supernatant was aspirated slowly in and out of the DPX tip three times before it was dispensed into a 2 mL vial for UHPLC-MS/MS analysis.

3.3. Instrumentation

A Nexera UHPLC system (Shimadzu, Kyoto, Japan) coupled to a QTRAP4500 MS/MS (Sciex, Framingham, USA) was used for the detection of CYCP. The hybrid system was controlled by Analyst 1.6.2 software (Sciex, Framingham, USA). The UHPLC was equipped with a Luna Omega 2.1 × 100 mM, 1.6 μm C_{18} column (Phenomenex, Torrance, USA) held at 40 °C. The mobile phase comprised 5 mM NH_4COOH (solvent A) and MeCN (solvent B) with a constant flow rate of 0.3 mL min^{-1}. The gradient elution used was 45%B (0.0 min) → 95%B (6.0 min) → 95%B (7 min) → 45%B (7.1 min) → 45%B (10.0 min). The injection volume was 5 μL.

The MS/MS was equipped with an electrospray ionization (ESI) source operating in the positive mode (ESI+). Qualification of CYCP was fulfilled by two ion transitions under multiple reaction monitoring (MRM) mode, and the one with the highest abundance was selected for quantification. Relevant MRM parameters are shown in Table 3. Other conditions used were as follows: ionspray voltage, 5500 V; source temperature, 550 °C; curtain gas pressure, 0.2 mPa; ion spray gas pressure, 0.3 mPa; auxiliary heating gas pressure, 0.3 mPa; dwell time for each transition, 100 ms. MultiQuant 3.0.3 software (Sciex, Framingham, USA) was employed for Data analysis.

Table 3. MS/MS parameters for detection of CYCP.

Compound	Molecular Formula	Retention Time (min)	Ion Transition (m/z)	DP (V)	CE (V)	CXP (V)	MRM Ratio
CYCP	$C_{21}H_{17}Br_2Cl_2N_5O_2$	3.94	599.9 > 283.8 [a,b]; 599d.9 > 514.8 [a]	90; 90	23; 32	9; 21	0.28

[a] For qualification, [b] For quantification.

3.4. Method Validation

CYCP content in different matrices was quantified by the external standard method. Since the influence of matrix effects (ME) on the quantification is inevitable during LC-MS analysis [32], a 7-point (0.001, 0.005, 0.01, 0.05, 0.1, 0.5, and 1 μg mL^{-1}) matrix-matched calibration curve instead of a curve comprising solvent standards was used to deal with any signal enhancement or suppression caused by ME. The extent of the ME on analyte ion abundance in different matrices was measured by the following equation [33]:

$$ME = \left(\frac{\text{Slope of matrix} - \text{matched calibration curve}}{\text{Slope of solvent calibration curve}} - 1\right) \times 100\%$$

The developed method was validated with respect to recovery (accuracy), and corresponding relative standard deviations (RSDs) (precision) by spiking blank samples at three levels of 0.01, 0.1, and 1 mg kg^{-1}. The recovery test at each spike level was conducted as intra-day (n = 5) and inter-day (n = 15). Based on SANTE/11312/2022 guidelines [30], the limit of quantification (LOQ) is identified as the lowest spike level of the analyte in matrix.

Ten samples of each kind of fruit and vegetables used in this study were purchased from 5 food markets in Changchun, Jilin, China. All collected samples (n = 60) were pretreated and analyzed using the developed method to assess the residue level of CYCP and further validate the reliability of this method.

4. Conclusions

The present study combined DPX sample cleanup with UHPLC-MS/MS detection to provide rapid and accurate determination of CYCP in six fruits and vegetables. Despite its excellent Lepidopteran control efficacy, CYCP is a possible carcinogen, so its contents in agricultural products must be monitored routinely to avoid any potential risks caused by its inappropriate use. The performance of this method satisfies relevant requirements of the EU analytical quality control regulation SANTE; thus, it can serve as a fast and reliable approach fulfilling the above purposes.

Supplementary Materials: The following supporting information can be downloaded at: https://www.mdpi.com/article/10.3390/molecules27196464/s1, Figure S1: Spectrum of CYCP using Q1 ESI+ scan (mass range 560–630 Da) (a). Spectrum of product ions of selected CYCP parent ion at 599.9 Da using ESI+ product ion mode (mass range 50–610 Da) (b). Spectrum of CYCP using Q1 ESI− scan (mass range 560–630 Da) (c). Spectrum of product ions of selected CYCP parent ion at 599.9 Da using ESI− product ion mode (mass range 50–610 Da) (d).

Author Contributions: Conceptualization, Z.L.; methodology, Z.L. and W.Y.; validation, Y.W.; investigation, Z.L., W.R. and W.Y.; resources, Z.L. and Y.L.; Data curation, Y.W.; writing—original draft preparation, Z.L.; writing—review and editing, Z.L. and Y.L.; supervision, Y.L.; project administration, Z.L.; funding acquisition, Z.L. All authors have read and agreed to the published version of the manuscript.

Funding: This study was financially supported by the National Natural Science Foundation of China (31901920).

Institutional Review Board Statement: Not applicable.

Informed Consent Statement: Not applicable.

Data Availability Statement: Not applicable.

Conflicts of Interest: The authors declare that there are no conflicts of interest.

Sample Availability: Samples of the compounds are not available from the authors.

References

1. IUPAC Pesticide Properties Database. Available online: http://sitem.herts.ac.uk/aeru/iupac/Reports/2632.htm (accessed on 1 May 2022).
2. Cordova, D.; Benner, E.A.; Sacher, M.D.; Rauh, J.J.; Sopa, J.S.; Lahm, G.P.; Selby, T.; Stevenson, T.; Flexner, L.; Gutteridge, S.; et al. Anthranilic diamides: A new class of insecticides with a novel mode of action, ryanodine receptor activation. *Pestic. Biochem. Physiol.* **2006**, *84*, 196–214. [CrossRef]
3. Hamamoto, T. Chemical Structure Advantage of Cyclaniliprole Against the Diamides Resistant Diamondback Moth, Plutella Xylostella. In Proceedings of the International Congress of Entomology, Orlando, FL, USA, 25–30 September 2016.
4. Pereira, R.M.; Neto, D.A.; Amado, D.; Durigan, M.R.; Franciscatti, R.A.; Mocheti, M.; Omoto, C. Baseline susceptibility and frequency of resistance to diamide insecticides in Helicoverpa armigera (Lepidoptera: Noctuidae) populations in Brazil. *Crop Prot.* **2020**, *137*, 105266. [CrossRef]
5. Lu, Y.; Wang, G.; Zhong, L.; Zhang, F.; Bai, Q.; Zheng, X.; Lu, Z. Resistance monitoring of Chilo suppressalis (Walker) (Lepidoptera: Crambidae) to chlorantraniliprole in eight field populations from east and central China. *Crop Prot.* **2017**, *100*, 196–202. [CrossRef]
6. Wang, X.; Wu, Y. High Levels of Resistance to Chlorantraniliprole Evolved in Field Populations of Plutella xylostella. *J. Econ. Entomol.* **2012**, *105*, 1019–1023. [CrossRef] [PubMed]
7. Oerke, E.C. Crop losses to pests. *J. Agric. Sci.* **2006**, *144*, 31–43. [CrossRef]
8. European Food Safety Authority (EFSA). Peer review of the pesticide risk assessment of the active substance cyclaniliprole. *EFSA J.* **2016**, *14*, 4452.
9. Anastassiades, M.; Lehotay, S.J.; Stajnbaher, D.; Schenck, F.J. Fast and easy multiresidue method employing acetonitrile extraction/partitioning and "dispersive solid-phase extraction" for the determination of pesticide residues in produce. *J. AOAC Int.* **2003**, *86*, 412–431. [CrossRef]
10. Lehotay, S.J.; Mastovská, K.; Lightfield, A.R. Use of buffering and other means to improve results of problematic pesticides in a fast and easy method for residue analysis of fruits and vegetables. *J. AOAC Int.* **2005**, *88*, 615–629. [CrossRef]
11. Anastassiades, M.; Scherbaum, E.; Tasdelen, B.; Stajnbaher, D. Recent Developments in QuEChERS Methodology for Pesticide Multiresidue Analysis. In *Pesticide Chemistry: Crop Protection, Public Health, Environmental Safety*; Ohkawa, H., Miyagawa, H., Lee, P.W., Eds.; Wiley: Weinheim, Germany, 2007; pp. 439–458.

12. Association of Official Analytical Chemists (AOAC). AOAC Official Method 2007.01 Pesticide Residues in Foods by Acetonitrile Extraction and Partitioning with Magnesium Sulfate. *J. AOAC Int.* **2007**, *90*, 485–520.
13. *EN 15662:2008*; Foods of Plant Origin—Determination of Pesticide Residues Using GC-MS and/or LC-MS/MS following Acetonitrile Extraction and Partitioning and Clean-Up by Dispersive SPE—QuEChERS-Method. European Committee for Standardization: Brussels, Belgium, 2008.
14. Shin, D.; Kim, J.; Kang, H.S. Simultaneous determination of multi-pesticide residues in fish and shrimp using dispersive-solid phase extraction with liquid chromatography-tandem mass spectrometry. *Food Control* **2020**, *120*, 107552. [CrossRef]
15. Acosta-Dacal, A.; Rial-Berriel, C.; Diaz-Diaz, R.; Bernal-Suarez, M.D.; Luzardo, O.P. Optimization and validation of a QuEChERS-based method for the simultaneous environmental monitoring of 218 pesticide residues in clay loam soil. *Sci. Total Environ.* **2021**, *753*, 142015. [CrossRef]
16. Gawel, M.; Kiljanek, T.; Niewiadowska, A.; Semeniuk, S.; Goliszek, M.; Burek, O.; Posyniak, A. Determination of neonicotinoids and 199 other pesticide residues in honey by liquid and gas chromatography coupled with tandem mass spectrometry. *Food Chem.* **2019**, *282*, 36–47. [CrossRef]
17. Cao, J.L.; Zheng, Y.X.; Kaium, A.; Liu, X.G.; Xu, J.; Dong, F.S.; Wu, X.H.; Zheng, Y.Q. A comparative study of biochar, multiwalled carbon nanotubes and graphitized carbon black as QuEChERS absorbents for the rapid determination of six triazole fungicides by UPLC-MS/MS. *Int. J. Environ. Anal. Chem.* **2019**, *99*, 209–223. [CrossRef]
18. Xu, X.; Xu, X.Y.; Han, M.; Qiu, S.T.; Hou, X. Development of a modified QuEChERS method based on magnetic multiwalled carbon nanotubes for the simultaneous determination of veterinary drugs, pesticides and mycotoxins in eggs by UPLC-MS/MS. *Food Chem.* **2019**, *276*, 419–426. [CrossRef]
19. Bordin, D.C.M.; Alves, M.N.R.; de Campos, E.G.; De Martinis, B.S. Disposable pipette tips extraction: Fundamentals, applications and state of the art. *J. Sep. Sci.* **2016**, *39*, 1168–1172. [CrossRef]
20. Lu, Z.; Fang, N.; Zhang, Z.; Hou, Z.; Lu, Z.; Li, Y. Residue analysis of fungicides fenpicoxamid, isofetamid, and mandestrobin in cereals using zirconium oxide disposable pipette extraction clean-up and ultrahigh-performance liquid chromatography-tandem mass spectrometry. *J. Chromatogr. A* **2020**, *1620*, 461004. [CrossRef]
21. Teixeira, R.A.; Flores, D.H.; Da Silva, R.C.S.; Dutra, F.V.A.; Borges, K.B. Pipette-tip solid-phase extraction using poly(1-vinylimidazole-co-trimethylolpropane trimethacrylate) as a new molecularly imprinted polymer in the determination of avermectins and milbemycins in fruit juice and water samples. *Food Chem.* **2018**, *262*, 86–93.
22. Mastrianni, K.R.; Metavarayuth, K.; Brewer, W.E.; Wang, Q. Analysis of 10 β-agonists in pork meat using automated dispersive pipette extraction and LC-MS/MS. *J. Chromatogr. B* **2018**, *1084*, 64–68. [CrossRef]
23. Li, Z.; Li, Y.; Liu, X.; Li, X.; Zhou, L.; Pan, C. Multiresidue Analysis of 58 Pesticides in Bean Products by Disposable Pipet Extraction (DPX) Cleanup and Gas Chromatography–Mass Spectrometry Determination. *J. Agric. Food Chem.* **2012**, *60*, 4788–4798. [CrossRef]
24. Guan, H.; Brewer, W.E.; Garris, S.T.; Morgan, S.L. Disposable pipette extraction for the analysis of pesticides in fruit and vegetables using gas chromatography/mass spectrometry. *J. Chromatogr. A* **2010**, *1217*, 1867–1874. [CrossRef]
25. Patring, J.D.; Jastrebova, J.A. Application of liquid chromatography–electrospray ionisation mass spectrometry for determination of dietary folates: Effects of buffer nature and mobile phase composition on sensitivity and selectivity. *J. Chromatogr. A* **2007**, *1143*, 72–82. [CrossRef] [PubMed]
26. Garcia, M. The effect of the mobile phase additives on sensitivity in the analysis of peptides and proteins by high-performance liquid chromatography–electrospray mass spectrometry. *J. Chromatogr. B* **2005**, *825*, 111–123. [CrossRef] [PubMed]
27. Hennion, M.-C. Solid-phase extraction: Method development, sorbents, and coupling with liquid chromatography. *J. Chromatogr. A* **1999**, *856*, 3–54. [CrossRef]
28. Perestrelo, R.; Silva, P.; Porto-Figueira, P.; Pereira, J.A.M.; Silva, C.; Medina, S.; Camara, J.S. QuEChERS-Fundamentals, relevant improvements, applications and future trends. *Anal. Chim. Acta* **2019**, *1070*, 1–28. [CrossRef]
29. Lehotay, S.J.; Majors, R.E.; Anastassiades, M. The QuEChERS revolution. *LCGC Asia Pac.* **2010**, *13*, 30–36.
30. European Commission SANTE/11312/2022. Guidance Document on Analytical Quality Control and Method Validation Procedures for Pesticide Residues and Analysis in Food and Feed. 2022. Available online: https://ec.europa.eu/food/system/files/2022-02/pesticides_mrl_guidelines_wrkdoc_2021-11312.pdf (accessed on 1 May 2022).
31. China Pesticide Information Network. Available online: http://www.chinapesticide.org.cn/hysj/index.jhtml (accessed on 1 May 2022).
32. Niessen, W.M.A.; Manini, P.; Andreoli, R. Matrix effects in quantitative pesticide analysis using liquid chromatography-mass spectrometry. *Mass Spectrom. Rev.* **2006**, *25*, 881–899. [CrossRef]
33. Stahnke, H.; Kittlaus, S.; Kempe, G.; Alder, L. Reduction of matrix effects in liquid chromatography-electrospray ionization-mass spectrometry by dilution of the sample extracts: How much dilution is needed? *Anal. Chem.* **2012**, *84*, 1474–1482. [CrossRef]

molecules

Article

Pesticide Residues in Vegetables and Fruits from Farmer Markets and Associated Dietary Risks

El-Sayed A. El-Sheikh [1,*], Mahmoud M. Ramadan [1], Ahmed E. El-Sobki [1], Ali A. Shalaby [1], Mark R. McCoy [2], Ibrahim A. Hamed [1], Mohamed-Bassem Ashour [1] and Bruce D. Hammock [2]

[1] Plant Protection Department, Faculty of Agriculture, Zagazig University, Zagazig 44511, Egypt
[2] Department of Entomology and Nematology, UC Davis Cancer Center, University of California, Davis, CA 95616, USA
* Correspondence: eaelsheikh@zu.edu.eg

Abstract: The use of pesticides leads to an increase in agricultural production but also causes harmful effects on human health when excessively used. For safe consumption, pesticide residues should be below the maximum residual limits (MRLs). In this study, the residual levels of pesticides in vegetables and fruits collected from farmers' markets in Sharkia Governorate, Egypt were investigated using LC-MS/MS and GC-MS/MS. A total number of 40 pesticides were detected in the tested vegetable and fruit samples. Insecticides were the highest group in detection frequency with 85% and 69% appearance in vegetables and fruits, respectively. Cucumber and apple samples were found to have the highest number of pesticide residues. The mean residue levels ranged from 7 to 951 μg kg^{-1} (in vegetable samples) and from 8 to 775 μg kg^{-1} (in fruit samples). It was found that 35 (40.7%) out of 86 pesticide residues detected in vegetables and 35 (38.9%) out of 90 pesticide residues detected in fruits exceeded MRLs. Results for lambda-cyhalothrin, fipronil, dimothoate, and omethoate in spinach, zucchini, kaki, and strawberry, respectively, can cause acute or chronic risks when consumed at 0.1 and 0.2 kg day^{-1}. Therefore, it is necessary for food safety and security to continuously monitor pesticide residues in fruits and vegetables in markets.

Keywords: pesticide residues; vegetables; fruits; dietary risk exposure

Citation: El-Sheikh, E.-S.A.; Ramadan, M.M.; El-Sobki, A.E.; Shalaby, A.A.; McCoy, M.R.; Hamed, I.A.; Ashour, M.-B.; Hammock, B.D. Pesticide Residues in Vegetables and Fruits from Farmer Markets and Associated Dietary Risks. *Molecules* **2022**, 27, 8072. https://doi.org/10.3390/molecules27228072

Academic Editor: Miguel Ángel González-Curbelo

Received: 15 October 2022
Accepted: 13 November 2022
Published: 21 November 2022

1. Introduction

Pests and diseases cause high losses in crop yields worldwide that can reach approximately 45% loss annually [1]. Due to the rapid growth of world population, increase in the agricultural productivity is urgent to meet rising food needs. Chemical pesticides are considered the main component in protecting agricultural products in the field and store to maintain crop yield and quality [2,3]. Pesticides usage in Egypt has increased, according to the Food and Agriculture Organization (FAO) report, from 4931 tons in 2000 to 13,178 tons in 2019 [4]. Globally, the total pesticides use in agriculture was 4.12 million tons in 2018. The worldwide application of pesticides was 2.63 kg ha^{-1} in 2018, which showed a more than doubled increase in pesticide usage in the 2010s compared with the 1990s [5]. Since banning of organochlorides, other groups (organophosphates (OPs), carbamates, and synthetic pyrethroids) were the most widely used classes of insecticides due to their high activity and relatively low persistence [6]. New groups of chemical insecticides have been also introduced in agriculture, including neonicotinoids, spinosyns, avermectins, and diamides [7].

Fruits and vegetables are important nutritional components in different societies. They are recommended to be eaten fresh, unpeeled, and unprocessed for their high nutritional value and content of minerals, vitamins, fibers, and antioxidants [8–10]. On the other hand, food (especially fruits and vegetables) is one of the main ways through which humans are exposed to pesticides, at a rate five times higher than other methods such as air and water [11]. Accordingly, efforts to ensure a sustainable use of chemical pesticides to avoid the increase of pesticide levels in the environment and food commodities are necessary.

Pesticide residues in fruit and vegetable samples have been reported in many countries including Croatia [12], South America [13], Turkey [14], Poland [9], China [15], Jordan [16], UAE [17,18], Kenya [19], and South Korea [20]. The chronic effects of exposure from contaminated food intake are mostly unknown. Studies have demonstrated that exposure to pesticides has dose-related chronic and acute toxicity in humans through different mechanisms including deregulation of transporters or enzymes involved in xenobiotic metabolism. This has effects on cell processes such as growth, differentiation, and survival, including reactive oxygen species, cell damage through subsequent oxidative stress, and DNA damage [21]. There is growing evidence of carcinogenicity and genotoxicity as well as endocrine disruption capacity attributed to the ingestion of contaminated food or direct exposure to pesticides [22]. Despite the fact that the use of certain organochlorides, Ops, and carbamates are prohibited in many countries [23], some of these compounds have been detected in the environment worldwide due to their persistent nature or illegal use of the banned chemical pesticides [24]. A large number of programs are being implemented to address this issue. For instance, to protect the Brazilian population from severe risks associated with food contaminated with pesticides, the Brazilian National Sanitary Agency has initiated a nationwide monitoring program for pesticide residues in fruits and vegetables since 2001. In 2009, 20 types of fruits and vegetables were analyzed and the results indicated that 23.2% were positive for insecticide residues, and 14.3% of the samples exceeded the European Union maximum residue levels (MRLs) [25]. In contrast, there is lack of data on contamination of the food available in the Egyptian market. Only a few studies have been published on this subject over the past 20 years, such as Tchounwou et al. [26]. Constant evolution of the pesticide industry requires closer surveillance and better assessment of factors including pesticides bioaccumulation, stability, widespread usage, and food quality and safety that impact directly on human health [27]. Hence, the aim of this study was to determine pesticide residues in vegetable and fruit samples in local markets in Egypt and to show the differences and frequencies in pesticides detection. The most common pesticides and the type of crops with the highest number of pesticide residues are also shown. This study will help understanding of the most applied pesticides on vegetables and fruits as well as the most common polluted crops locally. Risk assessment of pesticides exceeding MRLs in vegetable and fruit samples was also determined.

2. Results

2.1. Multi-Residues of Pesticides in Vegetable and Fruits

For pesticide residues in vegetables, 66 samples belonging to 13 types of vegetables collected from the farmer markets of Sharkia Governorate were analyzed. Pesticide residues were detected in 44 (67%) samples and 22 (33%) samples showed no pesticides detection. Regarding pesticide residues in fruits, it was found that out of 54 samples analyzed, 33 (61%) samples were positive for the presence of pesticides and 21 (39%) samples had no pesticide residues.

The number of pesticides that were detected in each vegetable sample ranged from 1 to 15 pesticides. Carrot was the sample that showed the lowest number of pesticide residues while cucumber was the highest sample with 15 pesticide residues. Pesticide residues in each fruit sample ranged from 1 to 20 pesticides. One pesticide residue appeared in banana while 20 pesticide residues were found in apples (Figure 1).

A total of 40 different pesticides were detected in the tested vegetable and fruit samples belonging to classes of insecticides, fungicides, and herbicides. It was shown that 12 pesticide residues (in vegetables) and 21 pesticide residues (in fruits) were detected one time only. The number of pesticides that were detected two times or more was 16 pesticides (in vegetables) and 16 pesticides (in fruits). The total number of pesticides detected in vegetable and fruits were 28 and 37, respectively (Figure 2).

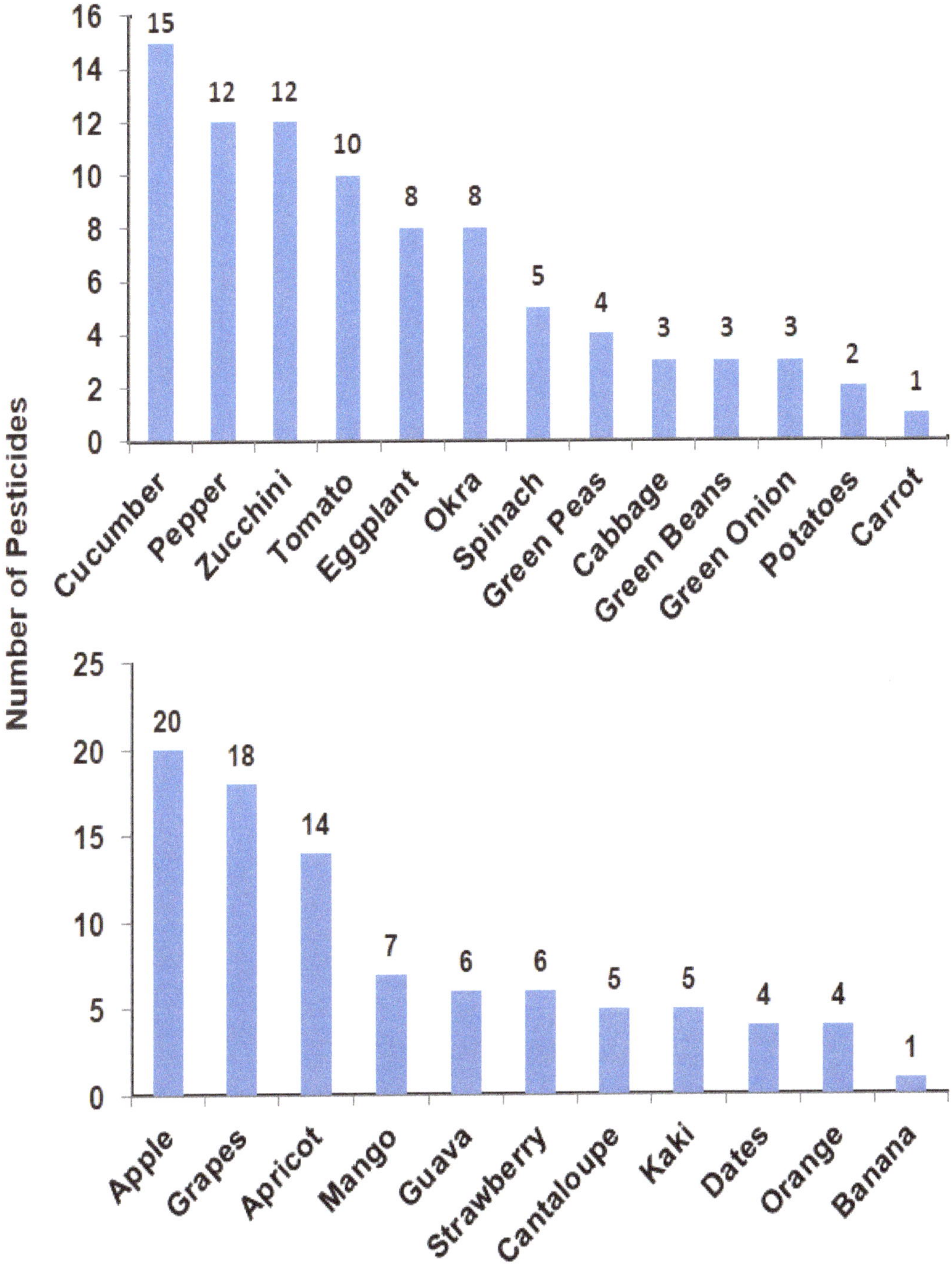

Figure 1. Number of pesticides detected in each vegetable (**upper**) or fruit (**lower**) sample.

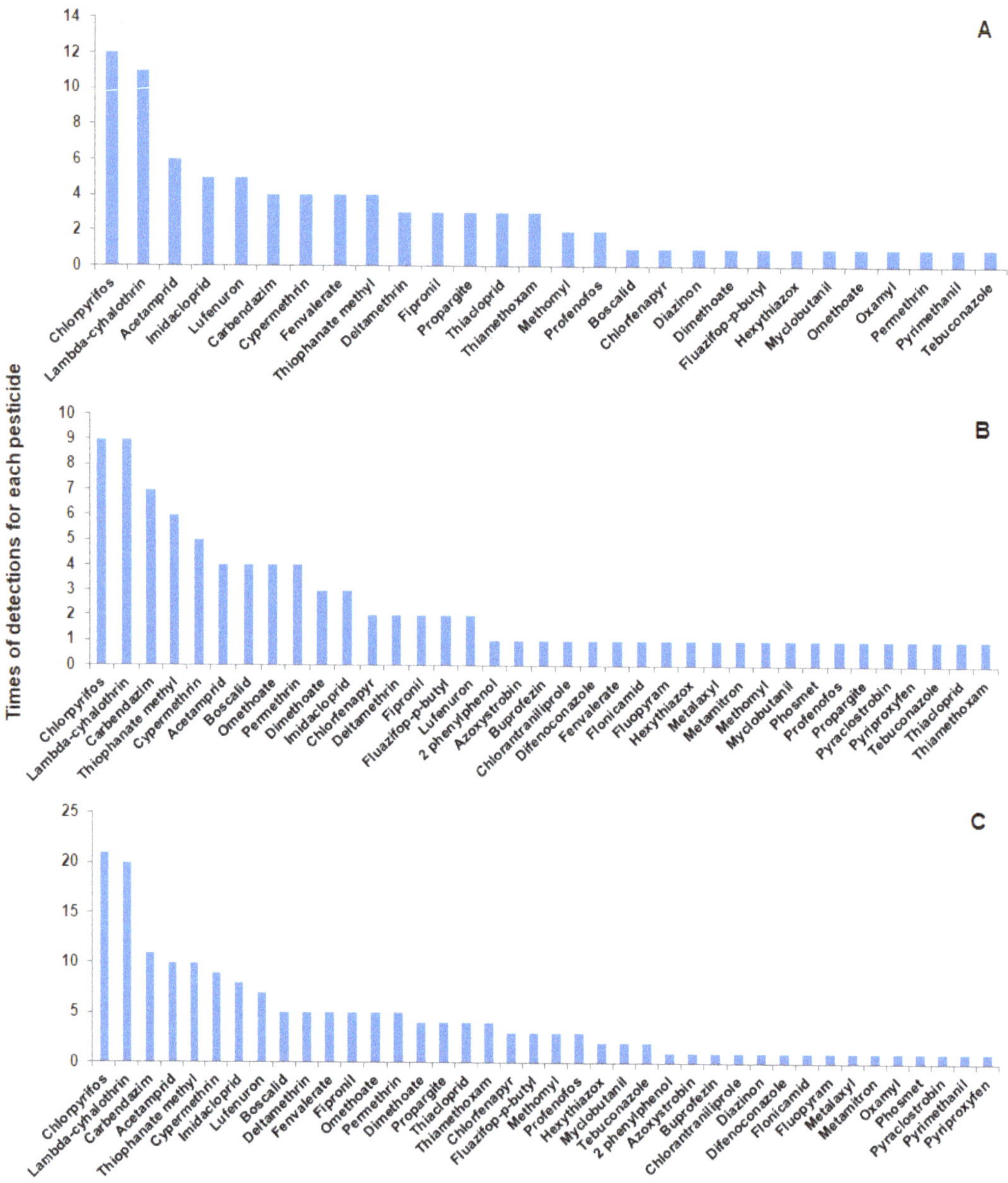

Figure 2. Pesticides detected in vegetables and fruits and times of detection for each pesticide in vegetables (**A**), fruits (**B**), and the total in both vegetables and fruits (**C**).

The total number of pesticides from each group that were detected in the tested vegetable or fruit samples collected from the markets of Sharkia Governorate in Egypt is presented in Figure 3. It was found that the insecticide group is the highest in detection in the vegetable samples with 73 insecticides (84.88%), compared to 62 insecticides (68.89%) in fruit samples. The percentage of fungicides and herbicides was recorded as 13.95 and 1.16% in the vegetable samples and 27.78 and 3.33% in the fruit samples, respectively (Figure 3).

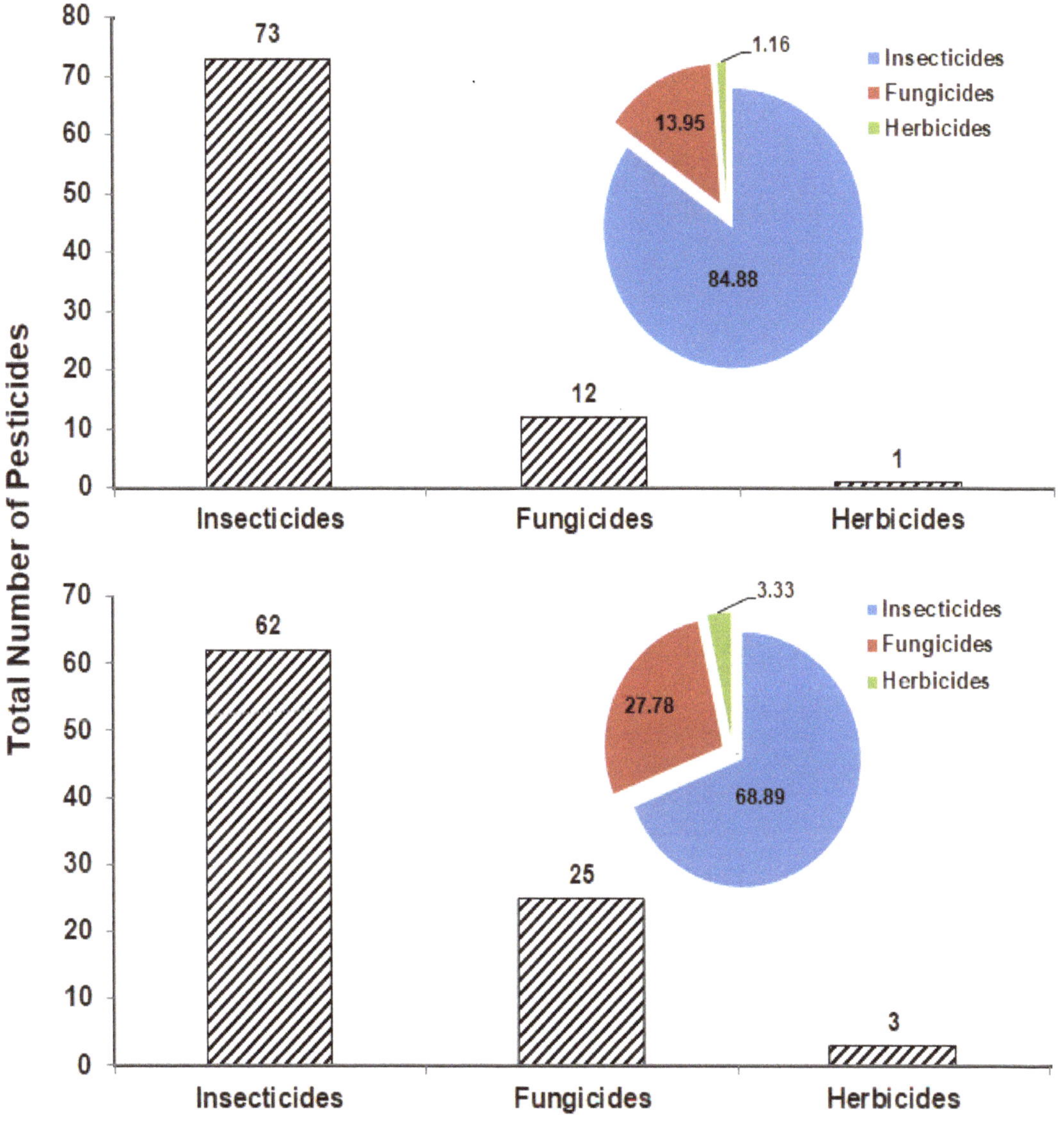

Figure 3. Total number of insecticides, fungicides, and herbicides and the frequency percentage detected in vegetable (**upper**) and fruit (**lower**) samples.

Data presented in Tables 1 and 2 show the pesticide residue levels in vegetables and fruits. The residue ranges, residue mean values, limits of detection (LODs), limits of quantification (LOQs), and registered MRLs from the European commission database for pesticide residues are shown. For residues detected in vegetable samples, 86 pesticide residues were detected in 13 types of vegetables (carrot, cabbage, cucumber, eggplant, green beans, green onion, green peas, okra, pepper, potatoes, spinach, tomato, and zucchini). Out of 86 pesticide residues, 35 (40.7%) residues exceeded MRLs. The lowest value detected was 7.33 μg kg^{-1} for chlorpyrifos in cabbage while the highest value detected was 951 μg kg^{-1} for profenofos in green onion (Table 1).

Table 1. Range and mean concentrations of pesticide residues (µg Kg^{-1}) found in vegetable sample collected from farmer markets in Sharkia Governorate, Egypt.

Samples	Pesticides	Type *	RT (Min.)	Range (µg Kg^{-1})	Mean ± SD	LOD (µg Kg^{-1})	LOQ (µg Kg^{-1})	MRL ** (µg Kg^{-1})
Carrot								
	Lufenuron	I	11.08	25–99	62.00 ± 37.00	0.5	5	10
Cabbage								
	Chlorpyrifos	I	11.02	4–10	7.33 ± 3.06	1.7	3	10
	Lambda-cyhalothrin	I	35.88	9–19	13.33 ± 5.13	0.5	10	150
	Thiacloprid	I	5.06	6–17	11.33 ± 5.51	0.5	10	300
Cucumber								
	Acetamprid	I	4.64	8–22	13.33 ± 7.57	0.3	10	300
	Carbendazim	F	4.92	8–28	18.33 ± 10.02	0.3	10	100
	Chlorfenapyr	I	10.70	22–36	28.67 ± 7.02	2.5	10	10
	Chlorpyrifos	I	11.02	36–100	59.33 ± 35.35	1.7	5	10
	Cypermethrin	I	41.95	29–45	35.00 ± 8.72	4	25	200
	Fenvalerate	I	45.07	55–80	65.67 ± 12.90	0.3	25	20
	Fipronil	I	9.25	11	11.00 ± 0.00	0.3	5	5
	Imidacloprid	I	4.93	12–32	21.67 ± 10.02	0.3	5	500
	Lambda-cyhalothrin	I	35.88	8–48	22.33 ± 22.28	0.5	10	50
	Methomyl	I	4.97	48	48.00 ± 0.00	5	10	10
	Oxamyl	I	1.51	17	17.00 ± 0.00	2.5	10	10
	Pyrimethanil	F	7.52	54	54.00 ± 0.00	1.7	5	800
	Thiacloprid	I	5.06	9–13	10.67 ± 2.08	0.5	10	500
	Thiamethoxam	I	4.78	17–40	26.67 ± 11.93	1.7	5	500
	Thiophanate methyl	F	5.44	66–100	85.00 ± 17.35	0.3	1	100
Eggplant								
	Acetamprid	I	4.64	10–14	12.00 ± 2.00	0.3	10	200
	Chlorpyrifos	I	11.02	11	11.00 ± 0.00	1.7	5	10
	Cypermethrin	I	41.95	11–21	16.00 ± 5.00	4	25	500
	Deltamethrin	I	35.01	44–74	57.33 ± 15.28	1.8	10	400
	Hexythiazox	I	11.05	13	13.00 ± 0.00	0.3	5	100
	Lambda-cyhalothrin	I	35.88	10–63	37.67 ± 26.58	0.5	10	300
	Lufenuron	I	11.08	17–34	24.33 ± 8.74	0.5	5	300
	Propargite	I	11.32	11	11.00 ± 0.00	2	5	10
Green Beans								
	Chlorpyrifos	I	11.02	8	8.00 ± 0.00	1.7	5	10
	Lambda-cyhalothrin	I	35.88	9–15	11.67 ± 3.06	0.5	10	400
	Lufenuron	I	11.08	233–453	328.33 ± 112.90	0.5	5	10
Green Onion								
	Chlorpyrifos	I	11.02	7–19	12.67 ± 6.03	1.7	5	10
	Lambda-cyhalothrin	I	35.88	12	12.00 ± 0.00	0.5	10	200
	Profenofos	I	10.46	951	951.00 ± 0.00	10	25	20
Green Peas								
	Chlorpyrifos	I	11.02	16	16.00 ± 0.00	1.7	5	10
	Lambda-cyhalothrin	I	35.88	24–101	63.00 ± 38.51	0.5	10	200
	Lufenuron	I	11.08	40–96	67.67 ± 28.01	0.5	5	10
	Thiamethoxam	I	4.78	9	9.00 ± 0.00	1.7	5	300

Table 1. *Cont.*

Samples	Pesticides	Type *	RT (Min.)	Range (μg Kg^{-1})	Mean ± SD	LOD (μg Kg^{-1})	LOQ (μg Kg^{-1})	MRL ** (μg Kg^{-1})
Okra								
	Chlorpyrifos	I	11.02	14–18	16.00 ± 2.83	1.7	5	10
	Cypermethrin	I	41.95	13–44	27.67 ± 15.57	4	25	500
	Deltamethrin	I	35.01	11	11.00 ± 0.00	1.8	10	10
	Fenvalerate	I	45.07	9–50	27.00 ± 20.95	0.3	25	20
	Fipronil	I	9.25	26	26.00 ± 0.00	0.3	5	5
	Imidacloprid	I	4.93	8–30	16.67 ± 11.72	0.3	5	500
	Lambda-cyhalothrin	I	35.88	39–99	68.33 ± 30.02	0.5	10	300
	Propargite	I	11.32	14	14.00 ± 0.00	2	5	10
Pepper								
	Acetamprid	I	4.64	99–221	151.67 ± 62.68	0.3	10	300
	Boscalid	F	8.03	50–106	79.67 ± 28.15	0.5	1	3000
	Carbendazim	F	4.92	98–330	232.67 ± 120.42	0.3	10	100
	Chlorpyrifos	I	11.02	24	24.00 ± 0.00	1.7	5	10
	Dimethoate	I	4.95	11	11.00 ± 0.00	0.3	1	10
	Fluazifop-p-butyl	H	10.74	60–90	77.33 ± 15.53	0.3	5	10
	Imidacloprid	I	4.93	9–22	16.00 ± 6.56	0.3	5	900
	Lambda-cyhalothrin	I	35.88	21–64	43.00 ± 21.52	0.5	10	300
	Methomyl	I	4.97	9–60	32.00 ± 25.87	5	10	40
	Myclobutanil	F	8.26	19	19.00 ± 0.00	1.7	5	3000
	Profenofos	I	10.46	35	35.00 ± 0.00	10	25	10
	Thiophanate methyl	F	5.44	17–60	38.33 ± 21.50	0.3	1	100
Potatoes								
	Chlorpyrifos	I	11.02	13–14	13.50 ± 0.71	1.7	5	10
	Lufenuron	I	11.08	60–120	87.00 ± 30.45	0.5	5	10
Spinach								
	Acetamprid	I	4.64	11	11.00 ± 0.00	0.3	10	600
	Chlorpyrifos	I	11.02	12	12.00 ± 0.00	1.7	5	10
	Lambda-cyhalothrin	I	35.88	340–521	434.33 ± 90.74	0.5	10	600
	Omethoate	I	7.38	12	12.00 ± 0.00	3.3	5	10
	Thiacloprid	I	5.06	9–31	17.00 ± 12.17	0.5	10	150
Tomato								
	Acetamprid	I	4.64	8–23	15.00 ± 7.55	0.3	10	500
	Carbendazim	F	4.92	16–109	62.50 ± 65.76	0.3	10	300
	Chlorpyrifos	I	11.02	52	52.00 ± 0.00	1.7	5	10
	Cypermethrin	I	41.95	26	26.00 ± 0.00	4	25	500
	Fenvalerate	I	45.07	12	12.00 ± 0.00	0.3	25	100
	Imidacloprid	I	4.93	11–30	20.50 ± 13.44	0.3	5	300
	Lambda-cyhalothrin	I	35.88	29–72	51.67 ± 21.59	0.5	10	70
	Permethrin	I	23.09	23	23.00 ± 0.00	0.3	1	50
	Propargite	I	11.32	88–219	152.00 ± 65.55	2	5	10
	Thiophanate methyl	F	5.44	120–210	163.33 ± 45.09	0.3	1	100
Zucchini								
	Acetamprid	I	4.64	7–40	20.33 ± 17.39	0.3	10	300
	Carbendazim	F	4.92	23	23.00 ± 0.00	0.3	10	100
	Chlorpyrifos	I	11.02	16–80	48.00 ± 45.25	1.7	5	10
	Deltamethrin	I	35.01	9–22	14.33 ± 6.81	1.8	10	200
	Diazinon	I	9.31	11	11.00 ± 0.00	1.7	5	10

Table 1. *Cont.*

Samples	Pesticides	Type *	RT (Min.)	Range (μg Kg^{-1})	Mean ± SD	LOD (μg Kg^{-1})	LOQ (μg Kg^{-1})	MRL ** (μg Kg^{-1})
	Fenvalerate	I	45.07	16	16.00 ± 0.00	0.3	25	20
	Fipronil	I	9.25	33	33.00 ± 0.00	0.3	5	5
	Imidacloprid	I	4.93	90–541	320.67 ± 225.68	0.3	5	400
	Lambda-cyhalothrin	I	35.88	68–174	121.00 ± 74.95	0.5	10	150
	Tebuconazole	F	8.59	9–70	37.33 ± 30.73	1.7	10	600
	Thiamethoxam	I	4.78	4–131	66.33 ± 63.53	1.7	5	500
	Thiophanate methyl	F	5.44	105	105.00 ± 0.00	0.3	1	100

* Types of pesticides detected: insecticide (I), fungicide (F), and herbicide (H). ** MRL: mean maximum residue limits obtained from European commission pesticide residue database.

Table 2. Range and mean concentrations of pesticide residues (μg Kg^{-1}) found in fruit samples collected from farmer markets in Sharkia Governorate, Egypt.

Samples	Pesticides	Type *	RT (Min.)	Range (μg Kg^{-1})	Mean ± SD	LOD (μg Kg^{-1})	LOQ (μg Kg^{-1})	MRL ** (μg Kg^{-1})
Apple								
	Acetamprid	I	4.64	9–48	31.00 ± 19.97	0.3	10	400
	Boscalid	F	8.03	11–23	16.67 ± 6.03	0.5	1	2000
	Carbendazim	F	4.92	41–125	82.00 ± 42.04	0.3	10	200
	Chlorantraniliprole	I	7.45	12–22	15.33 ± 5.77	0.3	1	400
	Chlorpyrifos	I	11.02	12–394	195.33 ± 191.46	1.7	5	10
	Cypermethrin	I	41.95	90–362	246.00 ± 140.34	4	25	1000
	Difenoconazole	F	10.11	10–13	11.33 ± 1.53	0.5	5	800
	Fipronil	I	9.25	17	17.00 ± 0.00	0.3	5	5
	Flonicamid	I	2.01	10	10.00 ± 0.00	3.3	10	300
	Fluazifop-p-butyl	H	10.74	16	16.00 ± 0.00	0.3	5	10
	Fluopyram	F	8.69	11–40	24.33 ± 14.64	0.3	5	800
	Hexythiazox	I	11.05	14	14.00 ± 0.00	0.3	5	400
	Imidacloprid	I	4.93	12–61	35.00 ± 24.64	0.3	5	10
	Lambda-cyhalothrin	I	35.88	23–25	24.00 ± 1.41	0.5	10	80
	Metalaxyl	F	7.01	9–20	13.00 ± 6.08	1.7	5	1000
	Permethrin	I	23.09	17	17.00 ± 0.00	0.3	1	50
	Phosmet	I	19.41	20–93	55.67 ± 36.53	17	50	500
	Propargite	I	11.32	16–20	18.00 ± 2.00	2	5	10
	Tebuconazole	F	8.59	31–99	66.33 ± 34.08	1.7	10	300
	Thiophanate methyl	F	5.44	20–44	31.33 ± 12.06	0.3	1	500
Apricot								
	Acetamprid	I	4.64	30–120	79.67 ± 45.72	0.3	10	800
	Azoxystrobin	F	7.95	13–76	43.67 ± 31.53	1.3	4	2000
	Boscalid	F	8.03	10	10.00 ± 0.00	0.5	1	5000
	Buprofezin	I	9.19	90–199	147.00 ± 54.67	1	5	10
	Carbendazim	F	4.92	131–340	231.67 ± 104.71	0.3	10	200
	Chlorpyrifos	I	11.02	86–311	183.00 ± 115.66	1.7	5	10
	Cypermethrin	I	41.95	109–261	177.67 ± 77.05	4	25	2000
	Deltamethrin	I	35.01	9–45	26.67 ± 18.01	1.8	10	150
	Dimethoate	I	4.95	14	14.00 ± 0.00	0.3	1	10
	Fenvalerate	I	45.07	23–29	26.00 ± 3.00	0.3	25	200
	Imidacloprid	I	4.93	11–41	25.67 ± 15.01	0.3	5	10
	Lambda-cyhalothrin	I	35.88	22–101	64.33 ± 39.80	0.5	10	150
	Metamitron	H	4.29	9–219	108.67 ± 105.41	3.3	10	10
	Profenofos	I	10.46	86–230	155.33 ± 72.15	10	25	10

Table 2. *Cont.*

Samples	Pesticides	Type *	RT (Min.)	Range ($\mu g\ Kg^{-1}$)	Mean ± SD	LOD ($\mu g\ Kg^{-1}$)	LOQ ($\mu g\ Kg^{-1}$)	MRL ** ($\mu g\ Kg^{-1}$)
Banana								
	Thiamethoxam	I	4.78	10	10.00 ± 0.00	1.7	5	20
Cantaloupe								
	Acetamprid	I	4.64	7–111	44.33 ± 57.87	0.3	10	200
	Cypermethrin	I	41.95	18–29	22.33 ± 5.86	4	25	200
	Lambda-cyhalothrin	I	35.88	9–31	19.00 ± 11.14	0.5	10	60
	Lufenuron	I	11.08	11	11.00 ± 0.00	0.5	5	400
	Permethrin	I	23.09	13–22	17.00 ± 4.58	0.3	1	50
Dates								
	Carbendazim	F	4.92	5–71	33.67 ± 33.84	0.3	10	100
	Chlorpyrifos	I	11.02	14	14.00 ± 0.00	1.7	5	10
	Omethoate	I	7.38	53	53.00 ± 0.00	3.3	5	10
	Thiophanate methyl	F	5.44	41–60	50.00 ± 9.54	0.3	1	100
Grapes								
	Acetamprid	I	4.64	9–28	16.07 ± 10.39	0.3	10	500
	Boscalid	F	8.03	13–60	36.67 ± 23.50	0.5	1	5000
	Carbendazim	F	4.92	40–569	277.00 ± 268.75	0.3	10	300
	Chlorfenapyr	I	10.70	30–172	99.67 ± 71.04	2.5	10	10
	Chlorpyrifos	I	11.02	18–24	20.67 ± 3.06	1.7	5	10
	Cypermethrin	I	41.95	10–50	30.00 ± 20.00	4	25	500
	Deltamethrin	I	35.01	14	14.00 ± 0.00	1.8	10	200
	Dimethoate	I	4.95	32–108	73.33 ± 38.44	0.3	1	10
	Imidacloprid	I	4.93	60–123	94.00 ± 31.80	0.3	5	700
	Lambda-cyhalothrin	I	35.88	10	10.00 ± 0.00	0.5	10	80
	Myclobutanil	F	8.26	20	20.00 ± 0.00	1.7	5	1500
	Omethoate	I	7.38	10–44	28.00 ± 17.09	3.3	5	10
	Permethrin	I	23.09	6–21	14.00 ± 7.55	0.3	1	50
	Pyraclostrobin	F	9.70	19	19.00 ± 0.00	0.3	5	300
	Pyriproxyfen	I	10.05	10	10.00 ± 0.00	1.7	5	50
	Thiacloprid	I	5.06	90–300	167.33 ± 115.42	0.5	10	10
	Thiamethoxam	I	4.78	10–96	53.67 ± 43.02	1.7	5	400
	Thiophanate methyl	F	5.44	86–635	368.00 ± 274.81	0.3	1	100
Guava								
	Carbendazim	F	4.92	14–61	36.33 ± 23.59	0.3	10	100
	Chlorpyrifos	I	11.02	18	18.00 ± 0.00	1.7	5	10
	Fipronil	I	9.25	17	17.00 ± 0.00	0.3	5	5
	Lambda-cyhalothrin	I	35.88	11–24	17.33 ± 6.51	0.5	10	10
	Methomyl	I	4.97	124	124.00 ± 0.00	5	10	10
	Thiophanate methyl	F	5.44	18	18.00 ± 0.00	0.3	1	100
Kaki								
	Chlorpyrifos	I	11.02	123	123.00 ± 0.00	1.7	5	10
	Dimethoate	I	4.95	520–990	775.00 ± 237.54	0.3	1	10
	Fluazifop-p-butyl	H	10.74	26	26.00 ± 0.00	0.3	5	10
	Lambda-cyhalothrin	I	35.88	44–126	86.33 ± 41.06	0.5	10	90
	Thiophanate methyl	F	5.44	8–25	16.00 ± 8.54	0.3	1	100

Table 2. *Cont.*

Samples	Pesticides	Type *	RT (Min.)	Range (µg Kg^{-1})	Mean ± SD	LOD (µg Kg^{-1})	LOQ (µg Kg^{-1})	MRL ** (µg Kg^{-1})
Mango								
	2 phenylphenol	F	5.21	40–59	46.67 ± 10.69	11	20	10
	Carbendazim	F	4.92	17–25	21.00 ± 5.66	0.3	10	500
	Chlorfenapyr	I	10.70	21	21.00 ± 0.00	2.5	10	10
	Chlorpyrifos	I	11.02	12	12.00 ± 0.00	1.7	5	10
	Cypermethrin	I	41.95	25–61	39.00 ± 19.29	4	25	700
	Lambda-cyhalothrin	I	35.88	10–70	39.33 ± 30.02	0.5	10	200
	Permethrin	I	23.09	18	18.00 ± 0.00	0.3	1	50
Orange								
	Chlorpyrifos	I	11.02	11	11.00 ± 0.00	1.7	5	10
	Lambda-cyhalothrin	I	35.88	8	8.00 ± 0.00	0.5	10	200
	Lufenuron	I	11.08	90–216	157.33 ± 63.45	0.5	5	300
	Omethoate	I	7.38	9	9.00 ± 0.00	3.3	5	10
Strawberry								
	Boscalid	F	8.03	10	10.00 ± 0.00	0.5	1	6000
	Carbendazim	F	4.92	65–269	150.33 ± 106.01	0.3	10	100
	Chlorpyrifos	I	11.02	111–301	193.33 ± 97.50	1.7	5	10
	Lambda-cyhalothrin	I	35.88	46–66	54.33 ± 10.41	0.5	10	200
	Omethoate	I	7.38	132–992	575.00 ± 430.59	3.3	5	10
	Thiophanate methyl	F	5.44	66–310	191.67 ± 122.17	0.3	1	100

* Types of pesticides detected: insecticide (I), fungicide (F), and herbicide (H). ** MRL: mean maximum residue limits obtained from European commission pesticide residue database.

For residues detected in 11 types of fruits, a total of 90 pesticide residues were recorded with 38.9% exceeding MRLs. The lowest and highest residue levels recorded were 8 and 775 µg kg^{-1} for lambda-cyhalothrin (in orange) and dimethoate (in kaki), respectively (Table 2).

2.2. Risk Analysis of Pesticide Residues

The results presented in Table 3 show the assessment of the acute and chronic risks of pesticide residues detected in vegetables or fruits that exceed the permissible MRLs, using two rates of consumption (0.1 kg day^{-1} for chronic risk and 0.2 kg day^{-1} for acute risk) for all the tested samples. Acute and chronic risks were determined for children, teenagers, and adults. The results showed existing acute risk with fipronil, lambad-cyhalothrin, dimethoate, and omethoate in the case of children consuming okra, zucchini, apples, guava (containing fipronil), spinach (containing lambada-cyhalothrin), kaki (containing dimethoate), and strawberries (containing omethoate). It was also found that acute risks appear in teenagers consuming spinach (containing lambada-cyhalothrin), kaki (containing dimethoate), and strawberries (containing omethoate), while the presence of acute risks appears in adults consuming kaki and strawberries contaminated with both dimethoate and omethoate, respectively. Regarding chronic risks, they appear in children consuming zucchini, spinach, kaki, and strawberries containing residues of fipronil, lambada-cyhalothrin, dimethoate, and omethoate, respectively, while chronic risks appear for teenagers when consuming kaki contaminated with dimethoate.

Table 3. Acute (% ARfD) and chronic (% ADI) risk assessment of pesticide residues in vegetable and fruit samples exceeding maximum residue levels using two consumption rates (0.1 kg for chronic risk and 0.2 kg for acute risk) in different population groups.

Pesticides/Samples	ARfD	Acute Dietary Exposure (%ARfD)			ADI	Chronic Dietary Exposure (%ADI)		
		Child [A]	Teenager [B]	Adult [C]		Child [A]	Teenager [B]	Adult [C]
Vegetables:								
Carrot								
Lufenuron	0.0150	8.80	3.77	2.20	0.015	2.76	1.18	0.69
Cabbage								
Chlorpyrifos	0.0100	1.33	0.57	0.33	0.01	0.49	0.21	0.12
Cucumber								
Chlorfenapyr	0.0150	3.20	1.37	0.80	0.015	1.27	0.55	0.32
Chlorpyrifos	0.0100	13.33	5.71	3.33	0.01	3.96	1.70	0.99
Fenvalerate	0.0125	8.53	3.66	2.13	0.0125	3.50	1.50	0.88
Fipronil	0.0090	1.63	0.70	0.41	0.0002	36.67	15.71	9.17
Imidacloprid	0.0800	0.53	0.23	0.13	0.06	0.24	0.10	0.06
Methomyl	0.0025	25.60	10.97	6.40	0.0025	12.80	5.49	3.20
Oxamyl	0.0010	22.67	9.71	5.67	0.001	11.33	4.86	2.83
Thiophanate methyl	0.2000	0.67	0.29	0.17	0.08	0.71	0.30	0.18
Eggplant								
Chlorpyrifos	0.0100	1.47	0.63	0.37	0.01	0.73	0.31	0.18
Propargite	0.0600	0.24	0.10	0.06	0.03	0.24	0.10	0.06
Green Beans								
Lufenuron	0.0150	40.27	17.26	10.07	0.015	14.59	6.25	3.65
Green Onion								
Chlorpyrifos	0.0100	2.53	1.09	0.63	0.01	0.84	0.36	0.21
Profenofos	1.0000	1.27	0.54	0.32	0.03	21.13	9.06	5.28
Green Peas								
Chlorpyrifos	0.0100	2.13	0.91	0.53	0.01	1.07	0.46	0.27
Lufenuron	0.0150	8.53	3.66	2.13	0.015	3.01	1.29	0.75
Okra								
Chlorpyrifos	0.0100	1.87	0.80	0.47	0.01	1.07	0.46	0.27
Deltamethrin	0.0100	1.47	0.63	0.37	0.01	0.73	0.31	0.18
Fenvalerate	0.0125	5.33	2.29	1.33	0.0125	1.44	0.62	0.36
Fipronil	0.0090	3.85	1.65	0.96	0.0002	86.67	37.14	21.67
Propargite	0.0600	0.31	0.13	0.08	0.03	0.31	0.13	0.08
Pepper								
Carbendazim	0.0200	22.00	9.43	5.50	0.02	7.76	3.32	1.94
Chlorpyrifos	0.0100	3.20	1.37	0.80	0.01	1.60	0.69	0.40
Dimethoate	0.0020	7.33	3.14	1.83	0.002	3.67	1.57	0.92
Fluazifop-p-butyl	0.0170	7.06	3.03	1.76	0.01	5.16	2.21	1.29
Methomyl	0.0025	32.00	13.71	8.00	0.0025	8.53	3.66	2.13
Profenofos	1.0000	0.05	0.02	0.01	0.03	0.78	0.33	0.19
Potatoes								
Chlorpyrifos	0.0100	1.73	0.74	0.43	0.01	0.90	0.39	0.23
Lufenuron	0.0150	10.67	4.57	2.67	0.015	3.87	1.66	0.97
Spinach								
Chlorpyrifos	0.0100	1.60	0.69	0.40	0.01	0.80	0.34	0.20
Lambda-cyhalothrin	0.0050	138.93	59.54	34.73	0.0025	115.82	49.64	28.96
Omethoate	0.0020	8.00	3.43	2.00	0.002	4.00	1.71	1.00

Table 3. *Cont.*

Pesticides/Samples	ARfD	Acute Dietary Exposure (%ARfD)			ADI	Chronic Dietary Exposure (%ADI)		
		Child A	Teenager B	Adult C		Child A	Teenager B	Adult C
Tomato								
Chlorpyrifos	0.0100	6.93	2.97	1.73	0.01	3.47	1.49	0.87
Lambda-cyhalothrin	0.0050	19.20	8.23	4.80	0.0025	13.78	5.90	3.44
Propargite	0.0600	4.87	2.09	1.22	0.03	3.38	1.45	0.84
Thiophanate methyl	0.2000	1.40	0.60	0.35	0.08	1.36	0.58	0.34
Zucchini								
Chlorpyrifos	0.0100	2.13	0.91	0.53	0.01	3.20	1.37	0.80
Diazenon	0.0250	0.59	0.25	0.15	0.0002	36.67	15.71	9.17
Fipronil	0.0090	4.89	2.10	1.22	0.0002	110.00	47.14	27.50
Imidacloprid	0.0800	9.02	3.86	2.25	0.06	3.56	1.53	0.89
Lambda-cyhalothrin	0.0050	18.13	7.77	4.53	0.0025	32.27	13.83	8.07
Thiophanate methyl	0.2000	0.70	0.30	0.18	0.08	0.88	0.38	0.22
Fruites:								
Apple								
Chlorpyrifos	0.0100	52.53	22.51	13.13	0.01	13.02	5.58	3.26
Fipronil	0.0090	2.52	1.08	0.63	0.0002	56.67	24.29	14.17
Fluazifop-p-butyl	0.0170	1.25	0.54	0.31	0.01	1.07	0.46	0.27
Imidacloprid	0.0800	1.02	0.44	0.25	0.06	0.39	0.17	0.10
Propargite	0.0600	0.44	0.19	0.11	0.03	0.40	0.17	0.10
Apricot								
Buprofezin	0.5000	0.53	0.23	0.13	0.01	9.80	4.20	2.45
Carbendazim	0.0200	22.67	9.71	5.67	0.02	7.72	3.31	1.93
Chlorpyrifos	0.0100	41.47	17.77	10.37	0.01	12.20	5.23	3.05
Dimethoate	0.0020	9.33	4.00	2.33	0.002	4.67	2.00	1.17
Imidacloprid	0.0800	0.68	0.29	0.17	0.06	0.29	0.12	0.07
Metamitron	0.1000	2.92	1.25	0.73	0.03	2.41	1.03	0.60
Profenofos	1.0000	0.31	0.13	0.08	0.03	3.45	1.48	0.86
Cantaloupe								
Dates								
Chlorpyrifos	0.0100	1.87	0.80	0.47	0.01	0.93	0.40	0.23
Omethoate	0.0020	35.33	15.14	8.83	0.002	17.67	7.57	4.42
Grapes								
Carbendazim	0.0200	37.93	16.26	9.48	0.02	9.23	3.96	2.31
Chlorfenapyr	0.0150	15.29	6.55	3.82	0.015	4.43	1.90	1.11
Chlorpyrifos	0.0100	3.20	1.37	0.80	0.01	1.38	0.59	0.34
Dimethoate	0.0020	72.00	30.86	18.00	0.002	24.44	10.48	6.11
Omethoate	0.0020	29.33	12.57	7.33	0.002	9.33	4.00	2.33
Thiacloprid	0.0200	20.00	8.57	5.00	0.01	11.16	4.78	2.79
Thiophanate methyl	0.2000	4.23	1.81	1.06	0.08	3.07	1.31	0.77
Guava								
Chlorpyrifos	0.0100	2.40	1.03	0.60	0.01	1.20	0.51	0.30
Fipronil	0.0090	2.52	1.08	0.63	0.0002	56.67	24.29	14.17
Lambda-cyhalothrin	0.0050	6.40	2.74	1.60	0.0025	4.62	1.98	1.16
Methomyl	0.0025	66.13	28.34	16.53	0.0025	33.07	14.17	8.27
Kaki								
Chlorpyrifos	0.0100	16.40	7.03	4.10	0.01	8.20	3.51	2.05
Dimethoate	0.0020	660.00	282.86	165.00	0.002	258.33	110.71	64.58
Fluazifop-p-butyl	0.0170	2.04	0.87	0.51	0.01	1.73	0.74	0.43
Lambda-cyhalothrin	0.0050	33.60	14.40	8.40	0.0025	23.02	9.87	5.76

Table 3. *Cont.*

Pesticides/Samples	ARfD	Acute Dietary Exposure (%ARfD)			ADI	Chronic Dietary Exposure (%ADI)		
		Child A	Teenager B	Adult C		Child A	Teenager B	Adult C
Mango								
2 phenylphenol	0.4000	0.20	0.08	0.05	0.4	0.08	0.03	0.02
Chlorfenapyr	0.0150	1.87	0.80	0.47	0.015	0.93	0.40	0.23
Chlorpyrifos	0.0100	1.60	0.69	0.40	0.01	0.80	0.34	0.20
Orange								
Chlorpyrifos	0.0100	1.47	0.63	0.37	0.01	0.73	0.31	0.18
Strawberry								
Carbendazim	0.0200	17.93	7.69	4.48	0.02	5.01	2.15	1.25
Chlorpyrifos	0.0100	40.13	17.20	10.03	0.01	12.89	5.52	3.22
Omethoate	0.0020	661.33	283.43	165.33	0.002	191.67	82.14	47.92
Thiophanate methyl	0.2000	2.07	0.89	0.52	0.08	1.60	0.68	0.40

Acute risk (% ARfD) and Chrinic risk (% ADI) were calculated with data of ARfD and ADI from European Commision pesticide residue database. Values of ADI were used when ARfD values were missing with pesticides lufenuron, chlorpyrifos, fenvalerate, dimethoate, omethoate, and 2-phenylphenol. The weight of different population groups used is (A) children (15 kg), (B) teenagers (35 kg), and (C) adults (60 kg).

3. Discussion

Taking into consideration that pesticides play a major role in increasing the production of agricultural products with high quality when moderately and safely applied in the control of crop pests, diseases, and weeds [28–30], their misuse may cause severe health problems. Pesticide residues' determination in food is an important action for monitoring contamination and ensuring food safety. This might help farmers and stakeholders in the proper handling of pesticides in terms of the applied dose, times of application, as well as the permissible level locally in each type of food for the health and safety of consumers. Our results showed that pesticide-free samples were 36% for both vegetables and fruits, while 64% of samples contained from one to 20 pesticide residues. Cucumber, pepper, zucchini, and tomato showed 15, 12, 12, and 10 pesticide residues, respectively. In fruit samples, apple, grapes, and apricot recorded 20, 18, and 14 pesticide residues, respectively (Figure 1). In agreement with our results, pesticide residue analyses in apples carried out by Pirsahib et al. [31] reported 26% of free-pesticide samples, 74% contained at least one pesticide, and 54%, 46%, and 26% of the samples had diazinon, chlorpyrifos, and both diazinon and chlorpyrifos residues, respectively.

Fruits and vegetables with multiple pesticide residues are widely observed globally, including 26% from Italy [32], 25% from China [33], 48% from Brazil [34], and 39% from Argentina [35]; the fruit and vegetable monitoring surveys found that carbendazim, pyrimethanil, imidacloprid, and procymidone had high detection frequency and showed wide use in fruits and vegetables in Colombia [36]. A survey in Poland and China found that strawberries had the highest frequency of multiple pesticide residues [37,38].

In the current study, insecticides were highly prevalent in vegetables and fruits (Figure 3). Some insecticides appeared one time and others were detected several times. Chlorantraniliprole is one of the insecticides detected one time only in apple. This insecticide is one of the diamide insecticides that are widely used against a variety of insect pests due to their selectivity and low mammalian toxicity [39–41]. Tian et al. [42] determined diamide insecticides in mushrooms and found that these insecticides can be effectively analyzed using HPLC-MS/MS with LOD and LOQ of 0.05 and 5 ug kg^{-1}, respectively, and recovery rates ranging from 73.5–110.2%. On the other hand, chlorpyrifos is an insecticide that was detected several times in the tested vegetable (12 times) and fruit (9 times) samples. Although this insecticide is recommended in Egypt against almond worms in cotton and termites in buildings according to the approved recommendations for agricultural pest control (Deposit No.: 13449/2022), it was detected in vegetables and fruits collected from farmers' markets (Tables 1 and 2). This insecticide is no longer approved by European

Commission [43] due to harmful effects on different organs [44]. In spite of that, it is still detected in a high percentage in many samples of fruits and vegetables [11,45], which is consistent with the results obtained in this study.

Fungicides were detected in fruits in a higher percentage than in vegetables (Figure 3). The fungicide pyraclostrobin was detected in grapes only (Table 2). The dissipation rate of this fungicide was studied in strawberry in Egypt when treated with the recommended field rate [46]. It was found that 82% of this fungicide degraded within 14 days of treatment with a half-life ($t1/2$) of 5 days. In contrast, the fungicide thiophanate-methyl was detected in four vegetable samples and in six fruit samples as recorded in Tables 1 and 2. As this fungicide is widely used in the control of a variety of pathogens pre- and post-harvest, it was detected in many vegetable and fruit samples [47–51], herbal medicine [52], raisins [45], salmon [53], beebread [54], and also in cow and human milk [55].

Samples of cucumber and apples were found to have ≥15 pesticide residues (Figure 1). Chlorpyrifos and lamda-cyhalothrin were detected in more than 15 samples (Figure 2) with some values higher than MRLs. In our study, pesticide residues exceeding MRLs in vegetables and fruits were 41 and 39%, respectively. Other studies showed the same results, i.e., in Mwanja et al. [56], pesticide residues were detected in 63.3% of the tested vegetable and fruit samples with residue levels exceeding MRLs of the codex Alimentarius in cabbage, tomato, and orange samples. Further, in the study of Hamed et al. [57], residues of pesticides in apples and grapes from Egypt were determined and they reported that 12.7 and 16.4% of pesticide residues exceeded the MRLs, which was slightly lower than what we found in the current study (25 and 33% exceeding MRLs for apple and grapes, respectively). Consistent with our findings, a study conducted by Parveen et al. [58] in Pakistan reported that pesticide residues in apple and grape samples exceeded MRLs with 28 and 20%, respectively.

Estimation of pesticide residues in imported food is necessary to know about food safety. A study in the United Kingdom for monitoring levels of pesticide residues in imported foods from different countries showed that 51.3% of Egypt samples, compared to 77% (Chile), 68.3% (Brazil), 55.1% (India), 46.1% (United States), and 45.7% (Kenya) [59] contained detectable pesticide residues. They recorded that India, Kenya, Brazil, Egypt, Chile, and the United States were countries with residue levels exceeding MRLs in 18.1%, 11.4%, 7.8%, 5.1%, 3.2%, and 2%, respectively. In the same context, Osaili et al. [18] determined pesticide residues in samples of imported vegetables to the United Arab Emirates. They found that 30.5% pesticide residues higher than MRLs in total imported samples and found 14% of the Egyptian samples compared to 47%, 33%, 13%, and 43% from India, United Kingdom, China, and Sri Lanka, respectively, contained residues higher than MRLs.

The results of monitoring pesticide residues in fruits and vegetables showed that some samples had residues that exceed the MRL standard, which may lead to risks when consuming food contaminated with these pesticides. In addition, some pesticides do not have corresponding residual limits, which make it difficult for farmers to safely use these pesticides and for the government to monitor their use. Therefore, identification of acute and chronic dietary risks is necessary to assess the risks associated with consuming vegetables or fruits that contain pesticide residues above the MRLs. In this regard, Chu et al. [3] evaluated the risks of food exposure to 26 insecticides on strawberries and found that despite the presence of high detection rates for these residues, they showed risks of acute and chronic exposure at a level of less than 100%.

In our results of risk assessment, residues of lambda-cyhalothrin, fipronil, dimothoate, and omethoate were found to have acute or chronic risks in consumers in the case of consuming 100 or 200 gm day^{-1} of spinach, zucchini, kaki, and strawberry, respectively (Table 3). In line with our findings, the results of Tao et al., 2021 showed that the fungicide carbendazim had a risk quotient value of 2.9 in wheat flour samples, indicating an unacceptable dietary risk. Furthermore, Tankiewicz and Berg [60] showed that pesticides of lambda-cyhalothrin in courgettes, captan in apples and cucumbers, dimethoate in courgettes, and linuron in carrots exceeded the MRLs and pose a health risk. In an Indian study conducted by Sinha et al. [61], they stated that excessive application of pesticides on grapes cause adverse health effects in developing countries as grapes and apples are contaminated with different classes of pesticides including organophosphate, which cause high health risks for consumers. The acute or chronic risk is dose-dependent and causes toxicity to humans through different mechanisms [21]. In this context, Javeres et al. [62] showed that the prolonged exposure to insecticides could lead to physiological disorders including high blood pressure, hyperglycemia, overweight or dyslipidemia, which may cause metabolic syndrome and other chronic diseases. For these adverse effects, it is important in each country to monitor pesticide residues in food for food safety and human health.

4. Materials and Methods

4.1. Sample Collection and Preparation

Samples of vegetables and fruits were collected from three different farmers' markets in Sharkia Governorate, Egypt. The weight of each sample, whether vegetable or fruit, was 3 kg purchased from 3 different sellers at the same farmer market (1 kg each). Samples were collected during the period from July 2020 to June 2021. Immediately after purchasing, the samples were transported to the laboratory, cut into pieces, packaged separately in marked plastic bags, and stored at −20 °C. On the next day, the samples were prepared for the extraction process [63] by mixing each sample (3 kg) separately in a laboratory blender (Warring laboratory blinder, model 8010S, USA) for two minutes. Ten g of the homogenized product of each sample was weighed in a 50 mL conical tube and then 10 mL of acetonitrile was added to each tube for the first extraction step and vortexed for 1 min. For the second extraction step, 4 g of magnesium sulphate ($MgSO_4$), 1 g of sodium chloride (NaCl), 1 g of trisodium citrate dehydrate, and 0.5 g of disodium hydrogen citrate sesquihydrate were added to each tube, vortexed for 1 min, and centrifuged at 4000 rpm for 10 min.

Four ml of the resulting supernatant was decanted into a 15 mL conical tube containing 300 mg $MgSO_4$, 50 mg primary secondary amine (PSA) for clean-up by dispersive solid phase extraction (dSPE). For samples with high content of chlorophyll and carotinoids, 5 mg graphitized carbon black (GCB) was used for dSPE, vortexed for 30 s, and centrifuged at 4000 rpm for 4 min. The supernatant was transferred into clean tubes following the clean-up process and acidified by adding a small amount of formic acid to improve the storage stability of certain base-sensitive pesticides, then employed for LC- and GC-MS/MS analysis.

4.2. LC-MS/MS Analysis

LC-MS/MS analysis of pesticide residues was determined using an Exion HPLC system (SCIEX) with a 6500+ QTRAP triple quadrupole mass spectrometer (AB SCIEX) and an electrospray ionization (ESI) source, operated in positive multiple reaction monitoring (MRM) mode for quantification. Chromatography was performed in a Zorbax XDB C18 column (Agilent Technologies, Santa Clara, CA, USA) with length, inner diameter, and particle size of 150 mm, 4.5 mm, and 5 μm, respectively. The column temperature was kept constant at 40 °C throughout the analysis and a consistent flow rate of 400 μL min^{-1} and injection volume of 5 μL. The mobile phase (A) consists of ammonium format (10 mM) solution at pH 4.0 in water (90/10; v/v) and methanol for phase (B). The gradient elution program of the mobile phase was as follows: 0 min, 100% phase (A); 13.0 min, 5% phase (A); 21.0 min, 5% phase (A); 28.0 min, 100% phase (A); 32.0 min, 100% phase (A). Data acquisition and processing for analyte confirmation and quantitative analysis were carried out using the analyst software (Version 1.8.1, Applied Biosystems). All studied analytes were detected in the positive ionization mode using MRM with MS/MS acquisition mode. Main ion source parameters were as follows: ion spray voltage, ion source temperature, and curtain gas were set as 5500 v, 400 °C, and 20 psi, respectively. Collision gas medium, nebulizer gas, and auxiliary gas were all set at 45 psi. Data were acquired in the positive ionization mode over the m/z range from 50 to 1100, with ESI using the following parameters derived from the flow rate used: capillary voltage, 4000 V; fragmentor voltage, 190 V; drying gas, 9 L/min; drying gas temperature, 325 °C.

4.3. GC-MS/MS Analysis

Pesticide residue analyses were performed using the Gas Chromatography (GC, 7890A; Agilent, Palo Alto, CA, USA) coupled with a triple-quadrupole tandem mass spectrometer (MS/MS) (7010B). Chromatographic separation was performed on an Agilent J & W HP-5MS capillary column (30 m × 0.25 mm × 0.25 µm; Agilent Technologies, USA) with helium as a carrier gas and electron impact (EI) ionization source. Sample volumes of 1.0 µL were injected in split/split less injection mode and a silica liner with a diameter of 2 mm was used. High purity helium (99.99%) was the carrier gas with a constant flow rate of 1 mL min^{-1}. High purity nitrogen was used as the collision cell gas with a flow rate of 1.5 mL min^{-1}, and the quench gas was helium at 4 mL min^{-1}. The temperature program of the oven was as follows: the initial temperature was set to 40 °C, which was held for 2 min before being increased to 220 °C at 30 °C min^{-1}. The oven temperature was then increased to 260 °C at 5 °C min^{-1} and then finally increased to 280 °C at 20 °C min^{-1} and held for 15 min. Other operating conditions: the split/splitless injector was set at a fixed temperature of 250 °C. The interface was set at 270 °C, manifold and trap temperatures were 50 and 210 °C, respectively, while MS1 and MS2 quadrupoles' temperature was set at 150 °C. The ion energy for electron impact was kept at 70 eV. For quantitative and qualitative analysis of the compounds, MRM transitions mode was used based on the most intensive precursor ion-product.

4.4. Dietary Risk Assessment

Risk assessment for the acute and chronic exposures was carried out to assess the exposure of the population to fruit and vegetable samples containing pesticide residues exceeding MRLs.

The acute reference dose percentage (% ARfD) was used to calculate the risk posed by the acute dietary intake [3]. If the calculated % ARfD is <100%, this indicates acceptable risk, while a value ≥ 100% indicates unacceptable risk and accordingly the lower risk is associated with the smaller % ARfD values. % ARfD was calculated through Formulas (1) and (2) as follows:

$$\mathrm{ESTI} = \frac{\mathrm{HPF} \times \mathrm{HRC}}{\mathrm{bw}} \quad (1)$$

$$\%\ \mathrm{ARfD} = \frac{\mathrm{ESTI}}{\mathrm{ARfD}} \times 100 \quad (2)$$

where ESTI is the estimated short-term intake (mg kg^{-1} day), HPF is the highest portion of food consumption in a day (kg), and HRC is the highest residual concentration detected for a pesticide (mg kg^{-1}).

The acceptable daily intake percentage (% ADI) was used to calculate the risk associated with chronic dietary intake [3] of each pesticide with residue exceeding MRL. The following Equations (3) and (4) were used in calculation:

$$\mathrm{NEDI} = \frac{\mathrm{APR} \times \mathrm{DFC}}{\mathrm{bw}} \quad (3)$$

$$\%\ \mathrm{ADI} = \frac{\mathrm{NEDI}}{\mathrm{ADI}} \times 100 \quad (4)$$

where NEDI is national estimated daily intake (mg kg^{-1} day), APR is average pesticide residue (mg kg^{-1}), DFC is the daily food consumption (kg), and ADI is the acceptable daily intake (mg kg^{-1} day). When the % ADI < 100%, it means the risk is acceptable; when it is ≥100%, the risk is unacceptable. Therefore, the risk is low whenever the value of the % ADI is low [64].

The toxicological values of ADI and ARfD were obtained from the European Pesticide Database of the European Commission [43]. The % ARfD and % ADI were calculated for children (bw: 15 kg), teenagers (bw: 35 kg), and adults (bw: 60 kg). The average food consumption and the highest portion of food consumption were used as 0.1 and 0.2 kg, respectively, for all samples of vegetables and fruits.

5. Conclusions

The presence of pesticide residues in food that exceed the permissible MRLs leads to significant environmental and health damages. To preserve the health of consumers, it is necessary to monitor pesticide residues in food on an ongoing basis to determine the dynamics of pesticide presence in food, especially vegetables and fruits that are freshly consumed. In the analyzed samples from the market, more than of 50% were found to contain pesticide residues, the highest of which were insecticides, followed by fungicides, while herbicides were the least detected. About 40% of the detected pesticide residues were higher than MRLs in vegetables and fruits, and 2.3% out of them may cause acute or chronic risks when eating contaminated vegetables or fruits in quantities equal to 0.1 or 0.2 kg per day.

Author Contributions: Design and conceptualization, E.-S.A.E.-S. and M.-B.A.; methodology, analysis and investigation, E.-S.A.E.-S., M.M.R., A.E.E.-S., A.A.S. and I.A.H.; data curation and supervision, E.-S.A.E.-S. and M.-B.A.; writing, reviewing, and editing the original draft, E.-S.A.E.-S., M.R.M., M.-B.A. and B.D.H.; project administration, E.-S.A.E.-S. All authors have read and agreed to the published version of the manuscript.

Funding: This study is based upon a project supported by the Science, Technology and Innovation Funding Authority (STDF), Ministry of Higher Education and Scientific Research, Egypt under grant USC-18 ID: 983.

Institutional Review Board Statement: Not applicable.

Informed Consent Statement: Not applicable.

Data Availability Statement: The data presented in this work are available on request from the corresponding author.

Conflicts of Interest: The authors declare no conflict of interest.

References

1. Kolani, L.; Mawussi, G.; Sanda, K. Assessment of organochlorine pesticide residues in vegetable samples from some agricultural areas in Togo. *Am. J. Anal. Chem.* **2016**, *7*, 332–341. [CrossRef]
2. Damalas, C.A.; Eleftherohorinos, I.G. Pesticide exposure, safety issues, and risk assessment indicators. *Int. J. Environ. Res. Public. Health* **2011**, *8*, 1402–1419. [CrossRef] [PubMed]
3. Chu, Y.; Tong, Z.; Dong, X.; Sun, M.; Gao, T.; Duan, J.; Wang, M. Simultaneous determination of 98 pesticide residues in strawberries using UPLC-MS/MS and GC-MS/MS. *Microchem. J.* **2020**, *156*, 104975. [CrossRef]
4. FAO. *World Food and Agriculture—Statistical Year Book 2021*; FAO: Rome, Italy, 2021.
5. FAO. *Pesticides Use: Global, Regional and Country Trends, 1990–2018*; FAOSTAT Analytical Brief Series No. 16; FAO: Rome, Italy, 2021.
6. Schimmel, S.C.; Garnas, R.L.; Patrick, J.M.; Moore, J.C. Acute Toxicity, Bioconcentration, and Persistence of Ac-222, 705, Benthiocarb, Chlorpyrifos, Fenvalerate, Methyl Parathion, and Permethrin in the Estuarine Environment. *J. Agric. Food Chem.* **1983**, *31*, 104–113. [CrossRef] [PubMed]
7. Cordova, D.; Benner, E.A.; Sacher, M.D.; Rauh, J.J.; Sopa, J.S.; Lahm, G.P.; Selby, T.P.; Stevenson, T.M.; Flexner, L.; Gutteridge, S.; et al. Anthranilic diamides: A new class of insecticides with a novel mode of action, ryanodine receptor activation. *Pestic. Biochem. Phys.* **2006**, *84*, 196–214. [CrossRef]
8. FAO; WHO. *Fruit and Vegetables for Health, Report of a Joint FAO/WHO Workshop*; WHO: Geneva, Switzerland; FAO: Rome, Italy, 2004.
9. Szpyrka, E.; Kurdziel, A.; Matyaszek, A.; Podbielska, M.; Rupar, J.; Słowik-Borowiec, M. Evaluation of pesticide residues in fruits and vegetables from the region of south-eastern Poland. *Food Control* **2015**, *48*, 137–142. [CrossRef]
10. Varela-Martínez, D.A.; Gonázlez-Curbelo, M.A.; González-Sálamo, J.; Hernandez-Borges, J. Analysis of multiclass pesticides in dried fruits using QuEChERS-gas chromatography tandem mass spectrometry. *Food Chem.* **2019**, *297*, 1–8. [CrossRef]

11. Mahdavi, V.; Eslami, Z.; Molaee-Aghaee, E.; Peivasteh-Roudsari, L.; Sadighara, P.; Thai, V.N.; Fakhri, Y.; Ravanlou, A.A. Evaluation of pesticide residues and risk assessment in apple and grape from western Azerbaijan Province of Iran. *Environ. Res.* **2022**, *203*, 111882. [CrossRef]
12. Knězevic, Z.; Serdar, M. Screening of fresh fruit and vegetables for pesticide residues on Croatian market. *Food Control* **2009**, *20*, 419–422. [CrossRef]
13. Hjorth, K.; Johansen, K.; Holen, B.; Andersson, A.; Christensen, H.B.; Siivinend, K.; Toome, M. Pesticide residues in fruits and vegetables from South America–A Nordic project. *Food Control* **2011**, *22*, 1701–1706. [CrossRef]
14. Bakırcı, G.T.; Acay, D.B.Y.; Bakırcı, F.; Otles, S. Pesticide residues in fruits and vegetables from the Aegean region, Turkey. *Food Chem.* **2014**, *160*, 379–392. [CrossRef] [PubMed]
15. Yu, R.; Liu, Q.; Liu, J.; Wang, Q.; Wang, Y. Concentrations of organophosphorus pesticides in fresh vegetables and related human health risk assessment in Changchun, Northeast China. *Food Control* **2016**, *60*, 353–360. [CrossRef]
16. Algharibeh, G.R.; AlFararjeh, M.S. Pesticide residues in fruits and vegetables in Jordan using liquid chromatography/tandem mass spectrometry. *Food Addit. Contam. Part B* **2019**, *12*, 65–73. [CrossRef] [PubMed]
17. Abd El-Mageed, N.M.; Abu-Abdoun, I.I.; Janaan, A.S. Monitoring of pesticide residues in imported fruits and vegetables in United Arab Emirates during 2019. *Int. Res. J. Pure Appl. Chem.* **2020**, *21*, 239–260.
18. Osaili, T.M.; Al Sallagi, M.; Dhanasekaran, D.K.; Wael, A.M.; Bani Odeh, W.; Al Ali, H.J.; Ahmed, A.S.A.; Al Ali, D.; Radwan, H.; Obaid, R.S.; et al. Pesticide residues in fresh vegetables imported into the United Arab Emirates. *Food Control* **2022**, *133*, 108663. [CrossRef]
19. Omwenga, I.; Kanja, L.; Zomer, P.; Louisse, J.; Rietjens, I.M.; Mol, H. Organophosphate and carbamate pesticide residues and accompanying risks in commonly consumed vegetables in Kenya. *Food Addit. Contam. Part B* **2021**, *14*, 48–58. [CrossRef]
20. Shin, J.; Choi, S.-J.; Park, Y.; Kwak, B.; Moon, S.H.; Yoon, Y.T.; Jo, S.A.; Yi, H.; Kim, S.; Park, S.K.; et al. Comparison of QuEChERS and Liquid–Liquid extraction methods for the simultaneous analysis of pesticide residues using LC-MS/MS. *Food Control* **2022**, *141*, 109202. [CrossRef]
21. Rizzati, V.; Briand, O.; Guillou, H.; Gamet-Payrastre, L. Effects of pesticide mixtures in human and animal models: An update of the recent literature. *Chem. Biol. Interact.* **2016**, *254*, 231–246. [CrossRef]
22. Hernandez, A.F.; Parron, T.; Tsatsakis, A.M.; Requena, M.; Alarcon, R.; Lopez-Guarnido, O. Toxic effects of pesticide mixtures at a molecular level: Their relevance to human health. *Toxicology* **2013**, *307*, 136–145. [CrossRef]
23. Jayaraj, R.; Megha, P.; Sreedev, P. Organochlorine pesticides, their toxic effects on living organisms and their fate in the environment. *Interdiscip. Toxicol.* **2016**, *9*, 90–100. [CrossRef]
24. Bayoumi, A.E. Deleterious effects of banned chemical pesticides on human health in developing countries. In *Pesticides-Updates on Toxicity, Efficacy and Risk Assessment*; Larramendy, M.L., Soloneski, S., Eds.; IntechOpen: London, UK, 2022. [CrossRef]
25. Ciscato, C.H.P.; Gebara, A.B.; Monteiro, S.H. Pesticide residue monitoring of Brazilian fruit for export 2006–2007. *Food Addit. Contam. Part B-Surveill.* **2009**, *2*, 140–145. [CrossRef]
26. Tchounwou, P.B.; Ashour, B.A.; Moreland-Young, C.; Ragheb, D.A.; Romeh, A.A.; Goma, E.; El-Sheikh, E.; Lidell, F.P.; Ibitayo, O.; Assad, J.-C. Health risk assessment of pesticide usage in Menia El-Kamh Province of Sharkia Governorate in Egypt. *Int. J. Mol. Sci.* **2002**, *3*, 1082–1094. [CrossRef]
27. Wahab, S.; Muzammil, K.; Nasir, N.; Khan, M.S.; Ahmad, M.F.; Khalid, M.; Ahmad, W.; Dawria, A.; Reddy, L.K.V.; Busayli, A.M. Advancement and new trends in analysis of pesticide residues in food: A comprehensive review. *Plants* **2022**, *11*, 1106. [CrossRef]
28. El-Sheikh, E.A.; Aamir, M. Comparative effectiveness and field persistence of insect growth regulators on a field strain of the cotton leafworm, Spodoptera littoralis, Boisd (Lepidoptera: Noctuidae). *Crop Prot.* **2011**, *30*, 645–650. [CrossRef]
29. El-Sheikh, E.A. Comparative toxicity and sublethal effects of emamectin benzoate, lufenuron and spinosad on Spodoptera littoralis Boisd. (Lepidoptera: Noctuidae). *Crop Prot.* **2015**, *67*, 228–234. [CrossRef]
30. Grimalt, S.; Dehouck, P. Review of analytical methods for the determination of pesticide residues in grapes. *J. Chromatogr. A* **2015**, *1433*, 1–23. [CrossRef]
31. Pirsaheb, M.; Fattahi, N.; Rahimi, R.; Sharafi, K.; Ghaffari, H.R. Evaluation of abamectin, diazinon and chlorpyrifos pesticide residues in apple product of Mahabad region gardens: Iran in 2014. *Food Chem.* **2017**, *231*, 148–155. [CrossRef] [PubMed]
32. Santarelli, G.; Marfoglia, C.; Centorame, P.; Agostino, A.; Aurelio, R.; Scattolini, S.; Aprea, G.; Iannetti, L.; Pomilio, F.; Migliorati, G. Survey on Pesticide Residues and Microbial Contamination in Leafy Vegetables Marketed in Italy. In Proceedings of the 18th International Symposium of the World Association of Veterinary Laboratory Diagnosticians, Sorrento, Italy, 7–10 June 2017.
33. Yang, X.; Luo, J.; Li, S.; Liu, C. Evaluation of nine pesticide residues in three minor tropical fruits from southern China. *Food Control* **2016**, *60*, 677–682. [CrossRef]
34. Jardim, A.N.O.; Caldas, E.D. Brazilian monitoring programs for pesticide residues in food—Results from 2001 to 2010. *Food Control* **2012**, *25*, 607–616. [CrossRef]
35. Loughlin, T.M.M.; Peluso, M.L.; Etchegoyen, M.A.; Alonso, L.L.; Castro, M.C.; Percudani, M.C.; Marino, D.J.G. Pesticide residues in fruits and vegetables of the Argentine domestic market: Occurrence and quality. *Food Control* **2018**, *39*, 129–138. [CrossRef]
36. Arias, L.A.; Bojaca, C.R.; Ahumada, D.A.; Schrevens, E. Monitoring of pesticide residues in tomato marketed in Bogota, Colombia. *Food Control* **2014**, *35*, 213–217. [CrossRef]

37. Chen, X.; Fan, X.; Ma, Y.; Hu, J. Dissipation behaviour, residue distribution and dietary risk assessment of tetraconazole and kresoxim-methyl in greenhouse strawberry via RRLC-QqQ-MS/MS technique. *Ecotoxicol. Environ. Saf.* **2018**, *148*, 799–804. [CrossRef]
38. Sójka, M.; Miszczak, A.; Sikorski, P.; Zagibajło, K.; Karlińska, E.; Kosmala, M. Pesticide residue levels in strawberry processing by-products that are rich in ellagitannins and an assessment of their dietary risk to consumers. *NFS J.* **2015**, *1*, 31–37. [CrossRef]
39. Dong, F.S.; Liu, X.G.; Xu, J.; Li, J.; Li, Y.B.; Shan, W.L.; Zheng, Y.Q. Determination of cyantraniliprole and its major metabolite residues in vegetable and soil using ultra-performance liquid chromatography/tandem mass spectrometry. *Biomed. Chromatogr.* **2012**, *26*, 377–383. [CrossRef] [PubMed]
40. Lu, Z.; Zhang, Z.B.; Fang, N.; Hou, Z.G.; Li, Y.R.; Lu, Z.B. Simultaneous determination of five diamide insecticides in food matrices using carbon nanotube multiplug filtration cleanup and ultrahigh-performance liquid chromatographytandem mass spectrometry. *J. Agric. Food Chem.* **2019**, *67*, 10977–10983. [CrossRef] [PubMed]
41. El-Sheikh, E.A.; Ashour, M.-B.A. Diamide insecticides: Efficacy, toxicity and analytical methods for residue monitoring in food samples. *Egypt. J. Chem.* **2022**, *65*, 165–177. [CrossRef]
42. Tian, F.; Qiao, C.; Luo, J.; Guo, L.; Pang, T.; Pang, R.; Li, J.; Wang, C.; Wang, R.; Xie, H. Development and validation of a method for the analysis of five diamide insecticides in edible mushrooms using modified QuEChERS and HPLC-MS/MS. *Food Chem.* **2020**, *333*, 127468. [CrossRef]
43. E.U. Pesticides Database 2020. Available online: https://ec.europa.eu/food/plant/pesticides/eu-pesticides-database/public (accessed on 15 May 2022).
44. Hassan, A.A.; Bel Hadj Salah, K.; Fahmy, E.M.; Mansour, D.A.; Mohamed, S.A.M.; Abdallah, A.A.; Ashkan, M.F.; Majrashi, K.A.; Melebary, S.J.; El-Sheikh, E.-S.A.; et al. Olive Leaf Extract Attenuates Chlorpyrifos-Induced Neuro- and Reproductive Toxicity in Male Albino Rats. *Life* **2022**, *12*, 1500. [CrossRef]
45. Constantinou, M.; Louca-Christodoulou, D.; Agapiou, A. Method validation for the determination of 314 pesticide residues using tandem MS systems (GC–MS/MS and LC-MS/MS) in raisins: Focus on risk exposure assessment and respective processing factors in real samples (a pilot survey). *Food Chem.* **2021**, *360*, 129964. [CrossRef]
46. Malhat, F.; Sabera, E.; Abd Elsalam, S.; Ahmed, M.; Amin, A. Consumer safety evaluation of pyraclostrobin residues in strawberry using liquid chromatography tandem mass spectrometry (LC-MS/MS): An Egyptian profile. *Regul. Toxicol. Pharmacol.* **2019**, *108*, 104450. [CrossRef]
47. Chen, J.-N.; Lian, Y.-J.; Zhou, Y.-R.; Wang, M.-H.; Zhang, X.-Q.; Wang, J.-H.; Wu, Y.-N.; Wang, M.-L. Determination of 107 pesticide residues in wolfberry with acetate-buffered salt extraction and sin-quechers nano column purification coupled with ultra performance liquid chromatography tandem mass spectrometry. *Molecules* **2019**, *24*, 2918. [CrossRef]
48. Malhat, F.; Abdallah, O.; Ahmed, F.; Abdel Salam, S.; Anagnostopoulos, C.; Ahmed, M.T. Dissipation behavior of thiophanate-methyl in strawberry under open field condition in Egypt and consumer risk assessment. *Environ. Sci. Pollut. Res.* **2021**, *28*, 1029–1039. [CrossRef]
49. Mozzaquatro, J.d.O.; César, I.A.; Pinheiro, A.E.B.; Caldas, E.D. Pesticide residues analysis in passion fruit and its processed products by LC–MS/MS and GC–MS/MS: Method validation, processing factors and dietary risk assessment. *Food Chem.* **2022**, *375*, 131643. [CrossRef]
50. Qin, G.; Chen, Y.; He, F.; Yang, B.; Zou, K.; Shen, N.; Zuo, B.; Liu, R.; Zhang, W.; Li, Y. Risk assessment of fungicide pesticide residues in vegetables and fruits in the mid-western region of China. *J. Food Compost. Anal.* **2021**, *95*, 103663. [CrossRef]
51. Xing, L.; Wang, Y.; Luo, R.; Li1, X.; Zou, L. Determination of 31 pesticide residues in wolfberry by LC-MS/MS and dietary risk assessment of wolfberry consumption. *Food Sci. Technol. Campinas* **2022**, *42*, e61921. [CrossRef]
52. Luo, L.; Dong, L.; Huang, Q.; Ma, S.; Fantke, P.; Li, J.; Jiang, J.; Fitzgerald, M.; Yang, J.; Jia, Z.; et al. Detection and risk assessments of multi-pesticides in 1771 cultivated herbal medicines by LC/MS-MS and GC/MS-MS. *Chemosphere* **2021**, *262*, 127477. [CrossRef]
53. Castilla-Fernández, D.; Moreno-González, D.; Bouza, M.; Saez-Gómez, A.; Ballesteros, E.; García-Reyes, J.F.; Molina-Díaz, A. Assessment of a specific sample cleanup for the multiresidue determination of veterinary drugs and pesticides in salmon using liquid chromatography/tandem mass spectrometry. *Food Control* **2021**, *130*, 108311. [CrossRef]
54. Kiljanek, T.; Niewiadowska, A.; Małysiak, M.; Posyniak, A. Miniaturized multiresidue method for determination of 267 pesticides, their metabolites and polychlorinated biphenyls in low mass beebread samples by liquid and gas chromatography coupled with tandem mass spectrometry. *Talanta* **2021**, *235*, 122721. [CrossRef] [PubMed]
55. Ramezani, S.; Mahdavi, V.; Gordan, H.; Rezadoost, H.; Conti, G.O.; Khaneghah, A.M. Determination of multi-class pesticides residues of cow and human milk samples from Iran using UHPLC-MS/MS and GC-ECD: A probabilistic health risk assessment. *Environ. Res.* **2022**, *208*, 112730. [CrossRef] [PubMed]
56. Mwanja, M.; Jacobs, C.; Mbewe, A.R.; Munyinda, N.S. Assessment of pesticide residue levels among locally produced fruits and vegetables in Monze district, Zambia. *Int. J. Food Contam.* **2017**, *4*, 11. [CrossRef]
57. Hamed, S.A.; EL-Ghanam, A.A.A.; Elhefny, D.E. Fast and easy method of 55 pesticide residues determination in commonly fruits and vegetables collected from Egyptian local markets. *J. Plant Protect. Pathol.* **2019**, *10*, 587–595. [CrossRef]
58. Parveen, Z.; Riazuddin, A.; Iqbal, S.; Bhutto, M.A.; Khuhro, M.I. Monitoring of multiple pesticide residues in some fruits in Karachi, Pakistan. *Pakistan J. Bot.* **2011**, *43*, 1915–1918.
59. Mert, A.; Qi, A.; Bygrave, A.; Stotz, H.U. Trends of pesticide residues in foods imported to the United Kingdom from 2000 to 2020. *Food Control* **2022**, *133*, 108616. [CrossRef]

60. Tankiewicz, M.; Berg, A. Improvement of the QuEChERS method coupled with GC–MS/MS for the determination of pesticide residues in fresh fruit and vegetables. *Microchem. J.* **2022**, *181*, 107794. [CrossRef]
61. Sinha, S.; Rao, M.; Vasudev, K.; Odetokun, M. A liquid chromatography mass spectrometry-based method to measure organophosphorous insecticide, herbicide and non-organophosphorous pesticide in grape and apple samples. *Food Control* **2012**, *25*, 636–646. [CrossRef]
62. Javeres, M.N.L.; Habib, R.; Laure, N.J.; Shah, S.T.A.; Valis, M.; Kuca, K.; Nurulain, S.M. Chronic exposure to organophosphates pesticides and risk of metabolic disorder in cohort from Pakistan and Cameroon. *Int. J. Environ. Res. Public Health* **2021**, *18*, 2310. [CrossRef]
63. Santana-Mayor, Á.; Socas-Rodríguez, B.; Herrera-Herrera, A.V.; Rodríguez-Delgado, M.Á. Current trends in QuEChERS method. A versatile procedure for food, environmental and biological analysis. *TrAC Trends Anal. Chem.* **2019**, *116*, 214–235. [CrossRef]
64. USEPA. *Baseline Human Health Risk Assessment*; Vasquez Boulevard and I-70 Superfund Site: Denver, CO, USA, 2001.

molecules

Article

Stable Isotope Analysis of Residual Pesticides via High Performance Liquid Chromatography and Elemental Analyzer–Isotope Ratio Mass Spectrometry

Hee Young Yun [1], Eun-Ji Won [1], Jisoo Choi [1], Yusang Cho [1], Da-Jung Lim [2], In-Seon Kim [2] and Kyung-Hoon Shin [1,*]

1 Institute of Marine and Atmospheric Sciences, Hanyang University, Ansan 15588, Republic of Korea
2 Department of Agricultural Chemistry, Chonnam National University, Gwangju 61186, Republic of Korea
* Correspondence: shinkh@hanyang.ac.kr; Tel.: +82-31-400-5536

Abstract: To broaden the range of measurable pesticides for stable isotope analysis (SIA), we tested whether SIA of the anthranilic diamides cyantraniliprole (CYN) and chlorantraniliprole (CHL) can be achieved under elemental analyzer/isotope ratio mass spectrometry with compound purification in high-performance liquid chromatography (HPLC). Using this method, carbon isotope compositions were measured in pesticide residues extracted from plants (lettuce) grown indoors in potting soil that were treated with 500 mg/kg CHL and 250 mg/kg CYN and were followed up for 45 days. Our results show that the CYN and CHL standard materials did not have significant isotope differences before and after clean-up processing in HPLC. Further, when applied to the CYN product and CHL product in soil, stable isotope differences between the soil and plant were observed at <1.0‰ throughout the incubation period. There was a slight increase in the variability of pesticide isotope ratio detected with longer-term incubation (CHL, on average 1.5‰). Overall, we measured the carbon isotope ratio of target pesticides from HPLC fraction as the purification and pre-concentration step for environmental and biological samples. Such negligible isotopic differences in pesticide residues in soils and plants 45 days after application confirmed the potential of CSIA to quantify pesticide behavior in environments.

Keywords: compound-specific isotope analysis; pollutant; agricultural application; soil; HPLC; SPE extraction

Citation: Yun, H.Y.; Won, E.-J.; Choi, J.; Cho, Y.; Lim, D.-J.; Kim, I.-S.; Shin, K.-H. Stable Isotope Analysis of Residual Pesticides via High Performance Liquid Chromatography and Elemental Analyzer–Isotope Ratio Mass Spectrometry. *Molecules* **2022**, *27*, 8587. https://doi.org/10.3390/molecules27238587

Academic Editor: Miguel Ángel González-Curbelo

Received: 8 November 2022
Accepted: 2 December 2022
Published: 6 December 2022

1. Introduction

Pesticides are used to prevent crop damage from pest insects and pathogens and to prolong the storage lives of agricultural products. However, persistence in pesticide-contaminated soils and repeated use can lead to an increase in the unintentional buildup of pesticide(s) residues that adversely affect non-target organisms and cause insecticide resistance, and negatively impact the environment and human health [1]. Many countries are seeking sustainable agriculture and farming practices by decreasing pesticide applications and by enforcing safe (or maximum acceptable) levels of pesticide concentrations detected in crops and agricultural products [2]. In this view, developing analytic techniques for pesticides in environmental and crop samples has received our attention. Current analytical approaches have focused on identifying pesticide compounds with a wide range of differences in physiochemical characteristics, polarity and thermal stability [3,4]. Although determining pesticide presence and their concentration from a complex matrix is an important mission for environmental pollution monitoring and for food safety issues, this concentration-based approach has difficulties in tracing the sources of pesticide contamination, where pesticides are directly given or sprayed in the environments [3].

Compound-specific stable isotope analysis (CSIA) is a standard way of pollutant risk assessment [3,5] for identifying their behavior and distribution patterns. It ultimately allows for characterizing pollutant sources (i.e., point- vs. nonpoint-contamination source) and their relative contributions. This applicability is based on the fact that naturally occurring stable isotope compositions of elements (e.g., $^{13}C/^{12}C$) in pesticide residues are closely related to stable isotope ratios of their source (parental compounds). Further, the carbon isotope ratio of the pesticide is changed only negligibly or slightly by plant absorption [6,7] and abiotic degradation processes (e.g., photolysis and hydrolysis) [8]. In contrast, soil type (sterile vs. non-sterile soil) and biodegradation processing fairly increase carbon isotope ratios in pesticides <13.5‰, e.g., throughout the incubation period (e.g., from days to months) [7]. Microorganisms tend to metabolize chemical substances with lighter isotopes of elements (e.g., ^{12}C) because of the lower energy required for breaking the bonds with the lighter isotope (i.e., the kinetic isotope effect). Thus, biodegradation leads to more compounds with heavier isotopes of elements (e.g., ^{12}C) remaining in the substrates than those with the lighter isotope, increasing the stable isotope variables [8,9]. Such distinctive isotopic variability in pesticides from abiotic or biodegradation processing helps detect the dynamic behavior of pesticide molecules in environments [3,6,10].

CSIA [3,6,10–12] is selectively accessible for several pesticides (insecticides, herbicides, and fungicides) (Figure 1) and pesticide metabolites, e.g., aminomethylphosphonic acid from glyphosate and desphenylchloridazon from chloridazon [13].

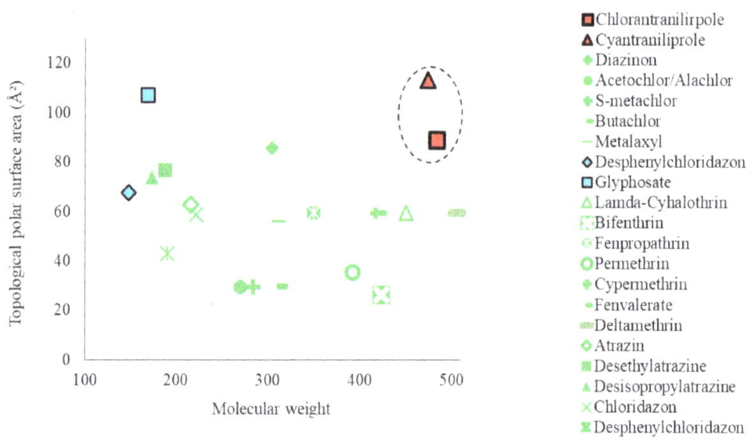

Figure 1. Distribution of diverse pesticides chemicals based on molecular characteristics and their analytic equipment, such as LC (or GC)–IRMS with blue symbols, GC–IRMS with green symbols [7–13], and EA–IRMS with red symbols (in this study) for stable isotope analysis.

Particularly, carbon is the most common element in pesticide substances and their metabolite compounds rather than other elements such as nitrogen [14]. The isotope ratios are mainly measured by an isotope ratio mass spectrometer (IRMS) system coupled to chromatographic systems such as gas chromatography (GC) and liquid chromatography (LC) (called GC–IRMS and LC–IRMS, respectively) depending on the characteristics of organic compounds such as polarities and molecular weight. As shown in Figure 1, GC–IRMS has generally been used for measuring the isotope composition of non-polar and volatile organic compounds, whereas LC–IRMS for polar and less volatile compounds (summarized in [3,5]). Additionally, an elemental analyzer (EA) directly connected to the IRMS system is capable of measuring solids or liquids (both types found in pesticide compounds) and is not restricted by characteristics of organic compounds such as polarities and molecular weights. EA–IRMS is generally used to determine the mass fractions of carbon, nitrogen, hydrogen, and sulfur of any organic compounds in diverse research fields (e.g., [3,15]). However, EA–IRMS does not inherently perform chromatographic separation for diverse organic compounds. To counter these issues, offline high-performance liquid chromatography (HPLC) methods were preceded prior to EA–IRMS (referred to as HPLC/EA–IRMS or the 'offline' CSIA method). HPLC/EA–IRMS were applied for measuring carbon isotopes in several biomolecules such as porphyrin [16] and amino acids [17,18] but rarely applied for pollutants (including pesticides).

Residual pesticides are commonly low in crops relative to topsoil and the CSIA approach in crop and environmental samples should collect a measurable amount of target analytes via significant extraction and pre-concentration steps. Moreover, traditional EA–IRMS systems require much higher amounts of analytes that produce reliable isotope variables compared to a GC–IRMS system. That is, the analyte mass required for EA–IRMS is approximately 20 μgC (or 50 μgN) [16,18,19], while the analyte mass for GC–IRMS via direct injection on the column is around 0.2 to 0.02 μgC (or 0.5 to 0.05 μgN). Fortunately, an improved capacity of the HPLC fraction collector (available injection volume up to 1 mL) with automated chromatographic separation might reduce the workload needed to detect target analytes over the detection limit. Additionally, practical HPLC/EA–IRMS methods should include the efficient removal of carbon-containing solvents (e.g., acetonitrile, dichloromethane, and methanol), since EA–IRMS does not have the ability to separate a pesticide compound from solvents based on molecular characteristics. Otherwise, the carbon isotope ratios of target analytes containing solvents might be shifted by carbon-containing solvents. Method optimization (i.e., extraction and clean-up procedure for complex matrices in soil and crop samples) of the EA–IRMS analysis is needed for broadening the measurable range of pesticides and for enhancing the CSIA applicability in environmental and agricultural issues.

The diamide pesticides such as cyantraniliprole (CYN) and chlorantraniliprole (CHL) used in this study are common due to their promising pest management effects via selective, uncontrolled calcium homeostasis in insects [20]. These pesticides were not applied in previous CSIA work, which has focused only on concentration determination in which CHL [21,22] and CYN [23] were used as parental compounds as well as metabolites [1]. Here, we sought to investigate an HPLC/EA–IRMS method for a CSIA approach in residual pesticides. First, the range of isotope ratio variability based on pesticide standards analyzed in EA–IRMS was compared before and after the analysis of HPLC fractions to verify the effects of isolation and pre-concentration procedures on pesticide isotope ratios. We also demonstrate the potential utility of this method by evaluating isotope compositions of the pesticide residue extracted from the soil and moved to lettuce in planted mesocosm set-ups with comparisons to the pesticide source. This CSIA approach will serve to monitor more diverse pesticide residues in the environment.

2. Results and Discussion

To our knowledge, this is the first study to measure stable isotope ratios of pesticides in EA–IRMS. Firstly, we investigated the possible uncertainties with isotopic measurements for pesticides CHL and CYN in HPLC/EA–IRMS. By combining a typical sorbent extraction method and chromatographic separations (time-based approach in HPLC chromatograms) with injection volume <200 μL, we collected considerable amounts of analytes (>3 nA particularly soil extract, corresponding to 0.15 mgC for CHL and 0.17 mgC for CYN) for obtaining reliable carbon isotopic compositions. Our HPLC/EA–IRMS approach helps to analyze stable isotope ratios of pesticides that had not been previously carried out, i.e., CHL and CYN in other studies, and this technique might broaden the range of measurable pesticides if LC–IRMS or GC–IRMS is not available.

2.1. Error Evaluation during Isotope Measurement

The effects from the tin capsule itself and/or solvents involved during the transfer of the analytes to the EA–IRMS system were tested to reveal the overall uncertainties associated with isotope analytical procedures. Our finding was that even a blank tin capsule (average weight = 65.80 mg) showed carbon contents of 61 ± 13 μgC with an amplitude of approximately 0.4 nA (Figure 2), while N content <0.1 nA was detected. Moreover, the tin capsule treated with 450 μL solvent (ACN, DCM and hexane) was used for transferring the target analyte from an extraction vial into a tin capsule. Subsequent drying for several hours in RT was consistent in providing the blank tin capsule with a peak height of approximately 0.4 nA (Figure 2). This indicates that N_2-drying might effectively remove organic solvents before isotopic measurement.

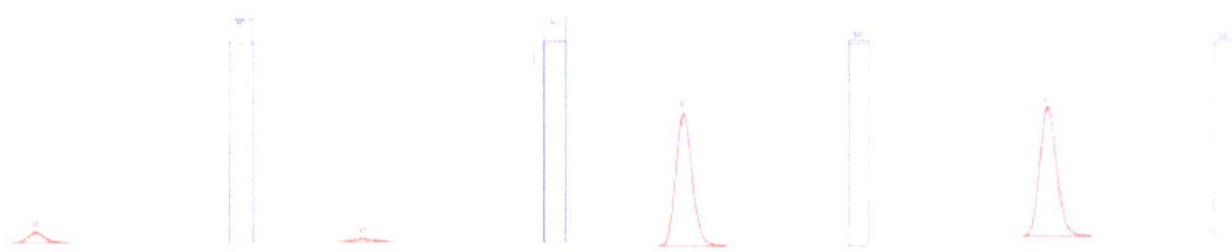

(a) (b) (c) (d)

Figure 2. Chromatograms of blank (empty) tin capsule (**a**), MeOH/DCM prewashed tin capsule (**b**), and standard chlorantraniliprole (**c**) and standard cyantraniliprole (**d**) in EA–IRMS. Square-shaped peaks indicate reference CO_2.

We assumed a very low carbon content in the blank tin capsules. To reveal the origin of unexpected carbon content, blank capsules were cleaned with DCM:MeOH (1:1, v:v) overnight, rinsed, and stored at 60 °C before use. Nonetheless, the cleaned blank tin showed a peak height of carbon <0.2 nA, which was much lower than the analytical blank (Figure 2). Overall, the unintended effect did not disappear completely during the isotope analysis. Similar to these results, previous studies showed that the interference effects of the blank tin capsules were consistent when using tins precleaned with an organic solvent mixture followed by combustion (400 °C, 5 h) [16,19]. Contamination may occur in the EA–IRMS system internally (e.g., from the gas line or autosampler) during isotope measurement. This indicates that the blank carbon content accounts for a non-negligible proportion of the analyte (Figure 2) and will not seriously contribute to isotope analysis. Thus, we suggested that blank correction should be considered when estimating the carbon contents of target compounds and $\delta^{13}C$ calculation [18]. To minimize the unintended contamination related to tin capsule use, the washed tin was used for further analysis.

2.2. $\delta^{13}C$ Variations in Standard Material: Before and after HPLC Purification

To validate the reliability of isotope information by HPLC/EA–IRMS, two aspects were considered in this study: (1) the relationship of amplitude (i.e., amount of analyte transferred to tin capsule) to the carbon isotope value and (2) the effects of HPLC performance on isotope variability. As shown in Figure 3, the $\delta^{13}C$ of CHL before HPLC was -26.18 ± 0.09‰ (average ± SD, n = 27), which was approximately 0.85‰ different than that after the HPLC fraction of -27.03 ± 0.30‰ (n = 18). Slightly lower isotope values after HPLC were also observed for CYN. That is, the carbon isotope composition of CYN after the HPLC fraction was on average 0.44% lower (average ± SD: -26.19 ± 0.65‰, n = 21) than that before the HPLC fraction (average ± SD: -25.75 ± 0.08‰, n = 24).

This isotopic difference might be related to the amplitude (or injection amount) of our target pesticides (Figure 3). The amplitude range of the CHL standard in our study was from 2.31 nA to 13.38 nA (the corresponding weights of standard material per tin capsule ranged from 0.01 mg to 0.13 mg), which was much broader than the amplitude from 1.52 nA to 2.48 nA (the estimated weight 0.03 mg to 0.08 mg) of CHL standard after the HPLC fraction. Thus, the amplitude of CHL from the HPLC fraction was close to the lowest level of analyte injection weight (mg) of powder CHL. The CYN standard results showed amplitudes per tin capsule from 4.32 nA to 14.15 nA (corresponding weight of the standard material from 0.01 mg to 0.15 mg), while the amplitudes were from 1.50 nA to 5.08 nA for CYN after the HPLC fraction (corresponding to their estimated weight from 0.03 mg to 0.15 mg per tin capsule). Moreover, SD was slightly different before and after the HPLC fraction (CHL: 0.09 vs. 0.30; CYN: 0.08 vs. 0.65). Higher precision in EA–IRMS relative to HPLC/EA–IRMS was also reported in other studies using other organic compounds, e.g., 0.08 vs. 0.16 in phenylalanine 11. Such isotope variability dependent on the sample amount (amplitude by IRMS) was widely reported in IRMS platforms. For instance, metabolite desphenylchloridazon formed by degradation of the herbicide chloridazon showed at least a 1‰ difference in $\delta^{13}C$ values depending on the injection amount of C, while a larger standard deviation is reported with a smaller injection amount [13]. The overall result confirms that HPLC performance after solvent removal by N_2-drying did not significantly change within a 1‰ difference for isotope measurements. To reduce the analytic errors from our EA–IRMS system, we also suggest a detection requirement of analyte per tin capsule (amplitude > 3 nA) in the HPLC/EA–IRMS method. Ultimately, securing the appropriate analyte amount in a tin capsule (approximately >0.03 mg) is important to produce reliable isotope compositions of specific compounds present in the environment by HPLC/EA–IRMS applications.

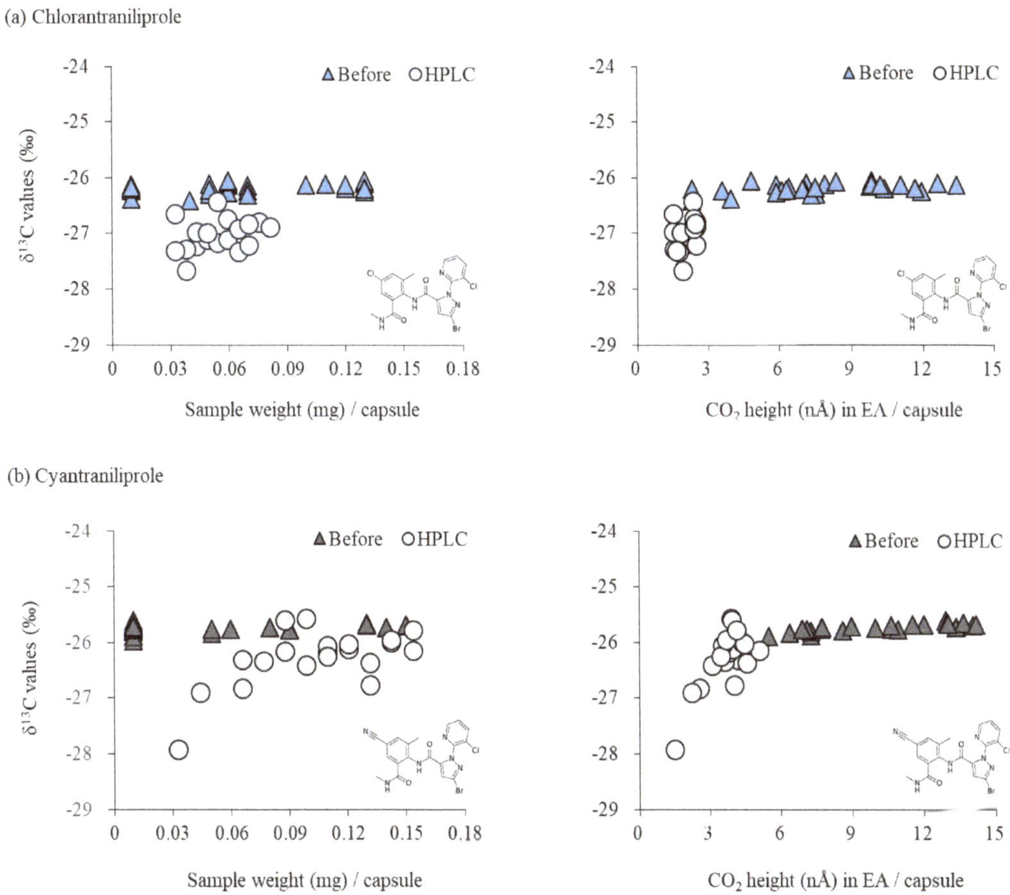

Figure 3. Effects of HPLC analytic procedure on determining isotope measurement in CHL standard (**a**) and CYN standard (**b**).

2.3. *Application of HPLC/EA–IRMS and $\delta^{13}C$ Determination of Target Compounds in a Soil-Crop System*

The pre-treatment procedure in samples of interest for CYN and CHL detection traditionally is known to follow solid phase extraction (SPE) based on silica [24]. However, our pre-treatment procedure involved SPE extraction combined with HPLC separation (see Section 3). To evaluate the needs of the complex pre-treatment procedures, the analytic procedure was applied to a CYN-based product (11 mg/mL), a CHL-based product (9 mg/mL), and pesticide-free samples (topsoil and plant parts from a local market). Briefly,

HPLC chromatograms showed that pesticides were present for CHL at 8 min and for CYN at 12 min under 45% ACN as a mobile phase, at least 2 min from the peaks of the pesticide-free sample matrix (interfering substances, present < 3.5 min) (Figure 4). This suggests that single SPE extraction (generally adopted for pesticide quantification) would experience unintended contamination from sample matrix effects when isotope values of pesticides are measured by an EA–IRMS system due to a lack of chromatographic separation ability. Consequently, SPE extraction after HPLC performance for collecting pesticide fractions would be essential to effectively exclude the interfering compound ('matrix' effect) and isolate residual pesticides, particularly from crop samples.

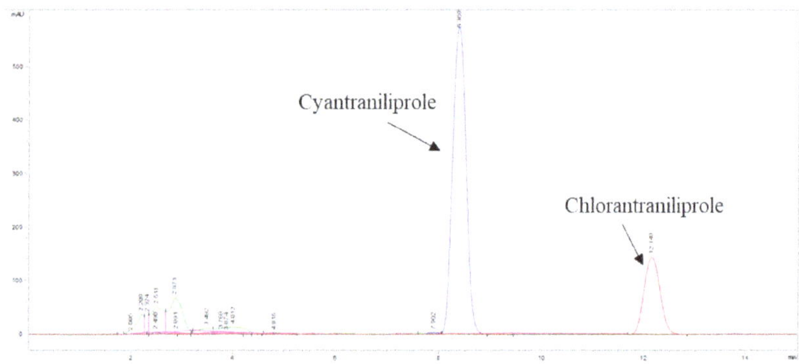

Figure 4. Chromatograms overlay in HPLC of two pesticides standards, cyantraniliprole (at 8.4 min) and chlorantraniliprole (12.3 min), and other pesticide-free samples of soil, leafy parts of lettuce, and root parts metrics (before 6 min) with constant flow (1mL/min) of 45% Acetonitrile in DW as mobile phase.

The percentage of CHL in soil decreased to 50.75 ± 0.02% (amount estimated as 0.25 ± 0.03 mg/g) and that of CYN decreased to 80.13 ± 1.22% (estimated concentration 0.80 mg/g). In contrast to the concentration-based results (Figure 5), there was no significant change in $\delta^{13}C$ in pesticides from soils, only a slight increase in $\delta^{13}C$ (on average 1.7‰ for CHL from the soil) at 45 days relative to the initial time (−29.90 ± 0.02‰, Table 1).

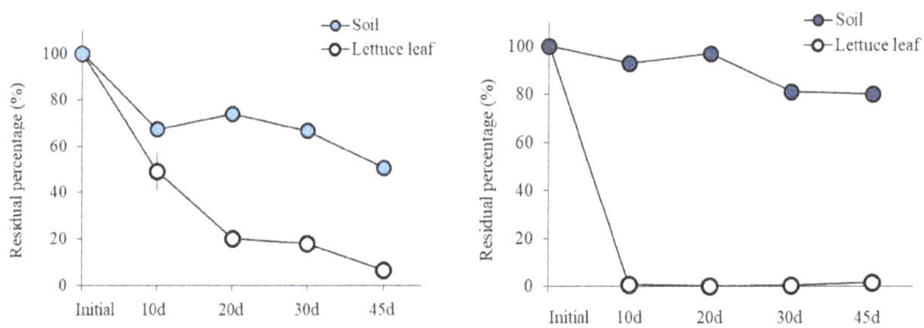

(**a**) Chlorantraniliprole (**b**) Cyantraniliprole

Figure 5. Estimate of amounts of residual pesticides in soil and lettuce leaves from indoor crops over a 45-day culture experiment (N = 3).

Table 1. Carbon isotope variability of chlorantraniliprole and cyantraniliprole extracted from samples.

Soil					Crop		
	$\delta^{13}C$	Estimated C (mg/Capsule)	Amount (mg/g)	N	$\delta^{13}C$	Estimated C (mg/Capsule)	N
(a) *Chlorantraniliprole*							
Initial	−29.90 ± 0.02	0.176 ± 0.010	0.5 *	2	Not applied		
10d	−29.24 ± 0.17	0.171 ± 0.004	0.337	3	−30.35 ± 0.06	0.268 ± 0.032	3
20d	−28.37 ± 0.13	0.222 ± 0.031	0.370	3	−30.27 ± 0.50	0.206 ± 0.017	3
30d	−28.85 ± 0.10	0.093 ± 0.012	0.334	3	−30.56 ± 0.40	0.085 ± 0.010	5
45d	−28.24 ± 0.28	0.110 ± 0.018	0.254	3	−31.11 ± 0.10	0.130 ± 0.037	3
Overall	−28.92	0.154			−30.57		
(b) *Cyantraniliprole*							
Initial	−29.42 ± 0.27	0.159 ± 0.023	1 *	3	Not applied		1
10d	−28.74 ± 0.17	0.212 ± 0.019	0.93	3	−29.03	0.043	1
20d	−29.15 ± 0.05	0.173 ± 0.051	0.97	3	−28.79	0.042	1
30d	−28.28 ± 0.58	0.142 ± 0.013	0.81	3	−28.54	0.070	1
45d	−28.80 ± 0.56	0.143 ± 0.026	0.80	3	−28.45	0.047	1
Overall	−28.88	0.166			−28.70		

* Amount of pesticides exposed to soil at initial time.

Moreover, CYN extracted from the soil at 0d was −29.41± 0.27‰, which changed slightly within <1.0‰ over 45 days (Table 1). Overall, the $\delta^{13}C$ values in CHL and CYN extracted from pesticide-treated soil were not significantly changed at 45 d (on average $\delta^{13}C < 1.7‰$ with overlapping SD of analytical uncertainty). The absence of significant isotope changes might be less influenced by biodegradation, which is assumed to induce significant isotope changes in pesticides based on kinetic isotope fractionation (e.g., [7]). Like our results, other studies reported that pesticide $\delta^{13}C$ compositions from sterilized soil (less dominated by microbial activity) showed a consistent pattern (on average $\delta^{13}C < 1.0‰$) for fenopropathrin, deltamethrin, α-cypermethrin [9], and lambda-cyhalothrin [8] under GC–IRMS. Even unsterilized soil (assuming active microbial biodegradation) showed a $\delta^{13}C$ increase of approximately <2‰ within 40 days [8,9]. Such a negligible isotope change was also reported in a short-term lab-scale experiment (similar to 45 days in this study) in other pesticides such as butachlor, S-metachlor, and metalaxyl as analyzed by GC–IRMS [7]. Other studies report that $\delta^{13}C$ did not change significantly for different pesticide amounts (2 mg/kg vs. 10 mg/kg) [9], planted or unplanted mesocosms, or soil types (forest soil vs. vineyard mesocosm) [7]. This indicates that degradation could trigger small, stable isotope changes, particularly for CHL and CYN in a soil environment. Other processes induce non-significant isotope fractionation such as sorption and leaching [6,8], which could be the major behavior of pesticides in the environment. The estimation of carbon isotope changes of CHL and CYN helped to broaden CSIA applications to monitor the short-term behavior of residual pesticides in environmental applications.

CHL and CYN residues from plants, transplanted to pesticide-treated soil for the experiment but previously grown in pesticide-free soil, were much less than those from soils (Figure 5). Indeed, only <2% of the initial pesticide treatment was detected in the plant grown in the CHL- and CYN-treated soils after 45 days, respectively. However, there were significant amounts (more than 55% of the initial treatment) of pesticides in top soils. These results suggest the pesticides might not persist well in crops. The residual CHL and CYN detected from plants suggest plant root uptake and transport of these pesticides, although their amounts in plants tended to decrease through the cultivation period. Similarly, the distribution of insecticides in plant parts decreased with an increased pesticide exposure period [1,25,26]. Although our study did not focus on the metabolites of CHL and CYN, there are reports that the amount of metabolites relative to the parental compound (CYN) increased [1,27]. The behavior may depend on plant tissue type, e.g., being highly abundant in leaves relative to fruits and flowers in tomato plants [1].

Compared to concentration changes, the overall CYN $\delta^{13}C$ value throughout the sampling period from the leaf part (−28.70‰) was similar to that from the soil (−28.88‰) (Table 1). Additionally, overall CHL $\delta^{13}C$ was 0.65‰ lighter in the plant than soil throughout the sampling period. In particular, the CHL $\delta^{13}C$ difference in the plant and soil was on average from 1.7‰ to 2.9‰. The weak carbon isotope fractionation in our target substances might be related to the transformation process of an insecticide (parent compound) to its metabolites. For instance, CYN has a structure very close to its metabolite IN-J9Z38 formed by ring closure, which is frequently formed as a result of environmental degradation or plant metabolism [28]. Such transformation processes may not involve chemical bond breakage, leading to a weak carbon isotope fractionation in our target pesticides. The less variable isotopic patterns from CHL and CYN suggest that carbon isotope ratios can be used as fingerprints to distinguish contamination sources in chemical products, particularly in environmental samples such as soil and groundwater [12,29–32].

CSIA approach of pesticide CHL and CYN was first addressed in our small-scale indoor incubation experiments. As long as chemicals of interest are extracted from the soil or crop samples, the pesticide stable isotope approach helps distinguish the source of the pesticide(s) released from direct pesticides used and/or unintended effects of residual pesticides in the environment. This is due to a negligible isotopic change in the pesticides in the environment within a short timeframe. However, the procedure for transferring the analyte and removing any solvents in the tin capsule might result in analyte loss and decrease the amplitude detected in the EA–IRMS system compared to the actual powder weight. In fact, when the commonly used CSIA system is employed directly via on-column injection in GC–IRMS, the analyte amount is <1 microgram of glyphosate [31] or <135 nmol C of desphenylchloridazon [13], which is demanding significantly lower than that in our EA–IRMS approach. Thus, the sensitivity-improved EA–IRMS system, referred to as nano-EA–IRMS [16], and highly effective preparation steps for extracting a large sample amount may improve HPLC/EA–IRMS applications for a wide range of pesticides.

Although the detection limit for our EA–IRMS system was not good enough to analyze other elements (such as ^{2}H,^{15}N and ^{37}Cl, [5]) in pesticides, the EA–IRMS-based method has potential advantages if the target compound is well isolated, purified, and collected beyond the detection limit. This is because the EA–IRMS platform is not affected by the molecular characteristics of organic compounds. In this regard, EA–IRMS is a reasonable tool to access chemical compounds with diverse polar/high molecular weight and their metabolites that are often more persistent and polar than their parental compounds in the environment [13]. Indeed, LC–IRMS can be used for a polar compound only but cannot be used for N isotope analysis, and GC–IRMS can be applied to midpolar or apolar compounds only but provides C and N in separated runs. Rather, EA–IRMS provides isotopic compositions of multi-elements (i.e., C and N) simultaneously in about 11 min. Additionally, EA–IRMS instrumentation is not expensive and is common in a stable isotope facility lab, and analytic services involving this technique are widely available. Multi-element CSIA leads to enhancing the discrimination power to verify sources of pesticide pollution in environments, rather than CSIA based on carbon only, as chloridazon standards are distinguished among suppliers [13]. Therefore, more analytic efforts should be involved in making more sensitive EA connected to IRMS (e.g., [19]) to produce dual isotopes more reliably in pesticides.

In conclusion, our HPLC/EA–IRMS approach was applied to improve the CSIA application availability for the pesticides, CYN and CHL with high polarity and molecular weight (Figure 1). Moreover, the HPLC performance with the solid phase extraction procedure reduces effectively the interference effects from sample matrices such as soil and crop samples but is also capable of obtaining reliable isotope measurements in residual pesticides. Overall, negligible changes in the carbon isotope value in pesticides propose that the pesticide remaining in the soil is directly related to the pesticide product applied to agricultural environments. Therefore, CSIA might successfully uncover the primary source of pesticides when severe CHL and CYN contamination events occur in the field.

3. Materials and Methods

3.1. Chemicals and Reagents

The analytical standards cyantraniliprole (CYN) and chlorantraniliprole (CHL) (>98% purity) were purchased from FUJIFILM Wako Pure Chemical Corporation (Japan). Acetonitrile, ethyl acetate, dichloromethane, and hexane, all HPLC grade, were supplied by Merck (Darmstadt, Germany). Anhydrous sodium sulfate (reagent grade: >97%) and NaCl were purchased from Sigma-Aldrich (St. Louis, MO, USA) and Junsei (Tokyo, Japan). Deionized water was prepared using an Aquapuri 5 series system with a resistivity level 18.2 MΩ cm (Young In Chromass, Republic of Korea). Pesticide products of CHL (5%, Altacoa®, FarmHannong, Kyungju, Republic of Korea) and CYN (10.26%, Benevia®, FarmHannong, Kyungju, Republic of Korea) were purchased from a local agricultural market in Republic of Korea.

3.2. Validation of HPLC/EA–IRMS Isotopic Measurement

To investigate possible bias during HPLC purification and subsequent pre-concentrating processes, isotopic compositions of pesticide standards before and after HPLC chromatographic separation were compared. Before HPLC, CYN and CHL standards from 0.01 mg to 0.15 mg were prepared in tin capsules (pressed capsule, 10 by 10 mm, Elemental Microanalysis, UK). After HPLC, the CYN and CHL were respectively dissolved in 45% and 55% acetonitrile (ACN, HPLC grade, purity > 99.9%) in deionized water (approximately 2.1 mg/mL). Then, 50 µL of dissolved standard substance was injected into the HPLC. Chromatographic separation was carried out by Zorbax Eclips XDB C18 column (Agilent) at 30 °C. The mobile phases at a flow rate 1 mL/min were 45% ACN for CYN and 55% ACN for CHL in Table 2. CYN and CHL eluted around 8 min and 6 min, respectively, and were fraction-collected. The HPLC system Agilent 1260 Infinity II series (Agilent, Santa Clara, CA) consisted of a quaternary pump (G7111B; Agilent), a column temperature controller (G7116A; Agilent), an autosampler (G7129A; Agilent), autosampler thermostat (G1330B; Agilent), an online-photodiode-array detector (DAD; G7115A; Agilent), and a fraction collector (G1364F; Agilent). The fractions were pooled and prepared from 0.4 mL to 1.5 mL in duplicate to determine the detection limit of our HPLC/EA–IRMS method.

Table 2. Carbon isotope variability of chlorantraniliprole and cyantraniliprole extracted from samples.

HPLC System	Cyantraniliprole	Chlorantraniliprole
Column	Zorbax Eclips XDB C18 column (4.6 × 250 mm, 5 µm)	Zorbax Eclips XDB C18 column (4.6 × 250 mm, 5 µm)
Mobile phase (duration time)	45% Acetonitrile in water (15 min)	55% Acetonitrile in water (15 min)
Flow rate	1 mL/min	1 mL/min
Column temperature	30 °C	30 °C
Detection	264 nm	254 nm
Injected volume	0.05mL to 0.50 mL	0.05mL to 0.50 mL

Subsequently, the fractions were dried under nitrogen gas, re-dissolved in hexane and DCM, filtered using a glass fiber GF-5 filter (estimated to 0.03mg to 0.15mg) and transferred to tin capsules. To avoid effect of washed solvents, the tin capsules were dried until they reached a constant weight (approximately >5 h at room temperature), and they were then introduced into EA–IRMS for isotopic measurements. To reduce putative contamination effects from the tin capsule itself [16,19], tin capsules were soaked with DCM:MeOH (1:1, v:v) overnight, rinsed and kept at 60 °C before use. The EA–IRMS system is composed of EA (Elementar Vario Isotope Select, Elementar, UK) coupled with IRMS (Isoprime vision, Elementar, UK). The EA–IRMS diluter function was not used during isotope measurement.

Carbon isotope values are reported in per mil (‰) using conventional delta notation relative to the international standards Vienna PeeDee Belemnite (V-PDB) and air, respectively:

$$\delta\, X\, (‰) = \left(\frac{R_{sample}}{R_{standard}} - 1 \right) \times 10^3 \quad (1)$$

where *R* denotes the $^{13}C/^{12}C$ ratio for carbon. Blank corrections using an analytical blank were applied for every $\delta^{13}C$ calculation. All reported isotope ratios are expressed as arithmetic means of replicate measurements with standard deviation in $\delta^{13}C$. Analytic precision was assessed as the SD of laboratory working standards, which were within 0.2 ‰ for $\delta^{13}C$ analysis.

3.3. Pesticide Application and Plant Growth

To illustrate the potential of the HPLC/EA–IRMS method in agricultural measurements, indoor container experiments were set up from mid-April 2021 to mid-June 2021 in the Isotope Ecology and Environmental Science Laboratory at Hanyang University. The containers were plastic and cuboid planters (52 cm × 14 cm × 15 cm) commonly used for home gardening. Potting soil mix was purchased from a local market (Ildungsangto®, Tosung, Republic of Korea) and was composed of 65–70% cocopeat, 10–15% zeolite, 5–10% perlite, and 1–4% biochar. The pH was 7.0, and the relative humidity was approximately 45%. Each container was filled with 4 kg (oven dry weight 2.3 kg) of the potting soil and reached 10 cm in depth. In total, 10 containers were prepared and allocated to either CYN or CHL treatment. Each pesticide product was mixed with 4L of water in a bucket and poured into the base of the container, allowing the container soil to absorb it. Final concentrations of CYN and CHL in the spiked soil container (pesticide mg/ soil kg) were 250mg/kg and 500mg/kg, respectively. This high concentration is not a recommended guideline amount for general pesticide users in Republic of Korea but was used for obtaining crop as well as soil samples with large amount of pesticide residue to reveal the unintended pesticide contamination sources in soil as well as crops.

Two days after pesticide treatment, fast-growing leafy vegetable lettuces (approximately one month old after seedling) that were purchased from a local farmer's market were transplanted into pesticide-treated containers. We assumed the pesticide was distributed evenly through soil in the container. Six plants were planted in each container, and 30 total plants were raised for each pesticide. The cuboid planters were incubated for two months in front of a sunroom-type window, and 4L of water every week were supplied to a base of the container. Then, the soil and crop samples (leaf parts from one plant) were collected at 10, 20, 30 and 45 days. Next, 150 g (wet weight) of soil was sampled with plastic spoons, and harvested leaves were rinsed with tap water to remove surface dust, chopped into small pieces, and then placed into polyethylene bags. Samples were immediately frozen and stored in a −20 °C freezer until analysis.

For two months, the average temperature was 19 °C with a maximum of 26 °C and a minimum of 15 °C. Additional light was not used in this study to maintain natural day and night conditions. Further, other external factors (e.g., rainfall) were not considered as the experiment was conducted under indoor conditions.

3.4. Sample Preparation: Isolation and Purification of Residual Pesticides In Situ Samples

The extraction method was based on previous protocols [23,24] with a slight modification of solvent volume. The overall workflow to extract residual pesticides from soil and crops is summarized in Figure 6. Samples were homogenized and extracted with 25 mL of ACN in a falcon tube and were centrifuged at 2500 rpm for 5 min. The supernatant was filtered through a syringe membrane filter (0.2 μm, PTEE-H). This ACN extraction procedure was conducted twice. The supernatant was transferred to a glass tube and concentrated to <1 mL in a turbo nitrogen evaporator with a water bath temperature of 40 °C.

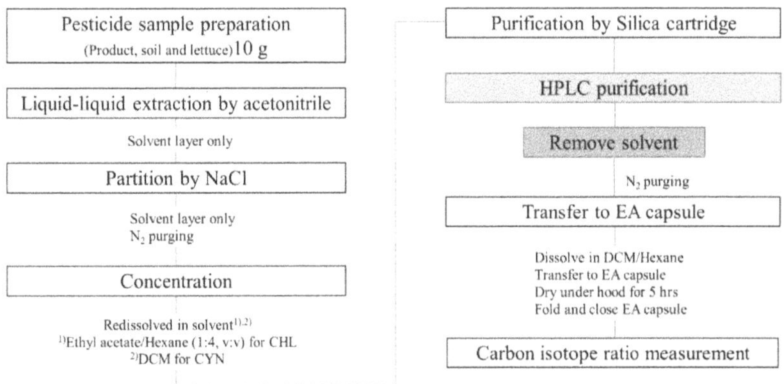

Figure 6. Schematic workflow to determine residual pesticides in samples for environmental and agricultural applications.

Further, the liquid–liquid partitioning process was different for CYN and CHL. For CYN, 20 mL distilled water, 5 mL saturated NaCl solution, and 20 mL hexane were added to the concentrated supernatants in the glass tube, and the glass tube was vigorously shaken. After complete layer separation, the lower layer was passed through 5 g anhydrous sodium sulfate. Then, the extract was collected and concentrated to dryness using a turbo nitrogen evaporator. The residue was dissolved in 3 mL DCM. The concentrated extract was loaded into a silica cartridge (6 cc Bond Elut, Agilent Technologies), which was pre-activated with 10 mL DCM. Then, the solution was washed with 3 mL 10% ethyl acetate in DCM. CYN was eluted with 3 mL 40% ethyl acetate in DCM. The eluted fraction was concentrated using a nitrogen evaporator, and the residue was dissolved in 2 mL 45% ACN in DW.

For extracting CHL, 10 mL distilled water, 10 mL saturated NaCl solution, and 20 mL DCM were added to the concentrated supernatant in a glass tube. After waiting until the layers were completely separated, the lower part was drained into 1.5 g anhydrous sodium sulfate. This partitioning was repeated with 10 mL DCM. Then, extracts were collected, pooled, and concentrated to dryness using a turbo nitrogen evaporator. The residue part was re-dissolved to 5 mL 20% ethyl acetate in hexane. The re-dissolved residue was loaded into a silica cartridge (1 g, 6 cc Bond Elut, Agilent Technologies), which was pre-washed using 10 mL hexane. Then, the cartridge was eluted with 6 mL 30% ethyl acetate in hexane and 6 mL 50% ethyl acetate in hexane. The eluted fractions were pooled and dried using a nitrogen evaporator, and then the residue was dissolved in 2 mL 55% ACN in DW and was stored at −20 °C prior to the HPLC procedure.

Then, pesticide extracts (<100 µL from soil and <900 µL from lettuce samples) were injected for HPLC. HPLC purification procedures were conducted with 10–12 repetitions to obtain pesticide fractions of CYN and CHL. The collected fractions were pooled and prepared in triplicate as described previously.

3.5. Standard Calibration Curve and Estimating Pesticide Amount

Standards of CHL and CYN were prepared by dissolving 20 mg of the compound in 20 mL solvent to obtain a 100 mg/L stock solution, respectively. From this stock solution, a working standard solution (20 mg/L) was prepared by dilution in solvent. This was then serially diluted to obtain standard solutions of 0.01, 0.05, 0.10, 0.20, 0.50, 1.00, 2.00, 3.00, 5.00, and 10.00 mg/L. An aliquot of 2.0 µL was injected into the HPLC, and a standard calibration curve was prepared based on the peak area. Limit of quantification (LOQ) was 0.082 mg/L for CHL and 0.0109 mg/L for CYN in our HPLC systems. LOQ was calculated as: LOQ (mg/ kg) = [minimum detectable amount (ng)/injection volume (uL)] × [final sample volume (mL)/sample amount (g)]. The instrumental conditions are shown in Table 1.

Author Contributions: Conceptualization, H.Y.Y., J.C., Y.C., D.-J.L., I.-S.K., E.-J.W. and K.-H.S.; methodology, H.Y.Y., J.C., Y.C., D.-J.L., I.-S.K., E.-J.W. and K.-H.S.; data curation, H.Y.Y., J.C. and Y.C.; writing—original draft preparation, H.Y.Y.; writing—review and editing, H.Y.Y., E.-J.W. and K.-H.S.; funding acquisition, K.-H.S. All authors have read and agreed to the published version of the manuscript.

Funding: This research was funded by the Rural Development Administration (Grant No. PJ0150530-12022), Republic of Korea.

Institutional Review Board Statement: Not applicable.

Informed Consent Statement: Not applicable.

Data Availability Statement: Not applicable.

Acknowledgments: The authors would like to thank the student interns (JS Son, TN Kim and S Sung) for their help in the HYU EABL lab. All authors have agreed to the acknowledgements.

Conflicts of Interest: The authors declare no conflict of interest.

References

1. Huynh, K.; Leonard, E.; Chong, J.-H.; Palmer, C.; Tharayil, N. Persistence and metabolism of the diamide insecticide cyantraniliprole in tomato plants. *Sci. Rep.* **2021**, *11*, 21570. [CrossRef]
2. Nicolopoulou-Stamati, P.; Maipas, S.; Kotampasi, C.; Stamatis, P.; Hens, L. Chemical Pesticides and Human Health: The Urgent Need for a New Concept in Agriculture. *Front. Public Health* **2016**, *4*, 148. [CrossRef] [PubMed]
3. Won, E.-J.; Yun, H.-Y.; Lee, D.-H.; Shin, K.-H. Application of Compound-Specific Isotope Analysis in Environmental Forensic and Strategic Management Avenue for Pesticide Residues. *Molecules* **2021**, *26*, 4412. [CrossRef] [PubMed]
4. González-Curbelo, M.Á.; Varela-Martínez, D.A.; Riaño-Herrera, D.A. Pesticide-Residue Analysis in Soils by the QuEChERS Method: A Review. *Molecules* **2022**, *27*, 4323. [CrossRef] [PubMed]
5. Höhener, P.; Guers, D.; Malleret, L.; Boukaroum, O.; Martin-Laurent, F.; Masbou, J.; Payraudeau, S.; Imfeld, G. Multi-elemental compound-specific isotope analysis of pesticides for source identification and monitoring of degradation in soil: A review. *Environ. Chem. Lett.* **2022**, *20*, 3927–3942. [CrossRef]
6. Cui, G.; Lartey-Young, G.; Chen, C.; Ma, L. Photodegradation of pesticides using compound-specific isotope analysis (CSIA): A review. *RSC Adv.* **2021**, *11*, 25122–25140. [CrossRef]
7. Pérez-Rodríguez, P.; Schmitt, A.-D.; Gangloff, S.; Masbou, J.; Imfeld, G. Plants affect the dissipation and leaching of anilide pesticides in soil mesocosms: Insights from compound-specific isotope analysis (CSIA). *Agric. Ecosyst. Environ.* **2021**, *308*, 107257. [CrossRef]
8. Shen, X.; Xu, Z.; Zhang, X.; Yang, F. Stable carbon isotope fractionation during the biodegradation of lambda-cyhalothrin. *Sci. Total Environ.* **2015**, *532*, 415–419. [CrossRef]
9. Xu, Z.; Liu, W.; Yang, F. A new approach to estimate bioavailability of pyrethroids in soil by compound-specific stable isotope analysis. *J. Hazard. Mater.* **2018**, *349*, 1–9. [CrossRef]
10. Elsner, M.; Imfeld, G. Compound-specific isotope analysis (CSIA) of micropollutants in the environment—Current developments and future challenges. *Curr. Opin. Biotechnol.* **2016**, *41*, 60–72. [CrossRef]
11. Torrentó, C.; Bakkour, R.; Glauser, G.; Melsbach, A.; Ponsin, V.; Hofstetter, T.B.; Elsner, M.; Hunkeler, D. Solid-phase extraction method for stable isotope analysis of pesticides from large volume environmental water samples. *Analyst* **2019**, *144*, 2898–2908. [CrossRef] [PubMed]
12. Kawashima, H.; Katayama, Y. Source Evaluation of Diazinon Using Stable Carbon Isotope Ratio. *Environ. Forensics* **2010**, *11*, 363–371. [CrossRef]
13. Melsbach, A.; Ponsin, V.; Torrentó, C.; Lihl, C.; Hofstetter, T.B.; Hunkeler, D.; Elsner, M. ^{13}C- and ^{15}N-Isotope Analysis of Desphenylchloridazon by Liquid Chromatography–Isotope-Ratio Mass Spectrometry and Derivatization Gas Chromatography–Isotope-Ratio Mass Spectrometry. *Anal. Chem.* **2019**, *91*, 3412–3420. [CrossRef] [PubMed]
14. Mogusu, E.O.; Wolbert, J.B.; Kujawinski, D.M.; Jochmann, M.A.; Elsner, M. Dual element (^{15}N/^{14}N, ^{13}C/^{12}C) isotope analysis of glyphosate and AMPA by derivatization-gas chromatography isotope ratio mass spectrometry (GC/IRMS) combined with LC/IRMS. *Anal. Bioanal. Chem.* **2015**, *407*, 5249–5260. [CrossRef]
15. Schürner, H.K.V.; Seffernick, J.L.; Grzybkowska, A.; Dybala-Defratyka, A.; Wackett, L.P.; Elsner, M. Characteristic Isotope Fractionation Patterns in s-Triazine Degradation Have Their Origin in Multiple Protonation Options in the s-Triazine Hydrolase TrzN. *Environ. Sci. Technol.* **2015**, *49*, 3490–3498. [CrossRef]
16. Isaji, Y.; Ogawa, N.O.; Boreham, C.J.; Kashiyama, Y.; Ohkouchi, N. Evaluation of δ^{13}C and δ^{15}N Uncertainties Associated with the Compound-Specific Isotope Analysis of Geoporphyrins. *Anal. Chem.* **2020**, *92*, 3152–3160. [CrossRef]
17. Phillips, A.A.; Wu, F.; Sessions, A.L. Sulfur isotope analysis of cysteine and methionine via preparatory liquid chromatography and elemental analyzer isotope ratio mass spectrometry. *Rapid Commun. Mass Spectrom.* **2021**, *35*, e9007. [CrossRef] [PubMed]

18. Broek, T.A.B.; Walker, B.D.; Andreasen, D.H.; McCarthy, M.D. High-precision measurement of phenylalanine $\delta^{15}N$ values for environmental samples: A new approach coupling high-pressure liquid chromatography purification and elemental analyzer isotope ratio mass spectrometry. *Rapid Commun. Mass Spectrom.* **2013**, *27*, 2327–2337. [CrossRef]
19. Ogawa, N.O.; Nagata, T.; Kitazato, H.; Ohkouchi, N. Ultra-sensitive elemental analyzer/isotope ratio mass spectrometer for stable nitrogen and carbon isotope analyse. In *Earth, Life, and Isotopes*; Ohkouchi, N., Tayasu, I., Koba, K., Eds.; Kyoto University Press: Kyoto, Japan, 2010; pp. 339–353.
20. Wang, D.; Lv, W.; Yuan, Y.; Zhang, T.; Teng, H.; Losey, J.E.; Chang, X. Mechanism of the different metabolome responses between *Plutella xylostella* and *Pieris rapae* treated with the diamide insecticides. *Ecotoxicol. Environ. Saf.* **2020**, *203*, 111033. [CrossRef]
21. Lee, J.W.; Kim, J.Y.; Kim, H.G.; Hur, K.J.; Kwon, C.H.; Hur, J.H. Residual Characteristics of Bistrifluron and Chlorantraniliprole in Strawberry (*Fragaria ananassa* Duch.) for Establishing Pre-Harvest Residue Limit. *Korean J. Environ. Agriculture. Korean Soc. Environ. Agric.* **2017**, *36*, 57–62. [CrossRef]
22. Zhang, J.-M.; Chai, W.-G.; Wu, Y.-L. Residues of chlorantraniliprole in rice field ecosystem. *Chemosphere* **2012**, *87*, 132–136. [CrossRef] [PubMed]
23. Yoon, J.H.; Lee, S.W.; Lim, D.; Kim, S.W.; Kim, I.S. Evaluation of Cyantraniliprole Residues Translocated by Lettuce, Spinach and Radish. *Korean J. Environ. Agric.* **2021**, *40*, 335–344. [CrossRef]
24. Park, J.-E.; Kwon, H.; Back, S.-J.; Ro, J.-H.; Kim, D.B.; Lee, H.-S.; Jin, Y. Development of multi-residue analytical method using QuEChERS and LC-MS/MS for 36 newly registered pesticides in Korea for major agricultural export commodities. *Korean J. Pestic. Sci.* **2018**, *22*, 378–392. [CrossRef]
25. Yoon, J.; Lim, D.; Lee, S.; Kim, I. Uptake of soil-residual diazinon by rotational lettuce under greenhouse conditions. *Foods* **2022**, *11*, 3510. [CrossRef]
26. Mei, J.; Ge, Q.; Han, L.; Zhang, H.; Long, Z.; Cui, Y.; Hua, R.; Yu, Y.; Fang, H. Deposition, Distribution, Metabolism, and Reduced Application Dose of Thiamethoxam in a Pepper-Planted Ecosystem. *J. Agric. Food Chem.* **2019**, *67*, 11848–11859. [CrossRef]
27. Melsbach, A.; Torrentó, C.; Ponsin, V.; Bolotin, J.; Lachat, L.; Prasuhn, V.; Hofstetter, T.B.; Hunkeler, D.; Elsner, M. Dual-Element Isotope Analysis of Desphenylchloridazon to Investigate Its Environmental Fate in a Systematic Field Study: A Long-Term Lysimeter Experiment. *Environ. Sci. Technol.* **2020**, *54*, 3929–3939. [CrossRef]
28. Lee, J.; Jung, M.W.; Lee, J.; Lee, J.; Shin, Y.; Kim, J.-H. Dissipation of the Insecticide Cyantraniliprole and Its Metabolite IN-J9Z38 in Proso Millet during Cultivation. *Sci. Rep.* **2019**, *9*, 11648. [CrossRef]
29. Schmidt, T.C.; Zwank, L.; Elsner, M.; Berg, M.; Meckenstock, R.U.; Haderlein, S.B. Compound-specific stable isotope analysis of organic contaminants in natural environments: A critical review of the state of the art, prospects, and future challenges. *Anal. Bioanal. Chem.* **2004**, *378*, 283–300.
30. Droz, B.; Drouin, G.; Maurer, L.; Villette, C.; Payraudeau, S.; Imfeld, G. Phase Transfer and Biodegradation of Pesticides in Water–Sediment Systems Explored by Compound-Specific Isotope Analysis and Conceptual Modeling. *Environ. Sci. Technol.* **2021**, *55*, 4720–4728. [CrossRef]
31. Limon, A.W.; Moingt, M.; Widory, D. The carbon stable isotope compositions of glyphosate and aminomethylphosphonic acid (AMPA): Improved analytical sensitivity and first application to environmental water matrices. *Rapid Commun. Mass Spectrom.* **2021**, *35*, e9017. [CrossRef]
32. Kuntze, K.; Eisenmann, H.; Richnow, H.-H.; Fischer, A. Compound-Specific Stable Isotope Analysis (CSIA) for Evaluating Degradation of Organic Pollutants: An Overview of Field Case Studies. In *Anaerobic Utilization of Hydrocarbons, Oils, and Lipids*; Boll, M., Ed.; Springer International Publishing: Cham, Switzerland, 2019; pp. 323–360.

molecules MDPI

Article

Monitoring and Exposure Assessment of Fosetyl Aluminium and Other Highly Polar Pesticide Residues in Sweet Cherry

Emrah Gormez [1], Ozgur Golge [2], Miguel Ángel González-Curbelo [3] and Bulent Kabak [4,*]

1 Pia Frucht Food Control Laboratory, Alaşehir 45600, Turkey
2 Department of Gastronomy and Culinary Arts, Faculty of Tourism, Alanya Alaaddin Keykubat University, Alanya 07450, Turkey
3 Departamento de Ciencias Básicas, Facultad de Ingeniería, Universidad EAN, Calle 79 nº 11-45, Bogotá 110221, Colombia
4 Department of Food Engineering, Faculty of Engineering, Hitit University, Corum 19030, Turkey
* Correspondence: bulentkabak@hitit.edu.tr

Abstract: Cherries are popular fruits due to their health benefits, organoleptic quality, and attractive appearance. Since highly polar pesticides are of low mass and amphoteric character, and are not amenable to traditional multi-residue extraction methods, they are more commonly not included in the pesticide monitoring program. This study aims to determine twelve highly polar pesticide residues in cherry samples intended for export from Turkey. A total of 16,022 cherry samples from 2018–2020 harvests in four production areas of Turkey were analyzed using a modification of the Quick Polar Pesticides method and liquid chromatography-tandem mass spectrometry. The method was validated at two fortification levels (0.01 and 0.05 mg kg^{-1}), and good recoveries (87.4–111.4%) and relative standard deviations (<6%) were achieved for all analytes. The limits of quantification were in the range of 1.08–2.55 µg kg^{-1}. Overall, 28.4% of the analyzed cherry samples were detected with phosphonic acid, calculated as fosetyl aluminium (fosetyl-Al) in amounts up to 77.7 mg kg^{-1}. For 2304 samples (14.4%), the residues exceeded the European Union maximum residue level of 2 mg kg^{-1}. There is no reason to be concerned about long-term exposure to phosphonic acid/fosetyl-Al, and the other highly polar pesticides through the consumption of sweet cherry.

Keywords: analytical method validation; chromatography; food safety; mass spectrometry; polar pesticides; risk assessment

Citation: Gormez, E.; Golge, O.; González-Curbelo, M.Á.; Kabak, B. Monitoring and Exposure Assessment of Fosetyl Aluminium and Other Highly Polar Pesticide Residues in Sweet Cherry. *Molecules* **2023**, *28*, 252. https://doi.org/10.3390/molecules28010252

Academic Editor: Jesus Simal-Gandara

Received: 28 November 2022
Revised: 20 December 2022
Accepted: 20 December 2022
Published: 28 December 2022

1. Introduction

Cherry is one of the most consumed fruit throughout the world, which belongs to the genus *Prunus*, under the *Rosaceae* family. While it has been identified more than thirty cherry spices, mainly native to Europe and West Asia, the sweet cherry (*Prunus avium*) and sour cherry (*Prunus cerasus*) are globally traded. Cherries are good sources of fibre, potassium, polyphenolics (anthocyanins and hydroxycinnamic acid), β-caratone, and vitamin C and possess a high antioxidant capacity. Sweet cherries are mostly consumed as fresh fruit, whereas sour cherries are most frequently incorporated in processed foods, such as juices, jam, jellies, pies, cakes, ice cream, and others [1,2]. Turkey is the world's leading producer of sweet cherries, producing 639,564 metric tonnes in 2018 and accounting for over 25% of total world production, followed by the United States (312,430 tonnes), Uzbekistan (172,035 tonnes), Chile (155,935 tonnes), and Iran (137,268 tonnes) [3]. Turkey exported cherries to the value of more than 150 million dollars in 2018 and accounting for 15% of global exports, the main importers being the Russian Federation and Germany [4].

Cherry is affected by insect and mite pests and by fungal, bacterial, and viral diseases during its growing process. It is susceptible to many diseases, including brown rot of stone fruit (*Monilinia fructigena*), brown rot of blossom (*Monilinia laxa*), *Armillaria* root rot (*Armillaria mellea*), *Phytophthora* root and crown rot (*Phytophthora* spp.), cherry leaf spot

(*Blumeriella jaapi*), cherry shot-hole (*Stigmina carpophila*), bacterial canker (*Pseudomonas syringae*), and crown gall (*Agrobacterium tumefaciens*) [5].

The most effective way to prevent, destroy or control harmful organisms and diseases in the crop is by using pesticides. Within the agricultural sector in Turkey, most pesticides utilized in 2018 were from fungicides/bactericides (42.5%), followed by insecticides (29.6%) and herbicides (27.3%) [3]. However, the residues in agricultural products are a growing concern because of their adverse acute and chronic health effects and environmental problems. The fungicides azadirachtin, cyprodinil, dithiocarbamates, fludioxonil, tebuconazole, and thiophanate, the insecticides deltamethrin, dimethoate, pirimicarb, 1pinosad, spirodiclofen, tau-fluvalinate and thiacloprid, and highly polar herbicides are the most used pesticides in the cultivation of cherries in Turkey [6].

The use of highly polar pesticides In agriculture and horticulture is widespread due to their low costs, high efficiency, low persistence in the environment, and relatively low toxicity in comparison with other pesticides towards mammals [7]. However, polar pesticides are more commonly not included in national pesticide monitoring programs as they have low mass, amphoteric character, and are not amenable to traditional multi-residue methods. Conventionally, a series of single residue methods are used to detect and quantify, which resulted in extra costs, time delays, and excluded from the surveillance program. For this reason, a fast and simple single analytical method that can analyze multi-residue polar pesticides in agricultural products and detect maximum residue level (MRL) violations with confidence is in great demand.

The Quick Polar Pesticides (QuPPe) method established by the European Reference Laboratory-Single Residue Methods (EURL-SRM) allows the simultaneous extraction of highly polar pesticides from a wide range of food commodities. This method involves extraction with acidified methanol without clean-up and liquid chromatography-tandem mass spectrometry (LC-MS/MS) measurement [8]. Several studies have been conducted in various countries to monitor multiple highly polar residues in vegetables and fruits [7,9–11], cereals [12], animal-derived products [13], honey [11], low alcoholic beverages [14] and human blood serum [15] in the last five years.

In this study, we aimed to determine twelve highly polar residues, namely aminomethylphosphonic acid (AMPA), N-acetyl-AMPA, chlorate, ethephon, ethephon-hydroxy (HEPA), fosetyl-aluminium (fosetyl-Al), glyphosate, glufosinate, N-acetyl-glufosinate, maleic hydrazide, 3-methylphosphinicopropionic acid (MPPA) and phosphonic acid in Turkish cherries intended for export to various countries mainly to Russian Federation and European countries. For that purpose, a modified method based upon the QuPPe extraction method followed by an LC-MS/MS measurement was validated.

2. Results and Discussion

2.1. Validation Data

The validation results for the cherry matrix compared favorably against the analytical performance described in the SANTE 11813/2017 guideline. As shown in Table 1, the coefficients of determination and the residuals were excellent ($R^2 > 0.99$ and residuals <20%). The limits of quantification (LOQs) ranged from 1.08 $\mu g\ kg^{-1}$ for phosphonic acid to 2.55 $\mu g\ kg^{-1}$ for glyphosate. The LOQs of target polar compounds were much lower than the European Union (EU) MRLs in cherry. According to Table 2, all recoveries were satisfactory, with mean values ranging from 87.4% to 111.4%, and relative standard deviations (RSD) values varying from 0.47 to 5.12% under repeatability conditions and from 1.68 to 5.04% under reproducibility conditions for target polar compounds, demonstrating good repeatability of the measurements in the absence/presence of ILIS. For all polar compounds, U_{exp} was greatly lower than the criteria of 50% specified in SANTE 11813/2017 guideline. The U_{exp} ranged between 7% for chlorate and fosetyl-Al and 27% for glyphosate.

Table 1. LOQs, EU MRLs and matrix-matched calibration data of the highly polar pesticides.

Analyte	LOQ (μg kg^{-1})	EU MRL (mg kg^{-1})	Linearity (Range: 5–250 μg kg^{-1})		Residual (%)
			Equation	R^2	
AMPA	1.76	None	$y = 4618x - 4087$	0.999	<12
N-acetyl-AMPA	2.19	None	$y = 21{,}781x - 19{,}397$	0.994	<16
Chlorate	1.46	0.05	$y = 30{,}155x - 14$	0.994	<16
Ethephon	1.10	5.0	$y = 43{,}871x - 54{,}173$	0.996	<14
HEPA	1.26	None	$y = 39{,}407x - 102{,}623$	0.999	<12
Fosetyl-Al	1.18	2.0 [b]	$y = 77{,}048x - 103{,}586$	0.999	<13
Glyphosate	2.55	0.1	$y = 9506x - 11{,}750$	0.998	<11
Glufosinate	2.23	0.15 [a]	$y = 3071x - 5666$	0.995	<20
N-acetyl-glufosinate	1.21	Part of glufosinate	$y = 13{,}979x - 29{,}314$	0.999	<12
Maleic hydrazide	2.14	0.2	$y = 2430x + 1373$	0.998	<18
MPPA	1.40	Part of glufosinate	$y = 26{,}002x - 26{,}203$	0.995	<13
Phosphonic acid	1.08	Part of fosetyl-Al	$y = 28{,}624x - 68{,}551$	0.998	<14

[a] Sum of glufosinate, its salts, MPPA and N-acetyl-glufosinate expressed as glufosinate equivalents. [b] Sum of fosetyl, phosphonic acid and their salts expressed as fosetyl.

Table 2. Recovery, precision and expanded uncertainty of the highly polar pesticides.

Analyte	Recovery (%)		Repeatability (%RSD, n = 5)		Reproducibility (%RSD, n = 10)		U_{exp} (%)
	0.01 (mg kg^{-1})	0.05 (mg kg^{-1})	0.01 (mg kg^{-1})	0.05 (mg kg^{-1})	0.01 (mg kg^{-1})	0.05 (mg kg^{-1})	
AMPA	98.0	95.3	3.94	2.98	5.04	3.74	16
N-acetyl-AMPA	92.2	95.7	3.36	2.38	2.47	1.77	17
Chlorate	101.1	102.9	1.53	1.35	2.31	1.68	7
Ethephon	94.7	96.0	3.58	2.21	1.97	2.30	14
HEPA	106.3	94.5	2.90	1.12	3.11	2.36	12
Fosetyl-Al	102.4	97.9	0.94	0.47	2.47	1.83	7
Glyphosate	89.4	95.7	5.12	2.00	3.90	3.42	27
Glufosinate	99.8	87.4	4.19	2.39	2.25	2.17	19
N-acetyl-glufosinate	100.6	92.3	4.08	2.05	2.77	2.13	16
Maleic hydrazide	103.2	111.4	3.25	2.37	3.55	2.49	20
MPPA	95.0	99.5	1.81	2.09	3.47	2.19	11
Phosphonic acid	96.8	90.6	2.69	3.05	3.39	2.27	17

2.2. *Pesticide Analysis in Sweet Cherry Samples and Exposure Assessment*

In total, 16,022 cherry samples from 2018, 2019 and 2020 harvests were monitored for the twelve highly polar pesticides. All the cherry samples were produced in four cherry production areas in Turkey. None of the target polar compounds was measured above the LOQs in cherries except for phosphonic acid, calculated as fosetyl-Al, sum. The frequency of cherry samples with fosetyl-Al residue produced in different areas of Turkey is shown in Figure 1. In the harvest years 2018, 2019, and 2020, 6.3–24.2%, 13.9–38.2%, and 17.5–43.4% of the samples, respectively, contained fosetyl-Al at different concentrations. In all three sampling years, Izmir samples had the most frequency of fosetyl-Al residues (24.2–43.4% frequency) in cherries, followed by Isparta samples (21.2–41.4%).

Figure 2 reveals the distribution of fosetyl-Al in sweet cherry samples, taking into consideration the harvest years. Overall, 78.3% of cherry samples harvested in 2018 were free from fosetyl-Al residue, while 12.2% of samples contained fosetyl-Al at levels not exceeding the respective EU MRL of 2 mg kg^{-1}. The level of fosetyl-Al exceeded the legal limit in 410 cherry samples (9.5%). No fosetyl-Al was found in 75.6% of cherry samples from the 2019 harvest, whereas 9.8% of the samples tested contained quantified residue of fosetyl-Al not exceeding the respective EU MRL. In 808 cherry samples (14.6%), fosetyl-Al residue levels exceeded the EU MRL. Cherry samples from the harvest of the year 2020 had a high frequency of fosetyl-Al (36.8%) compared to the other years. Out of 3354 quantified samples of cherries from the 2020 harvest, 1086 samples (17.6%) showed fosetyl-Al concentrations above the EU MRL. The concentrations of fosetyl-Al in cherries collected in three consecutive years, 2018–2020, varied from 0.013 to 18.6 mg kg^{-1} (mean = 0.862 mg kg^{-1}), from 0.005 to 16.9 mg kg^{-1} (mean = 0.308 mg kg^{-1}) and from 0.005 to 77.7 mg kg^{-1} (mean = 0.432 mg kg^{-1}), respectively. LC-MS/MS chromatograms of the extract of the cherry sample containing phosphonic acid (calculated as fosetyl-Al) at a level of 1.73 mg kg^{-1} are illustrated in Figure 3.

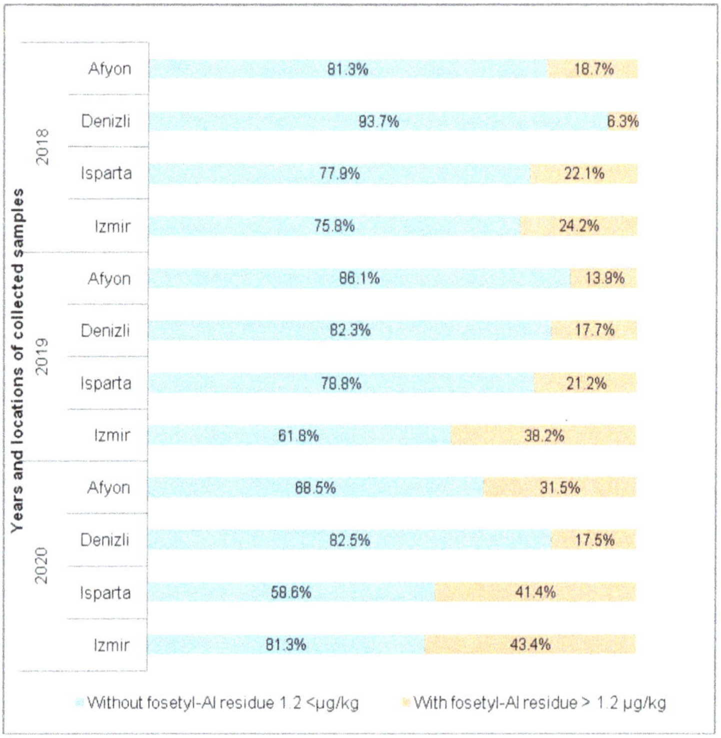

Figure 1. The frequency of cherry samples with fosetyl-Al per production area and harvest year.

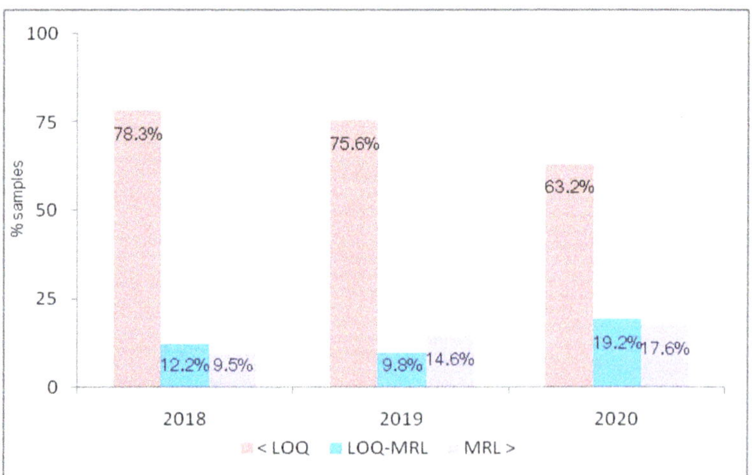

Figure 2. The distribution of fosetyl-Al content in cherry samples per harvest year.

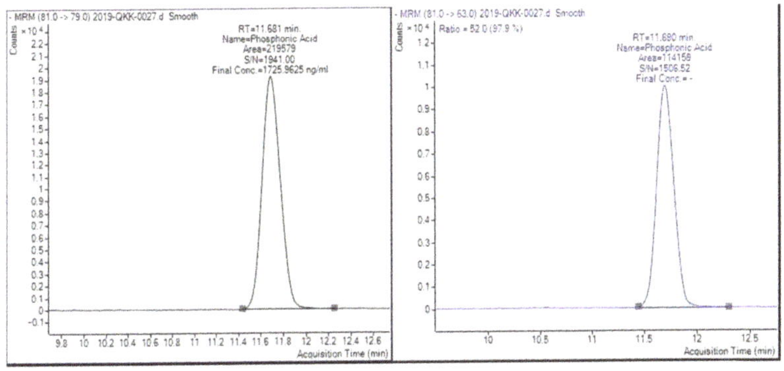

Figure 3. LC-MS/MS chromatograms of the extract of the cherry sample containing phosphonic acid (calculated as fosetyl-Al) at a level of 1.73 mg kg^{-1}.

The use of fosetyl-Al is not common among cherry farmers in Turkey. Farmers declared that they do not use fosetyl-Al contrary to foliar fertilizer. Residues of phosphonic acid, defined as fosetyl-Al, in cherries could occur as a result of applying foliar fertilizer containing phosphonic acid itself prior to harvest.

These results are inconsistent with the 2015 EU pesticide monitoring program findings performed by the EU Member States, Iceland, and Norway. Fosetyl-Al was present in 29.9% of 84,341 samples in quantifiable concentrations; 1.21% of them (59 samples) exceeded the EU MRL. Cherries were also found to contain different pesticides in 177 out of 719 samples analyzed (24.6%), 3.2% of which exceeded the respective MRLs [16]. In another extensive study, a total of 785 fresh fruit samples (including 23 sweet cherry samples) from conventional cultivation were analyzed by CVUA Stuttgart for over 750 different pesticides. Fosetyl, a sum fungicide was found to be the predominant pesticide detected in fresh fruits (47.4% of the samples analyzed) from 40 different countries, up to a level of 47.7 mg kg^{-1}. Cherries had fosetyl, sum, at concentrations varying from 0.083 to 2.1 mg kg^{-1} [17]. In contrast to our results, fosetyl-Al was not determined in any 225 sweet cherry samples consumed domestically in Turkey [18]. In a study by Da Silva et al. [7], ethephon was found in 547 out of 1048 fruits (53%) intended for export from Brazil; 17 of them (2%) had residues higher than the legal limit. Fosetyl was also detected in 20 out of 109 mango samples (18%) in measurable concentrations. The level of fosetyl exceeded the respective MRL in 4.6% of the mango samples. During the years 2004–2011, the Danish Veterinary and Food Administration monitored 17,309 food commodities, including fruits, vegetables, cereals, and animal origin products, for about 250 pesticides, but polar pesticides were not included in the monitoring program. Cherry samples (n = 24) were found to contain various pesticides, including bifenthrin (4.2% of the cherries), carbendazim (20.8%), lambda-cyhalothrin (8.3%), cypermethrin (16.7%), cyprodinil (4.2%), diazinon (8.3%), dimethoate (8.3%), iprodione (4.2%), monocrotophos (4.2%), myclobutanil (16.7%), and tebuconazole (12.5%) [19].

Fosetyl-Al is a systematic fungicide that has been used to protect many fruits and vegetables against plant pathogens such as *Phytophthora*, *Pythium*, *Plasmopara*, *Bremia* spp. as well as bacteria such as *Xanthomonas* and *Erwinia* spp. [20]. Fosetyl-Al does not show carcinogenic, genotoxic, or mutagenic properties in laboratory animals, and it does not pose developmental or reproductive effects of concern. An acceptable daily intake (ADI) of 3 mg kg^{-1} body weight (b.w.) per day and an acceptable operator exposure level (AOEL) of 5 mg kg^{-1} b.w. per day for fosetyl-Al has been established. The ADI of 2.52 mg kg^{-1} b.w. has also been set for phosphonic acid, expressed as fosetyl [21].

The mean long-term exposure to phosphonic acid/fosetyl from sweet cherry for adults ranged from 1.55×10^{-5} to 1.56×10^{-5} mg kg^{-1} b.w. day^{-1} (LB to UB). This is the first data on long-term exposure to phosphonic acid/fosetyl through the consumption of sweet cherries for adults. Since all sweet cherry data were left-censored for other highly polar substances analyzed, those exposure estimates were not included in the analysis. The HQ of fosetyl for adults was 0.0006% (LB/MB/UB). Applying the long-term exposure assessment method, none of the samples exceeded the toxicological reference value (max. 0.34% of the ADI) for fosetyl. There is, therefore, no reason to be concerned about long-term exposure to residues, phosphonic acid/fosetyl, and other highly polar substances through the consumption of sweet cherries.

3. Materials and Methods

3.1. Chemicals and Materials

LC-MS grade acetonitrile and methanol were supplied by J.T. Baker (Gliwice, Poland) and VWR Chemicals BDH® (Gdansk, Poland), respectively. Formic acid and glacial acetic acid were ordered from Merck KGaA (Darmstadt, Germany).

The analytical standards of AMPA (purity of 99.9%), ethephon (96.0%), HEPA (89.5%), fosetyl-Al (95.0%), glyphosate (98.7%), glufosinate (97.9%), N-acetyl-glufosinate (94.3%), maleic hydrazide (99.0%), MPPA (99%) and phosphonic acid (97.5%) were obtained from Dr. Ehrenstorfer GmbH (Augsburg, Germany). N-acetyl-AMPA (94.4%) and chlorate (99.0%) were from HPC Standards GmbH (Cunnersdorf, Germany). Isotopically labelled internal standards (ILISs) ethephon D4 (94.3%) and fosetyl-Al D15 (96.4%) were supplied from Dr. Ehrenstorfer GmbH (Augsburg, Germany). The ILISs of glyphosate-$^{13}C_2$,^{15}N (>95%) and $^{18}O_3$-phosphonic (≥95%) were obtained from EURL-SRM (Stuttgart, Germany) and Toronto Research Chemicals (Toronto, ON, Canada), respectively.

3.2. Samples

A total of 16,022 cherry samples, each weighing 2 kg harvested for export, were collected from Turkey for the analysis of the twelve highly polar residues. The samples were originating from four Turkish cherry production areas, namely İzmir-Kemalpaşa, Denizli, Isparta, and Afyon. Sampling was carried out for three consecutive years, 2018–2020, and yearly size varied between 4319 and 6170 samples. Each analytical result was derived from one laboratory sample taken from each lot.

3.3. Sample Preparation

Sweet cherry samples were extracted using the EURL-SRM QuPPe method [8], with slight modifications. The extraction procedures were schematically depicted in Figure 4. Briefly, ten grams of homogenized cherry samples were placed into 50 mL polypropylene centrifuge tubes, and 1.5 g of water was added and spiked with 50 µL of ILIS solution. Then, 10 mL of acidified MeOH (containing 1% formic acid, *v*/*v*) were added, shaken for 2 min in a Collomix shaker (VIBA 330, Gaimersheim, Germany), and the tubes were centrifuged (Rotofix 32 A, Hettich, Tuttlingen, Germany) for 5 min at 4000 rpm at room temperature. Formic acid was used for the adjustment of pH. Finally, 1 mL of supernatant was filtered through a regenerated cellulose syringe filter (0.20 µm) and collected in plastic autosampler vials.

3.4. LC-MS/MS Analysis

The LC-MS/MS system comprised of an Agilent 1290 LC coupled to an Agilent 6470 triple quadrupole (QQQ) mass spectrometer (Agilent, Santa Clara, CA, USA) equipped with a Jet Stream electrospray ionization (ESI) source. Instrument control, data acquisition, and quantitative analysis were performed using the Agilent MassHunter workstation software. Separation of highly polar compounds was achieved using a porous graphitic carbon-based Thermo Scientific™ Hypercarb column (100 × 2.1 mm, 5 µm particle size) at 40 °C. Eluent A composed of water containing 5% methanol and 1% acetic acid, and eluent B is composed of methanol containing 1% acetic acid. Gradient elution was performed as follows: 0–11 min 100–70% A, 0.2 mL min^{-1}; 11–19 min 70% A, 0.4 mL min^{-1}; 19–22 min 10% A, 0.4 mL min^{-1}; 22.1–30 min 100% A, 0.2 mL min^{-1}. The injection volume was 10 µL.

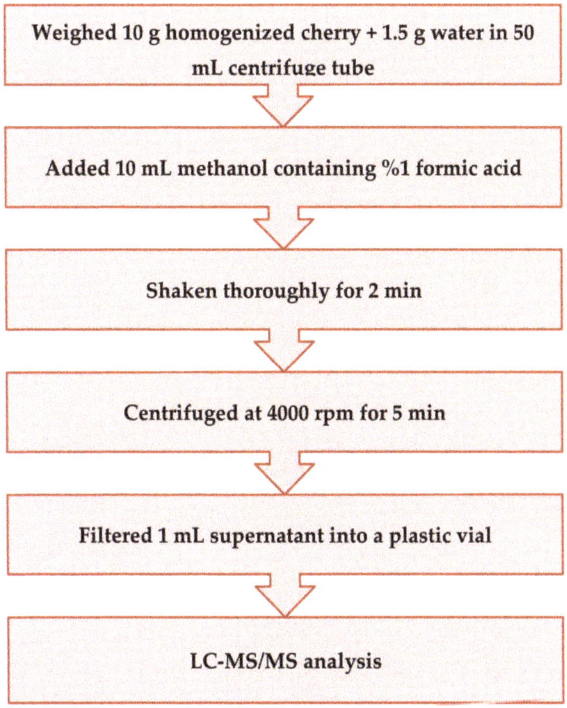

Figure 4. The schematic diagram of the QuPPe method.

Electrospray negative ionization (ESI-) was used for the monitoring of the twelve highly polar compounds and the four ILISs. The multiple reaction monitoring (MRM) settings for each polar compound were optimized by infusing neat standard solutions. The parameters for each target analyte's MRM transition are given in Table 3.

Table 3. MS/MS parameters for the analysis of target polar compounds in the MRM ESI-negative mode.

Analyte	Type of Pesticide [a]	Molecular Formula	t_R (min)	Quantifier (*m/z*)	CE [b] (V)	Qualifier (*m/z*)	CE (V)	Fragmentor (V)
AMPA	HB	CH_6NO_3P	3.04	110 → 63	21	110 → 79	35	116
N-acetyl-AMPA	HB	$C_3H_8NO_4P$	6.81	152 → 110	10	152 → 63	35	94
Chlorate	HB	$ClNaO_3$	5.75	85 → 69	21	83 → 67	21	74
Ethephon	PG	$C_2H_6ClO_3P$	7.87	143 → 107	10	143 → 79	10	72
HEPA	PG	$C_2H_7O_4P$	6.20	125 → 95	14	125 → 79	28	98
Fosetyl-Al	FU	$C_6H_{18}AlO_9P_3$	3.24	109 → 81	12	109 → 63	34	90
Glyphosate	HB	$C_3H_8NO_5P$	8.96	168 → 150	8	168 → 124	10	96
Glufosinate	HB	$C_5H_{15}N_2O_4P$	3.18	180 → 136	16	180 → 63	48	108
N-acetyl-glufosinate	HB	$C_7H_{14}NO_5P$	7.86	222 → 136	23	222 → 59	13	116
Maleic hydrazide	PG	$C_4H_4N_2O_2$	3.64	111 → 83	12	111 → 82	18	114
MPPA	HB	$C_4H_9O_4P$	8.40	151 → 133	12	151 → 107	14	104
Phosphonic acid	FU	H_3PO_3	11.68	81 → 79	15	81 → 63	35	54
Ethephon D4 (ILIS)		$C_2H_2ClO_3PD_4$	7.87	147 → 111	4			60
Fosetyl-Al D15 (ILIS)		$_3C_2D_5HO_3P.Al$	3.20	114 → 82	14			66
Glyhosate-$^{13}C_2$, ^{15}N (ILIS)		$C^{13}C_2H_8{}^{15}NO_5P$	17.9	171 → 63	33			102
$^{18}O_3$-Phosphonic acid (ILIS)		$H_3P^{18}O_3$	7.88	87 → 85	19			60

[a] FU: Fungicide; PG: Plant growth regulator; HB: Herbicide. [b] CE: Collision energy.

3.5. Validation Studies

The performance of the modified QuPPe method was assessed using SANTE/11813/2017 guideline [22]. Matrix-matched multi-residue calibration standards were constructed by adding six different concentrations (5, 10, 25, 50, 50, 100, and 250 $\mu g\ kg^{-1}$) of each polar compound in the blank cherry extract. The calibration curves for each target analyte were prepared by running matrix-matched calibration standards, and R^2 values of >0.99 were acceptable. The method's precision (repeatability and within-laboratory reproducibility) and accuracy were assessed by analyzing blank cherry samples fortified with 0.01 and 0.05 mg kg^{-1} for analytes. The analysis was performed in five replicates (n = 5) at each level. The LOQs were determined as the lowest concentration that provided an accuracy rate of 70–120% and RSD of $\leq$20%. Two sources of uncertainty (uncertainty associated with trueness (bias) and within-laboratory reproducibility) were considered in the determination of expanded measurement uncertainty (U_{exp}) for each analyte, as described in detail previously [23].

3.6. Exposure Analysis and Risk Assessment

The long-term dietary exposure to highly polar substances from the consumption of sweet cherries was calculated by multiplying the residue concentration by sweet cherry consumption data (Equation (1)) [24].

$$Dietary\ exposure = \frac{Concentration\ of\ residue\ in\ food\ \left(\frac{mg}{kg}\right) x\ Food\ consumption\ (kg/day)}{Body\ weight\ (kg)} \quad (1)$$

The non-detect results were treated by the substitution method as described in the EFSA Scientific Report [25]. The left-censored results were input as "zero", "a value of the respective LOQ", and "LOQ/2" according to Lower Bound (LB), Upper Bound (UB), and Middle Bound (MB) scenarios, respectively.

The consumption rate of sweet cherry (0.1107 g kg^{-1} b.w. day^{-1}) from the GEMS/Food G06 cluster diets and a standard body weight of 60 kg have been assumed to calculate dietary exposure to highly polar residues for adults [26].

To assess the health risks of polar residues, the Hazard Quotient (HQ), which was calculated by dividing the potential exposure to a chemical hazard by the reference dose (Equation (2)) as described by Reffstrup et al. [27].

$$Hazard\ Quotient\ (HQ) = \frac{Exposure\ of\ the\ concerned\ residue}{Reference\ value\ (ADI)} \quad (2)$$

4. Conclusions

This study was conducted to monitor twelve highly polar pesticides in sweet cherries intended for export from Turkey to various countries, mainly Russia and European countries. A modified QuPPe method was successfully validated and applied for the analysis of 16,022 cherry samples from 2018–2020 harvests. Among the polar compounds, only phosphonic acid residues, calculated as fosetyl-Al, sum, were detected in cherry samples. Fosetyl-Al was measured in 28.4% of the cherry samples in quantifiable concentrations; 2304 of these samples (14.4%) had fosetyl-Al above the MRL. There is no health risk in the consumption of sweet cherries intended for export from Turkey.

Author Contributions: Conceptualization, E.G. and B.K.; methodology, O.G. and B.K.; validation, E.G. and O.G.; formal analysis, E.G.; investigation, E.G. and B.K.; writing—review and editing, M.Á.G.-C. and B.K.; supervision, B.K.; funding acquisition, M.Á.G.-C. and B.K. All authors have read and agreed to the published version of the manuscript.

Funding: This research received no external funding. The APC was funded by Universidad EAN.

Institutional Review Board Statement: Not applicable.

Informed Consent Statement: Not applicable.

Data Availability Statement: Not applicable.

Acknowledgments: The authors thank A. G. Gozenc for her technical assistance.

Conflicts of Interest: The authors declare no conflict of interest.

Sample Availability: Samples of the compounds are not available from the authors.

References

1. McCune, L.M.; Kubota, C.; Stendell-Hollis, N.R.; Thomson, C.A. Cherries and health: A review. *Crit. Rev. Food Sci. Nutr.* **2010**, *51*, 1–12. [CrossRef] [PubMed]
2. Blando, F.; Oomah, B.D. Sweet and sour cherries: Origin, distribution, nutritional composition and health benefits. *Trends Food Sci. Technol.* **2019**, *86*, 517–529. [CrossRef]
3. FAO (Food and Agricultural Organization). *FAO Statistical Databases and Data Sets*; FAO (Food and Agricultural Organization): Rome, Italy, 2022; Available online: http://www.fao.org/faostat/en/#data/QC (accessed on 26 November 2022).
4. GAIN (Global Agricultural Information Network). Turkey Stone Fruit Annual 2019. GAIN Report Number TR9021. USDA Foreign Agricultural Service. Available online: https://agfstorage.blob.core.windows.net/misc/FP_com/2019/08/21/TurkStone_Fruit_Annual_Ankara_Turkey_8_15_2019.pdf (accessed on 26 November 2022).
5. GTHB (Republic of Turkey Ministry of Food, Agriculture and Livestock). Control of Sweet and Sour Cherries Diseases and Pests. Available online: https://www.tarimorman.gov.tr/GKGM/Belgeler/Uretici_Bilgi_Kosesi/Dokumanlar/kiraz.pdf (accessed on 26 November 2022).
6. GTHB (Republic of Turkey Ministry of Food, Agriculture and Livestock). Kiraz Ve Vişne Entegre Mücadele Teknik Talimatı. General Directorate of Agricultural Research and Policies. 2017. Available online: https://www.tarimorman.gov.tr/ (accessed on 26 November 2022).
7. Da Silva, H.C.M.P.; Bedor, D.C.G.; Cunha, A.N.; Rodrigues, H.O.S.; Telles, T.L.; Araújo, A.C.P.; Santana, D.P. Ethephon and fosetyl residues in fruits from São Francisco Valley, Brazil. *Food Addit. Contam. B* **2020**, *13*, 16–24. [CrossRef] [PubMed]
8. Anastassiades, M.; Kolberg, D.I.; Benkenstein, A.; Eichhorn, E.; Zechmann, S.; Mack, D. Quick method for the analysis of numerous highly polar pesticides in foods of plant origin via LC-MS/MS involving simultaneous extraction with methanol (QuPPe-method)—version 9.3. In *EU Reference Laboratory for Pesticides Requiring Single Residue Methods (EURL-SRM)*; CVUA: Stuttgart, Germany, 2017.
9. Lara, F.; Chan, D.; Dickinson, M.; Lloyd, A.S.; Adams, S.J. Evaluation of direct analysis in real time for the determination of highly polar pesticides in lettuce and celery using modified Quick Polar Pesticides Extraction method. *J. Chromatogr. A* **2017**, *1496*, 37–44. [CrossRef] [PubMed]
10. Melton, L.M.; Taylor, M.J.; Flynn, E.E. The utilisation of ion chromatography and tandem mass spectrometry (IC-MS/MS) for the multi-residue simultaneous determination of highly polar anionic pesticides in fruits and vegetables. *Food Chem.* **2019**, *298*, 125028. [CrossRef] [PubMed]
11. Gasparini, M.; Angelone, B.; Ferretti, E. Glyphosate and other highly polar pesticides in fruit, vegetables and honey using ion chromatography coupled with high resolution mass spectrometry: Method validation and its applicability in an official laboratory. *J. Mass Spectrom.* **2020**, *55*, 4624. [CrossRef] [PubMed]
12. Francesquett, J.Z.; Rizzetti, T.M.; Cadaval, T.R.S., Jr.; Perestes, O.D.; Adaime, M.B.; Zanella, R. Simultaneous determination of the quaternary ammonium pesticides paraquat, diquat, chlormequat, and mepiquat in barley and wheat using a modified quick polar pesticides method, diluted standard addition calibration and hydrophilic interaction liquid chromatography coupled to tandem mass spectrometry. *J. Chromatogr.A* **2019**, *10*, 101–111.
13. Herrera López, S.; Dias, J.; de Kok, A. Analysis of highly polar pesticides and their main metabolites in animal origin matrices by hydrophilic interaction liquid chromatography and mass spectrometry. *Food Control* **2020**, *115*, 107289. [CrossRef]
14. Herrera López, S.; Dias, J.; Mol, H.; de Kok, A. Selective multiresidue determination of highly polar anionic pesticides in plant-based milk, wine and beer using hydrophilic interaction liquid chromatography combined with tandem mass spectrometry. *J. Chromatogr. A* **2020**, *1625*, 461226. [CrossRef] [PubMed]
15. López Ruiz, R.; Romero-González, R.; Frenich, A.G. Simultaneous determination of polar pesticides in human blood serum by liquid chromatography coupled to triple quadrupole mass spectrometer. *J. Pharm. Biomed. Anal.* **2020**, *190*, 113492. [CrossRef] [PubMed]
16. EFSA (European Food Safety Authority). The 2015 European Union report on pesticide residues in food. *EFSA J.* **2017**, *15*, 4791.
17. CVUA Stuttgart (Chemisches und Veterinäruntersuchungsamt). Residues and Contaminants in Fresh Fruit from Conventional Cultivation. 2018. Available online: https://www.ua-bw.de/pesticides/beitrag_en.asp?subid=1&Thema_ID=5&ID=2931&Pdf=No&lang=EN (accessed on 26 November 2022).
18. Golge, O. Validation of quick polar pesticides (QuPPe) method for determination of eight polar pesticides in cherries by LC-MS/MS. *Food Anal. Methods* **2021**, *14*, 1432–1437. [CrossRef]
19. Petersen, A.; Jensen, B.H.; Andersen, J.H.; Poulsen, M.E.; Christensen, T.; Nielsen, E. *Pesticide Residues. Results from the Period 2004–2011*; National Food Institute, Technical University of Denmark: Kongens Lyngby, Denmark, 2013; p. 138.

20. FAO (Food and Agricultural Organization). Fosetyl-Aluminium, Aluminium Tris (Ethylphosphonate). FAO Specifications and Evaluations for Agricultural Pesticides. Available online: http://www.fao.org/agriculture/crops/core-themes/theme/pests/jmps/ps-new/en/ (accessed on 26 November 2022).
21. EFSA (European Food Safety Authority). Conclusion on the peer review of the pesticide risk assessment of the active substance potassium phosphonates. *EFSA J.* **2012**, *10*, 2963.
22. European Commission. *Guidance Document on Analytical Quality Control and Method Validation Procedures for Pesticides Residues and Analysis in Food and Feed*; Document No. SANTE/11813/2017; EC Directorate-General for Health and Food Safety: Brussels, Belgium, 2017.
23. Golge, O.; Hepsag, F.; Kabak, B. Health risk assessment of selected pesticide residues in green pepper and cucumber. *Food Chem. Toxicol.* **2018**, *121*, 51–64. [CrossRef]
24. IPCS (International Programme on Chemical Safety). *Principle and Methods for the Risk Assessment of Chemicals in Food. Environmental Health Criteria 240*; World Health Organization: Stuttgart, Germany, 2009.
25. EFSA (European Food Safety Authority). Management of left-censored data in dietary exposure assessment of chemical substances. *EFSA J.* **2010**, *8*, 1557.
26. EFSA (European Food Safety Authority). Guidance on the use of EFSA pesticide residue intake model (EFSA PRIMO revision 3). *EFSA J.* **2018**, *16*, 5147.
27. Reffstrup, T.K.; Larsen, C.J.; Meyer, O. Risk assessment of mixtures of pesticides. Current approaches and future strategies. *Regul. Toxicol. Pharmacol.* **2010**, *56*, 174–192. [CrossRef] [PubMed]

molecules

MDPI

Concept Paper

Quality Control of Pesticide Residue Measurements and Evaluation of Their Results

Árpád Ambrus [1,*], Vy Vy Ngoc Doan [2], Júlia Szenczi-Cseh [3], Henriett Szemánné-Dobrik [4] and Adrienn Vásárhelyi [5]

[1] Doctoral School of Nutrition and Food Sciences, University of Debrecen, 4032 Debrecen, Hungary
[2] Southern Pesticide Control and Testing Center, Plant Protection Department, 71007 Ho Chi Minh City, Vietnam
[3] Freelancer Food Safety Adviser, 1116 Budapest, Hungary
[4] Food Chain Safety Centre, Non-profit Ltd., Pesticide Residue Analytical Laboratory, 3529 Miskolc, Hungary
[5] Food Chain Safety Laboratory Directorate, 1095 Budapest, Hungary
* Correspondence: ambrusadr@yahoo.co.uk

Abstract: Pesticide residues are monitored in many countries around the world. The main aims of the programs are to provide data for dietary exposure assessment of consumers to pesticide residues and for verifying the compliance of the residue concentrations in food with the national or international maximum residue limits. Accurate residue data are required to reach valid conclusions in both cases. The validity of the analytical results can be achieved by the implementation of suitable quality control protocols during sampling and determination of pesticide residues. To enable the evaluation of the reliability of the results, it is not sufficient to test and report the recovery, linearity of calibration, the limit of detection/quantification, and MS detection conditions. The analysts should also pay attention to and possibly report the selection of the portion of sample material extracted and the residue components according to the purpose of the work, quality of calibration, accuracy of standard solutions, and reproducibility of the entire laboratory phase of the determination of pesticide residues. The sources of errors potentially affecting the measured residue values and the methods for controlling them are considered in this article.

Keywords: pesticide residues; quality control procedures; sources of errors of residue analyses; reproducibility of results

Citation: Ambrus, Á.; Doan, V.V.N.; Szenczi-Cseh, J.; Szemánné-Dobrik, H.; Vásárhelyi, A. Quality Control of Pesticide Residue Measurements and Evaluation of Their Results. *Molecules* **2023**, *28*, 954. https://doi.org/10.3390/molecules28030954

Academic Editor: Miguel Ángel González-Curbelo

Received: 29 November 2022
Revised: 10 January 2023
Accepted: 12 January 2023
Published: 18 January 2023

1. Introduction

A sufficient amount of safe food cannot be provided for the continuously growing population of the world without the use of pesticides at the current technological level. The global demand for, and the production as well as the use of pesticides have increased steadily during the past decades and are projected to continue growing [1,2]. Pesticides are chemical substances with various degrees of toxicity and modes of action [3,4]. To control the target pests certain concentrations of pesticide residues must remain in/on the treated species. Consumers are generally concerned about the toxic chemicals in their food. According to the survey conducted by the European Food Safety Authority, pesticide residues in food (40%) and antibiotic, hormone, or steroid residues in meat (39%) are the main food safety-related concerns among Europeans [5].

To protect consumers and the environment, the national authorities authorize the use of pesticides only after the critical evaluation of their toxicity, biological efficacy and residues remaining in/on food as well as in the environment [6–11]. The OECD Guidelines for Testing of Chemicals are a collection of the most relevant internationally agreed testing methods used by government, industry, and independent laboratories [12]. They are intended to enhance the validity and international acceptance of test data and reduce unnecessarily repeated tests [13–15]. Many non-OECD member countries adopt the same principles [12] or give permission for use only after [16–18] a pesticide active ingredient has been authorized by countries having an advanced registration system [8,10,11]. To facilitate international trade and assist the national registration authorities to establish their own limits, the CODEX maximum residue limits, MRLs, are elaborated by the FAO/WHO Joint Meeting on Pesticide Residues, JMPR, [19], further considered in a stepwise procedure by the Codex Committee on Pesticide Residues, and approved by the Codex Alimentarius Commission [20,21].

To control the safe and efficient use of pesticides, their residues are regularly monitored in food and environmental samples in many countries according to risk-based sampling plans [22–30] or targeted surveillance with limited scope and sampling targets. For example, the world-wide activities are demonstrated with some selected publications from Argentina to Vietnam [31–46]. In the European Union the largest number of residues tested within the EU-coordinated and national pesticide residue monitoring programs were reported in 2020 by Luxembourg (659), Malta (643), Germany (626), France (619) and Belgium (617) [47]. Concerning all 30 countries reporting their monitoring results to EFSA, multiple residues were detected in 27.2% of the samples, and 30%, 22.3%, 4.1%, 0.5%, and 0.02% of samples contained 0, 1, 5, 10 and 15 different residues, respectively. However, in extreme cases 18 and 31 residues were detected in single strawberry [47] and honeysuckle samples [48]. These results underlined the importance of applying screening methods of the widest possible scope with low limit of detection/limit of quantification (LOD/LOQ) values. For this purpose, good progress has been made in expanding the scope of the methods [49–51].

Most publications referenced above [30–50] mainly reported minor modifications in sample preparation procedures of the original QuEChERS (Quick, Easy, Cheap, Effective, Rugged, and Safe) method [52] and often provided details of the conditions of the MS mass spectrometry (MS) detection. Other authors reported various combinations of sample preparation [53–61]. The authors typically stated the recoveries, linearity, LOD and LOQ values, matrix effects and compared them to the acceptance criteria specified in the major guidance documents [62–64]. On the other hand, none of them provided information on the details of sampling, efficiency of subsampling and comminution affecting the reproducibility of the results, or accuracy of reference standard solutions, albeit these steps can be major hidden sources of random and systematic errors [65–70].

Drawing realistic conclusions and making appropriate corrective actions can only be done if the monitoring results are accurate and derived from the analyses of samples taken according to the specific objectives of the program. That can only be achieved by implementing rigorous internal quality control of the whole process of the determination of pesticide residues. The basic quality requirements for the monitoring results are defined in five major guidance documents [62–64,70,71]. However, several potential hidden errors are not explicitly addressed in these documents. Although over the last two decades several scientific publications have highlighted the effect of these errors on the accuracy and uncertainty of the measurement results [65,67,68,72–75], the actions for limiting them have rarely been reported in the monitoring studies. Therefore, the reliability of these study results cannot be assessed. Table 1 summarizes the main steps of residue analyses and gives examples for the sources of potential errors.

Every laboratory should introduce and implement appropriate quality control procedures to assure that the results of the analyses are as accurate as possible, and that their uncertainties are kept as low as practical. The random error indicated by the combined relative uncertainty of the results (CV_R) is influenced by four main factors (Equation (1)): sampling (S), laboratory sample handling including subsampling of large crops (CV_{SS}), comminution (CV_{Sp}), test portion selection and analyses of sample extracts (CV_A) [66].

$$CV_R = \sqrt{CV_S^2 + CV_{SS}^2 + CV_{Sp}^2 + CV_A^2} \quad (1)$$

The CV_R incorporates the relative precision of all steps of the determination of pesticide residues including sampling.

Table 1. Examples for the sources of errors in the results of pesticide residues determination.

Operation	Potential Sources of Errors Random	Systematic
Sampling	Sample size; heterogeneous distribution of analyte; varying temperature during shipping and storage	Sampling target selection; sampling plan and method; degradation, evaporation of analyte; contamination of the sample; mislabeling
Selection of the portion of commodity to be analyzed	Inconsistent preparation of sample portion	Wrong part of the sample selected for extraction,
Sample size reduction, subsampling	Subsample does not represent the composition of the laboratory sample	primary samples are not proportionally represented
Comminution of selected sample portions	Particle size distribution in the homogenate; varying temperature and duration of comminution	Decomposition, evaporation of analytes
Test portion selection	Test portion does not represent the comminuted sample matrix	
Extraction	Varying intensity and temperature of extraction	Efficiency of extraction
Clean-up	Variation in the composition (e.g., water, fat, and sugar content) of sample materials;	Loss of analyte
Qualitative/quantitative determination of residues	Changing the retention time—shifting mass acquisition window; linearity and confidence intervals of calibration	Deviation from residue definition; missing analytes present in targeted or non-targeted analyses; high LOD; inaccurate standard solutions; matrix effect

The analysts usually only report the within-laboratory repeatability/reproducibility of steps from the extraction of the test portions (CV_A). On the other hand, the reproducibility (CV_L) is the parameter that realistically characterizes the laboratory measurements including all steps from subsampling to the quantitative determination of residues.

$$CV_L = \sqrt{CV_{SS}^2 + CV_{Sp}^2 + CV_A^2} \quad (2)$$

The analyses phase can be further subdivided into extraction (Ex), clean-up (Cl), evaporation (Ev), and chromatographic determination (Ch):

$$CV_A = \sqrt{CV_{Ex}^2 + CV_{Cl}^2 + CV_{Ev}^2 + CV_{Ch}^2} \quad (3)$$

However, the individual quantification of the contributions of the steps affecting CV_A can only be done in practice with applying isotope labelled compounds in specialized laboratory conditions with specific detection instruments [74]. Therefore, their combined effect should be determined in practice with repeated recovery tests (CV_A) performed at the concentration range that is expected to occur in the samples. Such tests reflect only the effect of operations carried out after spiking the test portions. If the tests are carried out on different days by different analysts the calculated relative standard deviation of the results will only indicate an interim reproducibility of the analyses step, but it is not equivalent to CV_L as defined by Equation (2). The results of recovery tests can be used to characterize the within-laboratory reproducibility (CV_L) only if they are performed with samples containing incurred residues derived from the prior application of a pesticide [65,72,74].

Our objectives are to call attention to the hidden errors in the analyses of pesticide residues that can significantly affect the accuracy and reliability of the results. Without aiming for a full review of the vast amount of published data, we describe some practical options for the quality control actions that the program managers can get implemented by the laboratory staff to obtain accurate results with quantified uncertainty.

2. Methods

2.1. Sampling

The main objectives of the monitoring program, and in general the analyses of samples, are to obtain correct information with known uncertainty on the pesticide residue levels in the sampling targets and not only in the sample. It is generally recognized that the accuracy and validity of analytical results cannot be better than that of the samples analyzed. The sampling designs and methods are widely described in the scientific literature. Their coverage is beyond the scope of this paper. Briefly, for the monitoring of the pesticide residues in plant commodities and soil, stratified random sampling is the best choice. The sampling target (the area from where the samples are to be collected) can be stratified for instance according to crop, cultivation mode, growing season, soil type, etc. Random samples should be separately taken from each stratum. The minimum number of primary samples to be collected for one composite sample (sample size) depends on the objectives of the program. For instance, the provisions of the Codex sampling standard [70] for the minimum number of primary samples and total mass of a composite laboratory sample should be satisfied where the compliance with MRLs is assessed in goods offered for sale. It is not sufficient to collect [76], for instance, 4 pieces of head cabbages or Chinese cabbages (instead of the minimum five specified in the Codex GL) even though their total mass may be well over 10 kg and 4–5 kg, respectively, which are much larger than the specified minimum of 2 kg. A larger number of primary samples may be collected than the minimum, provided that their representative part can be effectively comminuted with the available laboratory equipment.

The sampling uncertainty is inversely proportional to the sample size (n, the number of primary samples) and depends on the variability of residues in crop units or in single sample increments (CV_1):

$$CV_S = \frac{CV_1}{n} \quad (4)$$

The variability of residues in individual crop units derived from a single field (called within field variability) is close to 80–100-fold [77,78], therefore increasing 'n' will decrease the uncertainty of the results and improve the accuracy of the estimated average residue in the sample. Under typical growing conditions the relative uncertainty of sampling is in the range of 25–40% for samples of size 10 and 5, respectively [79,80]. These uncertainties shall be considered when a product is tested before export.

2.2. Selection of Portion of Sample to Be Analysed

For testing compliance with MRLs, the portion of commodities specified in the Codex CAC/GL-41-1993 standard should be considered [81]. However, for providing data to estimate the dietary exposure of consumers, the edible portion of commodities should be analyzed. Since the edible portion varies and for instance depends on the variety, maturity of the crop, and local practices for its consumption. Consequently, the specific way of selecting the edible portions should be precisely described in the publications to enable the comparison of the results with other studies.

Most of the operations required for the preparation of the test portions depend very much on the actual condition of the test item and cannot be generally standardized. The laboratory assistants should be well trained on the principles enabling them to perform the tasks properly. Inconsistent operation may lead to high, uncontrollable variability (unquantifiable uncertainty) of the results. For instance, the way of removing adhering soil from root vegetables or outer, withered leaves from leafy vegetables can substantially influence the residues measured. The outer leaves usually contain much higher residues than the inner leaves. Therefore, only the loose leaves should be removed (Figure 1) otherwise the measured residues will not correctly reflect the residue content and may lead to dispute if the lot is repeatedly sampled along the commercial chain.

Figure 1. Phases of preparation of head cabbage for analysis: (**a**) collecting head cabbage; (**b**) removing outer leaves; (**c**) obtaining portion of commodity to be analysed after cutting off the stalk.

To prepare samples for the analyses of the edible portion the peeling of fruits with inedible peel should be made in a way that the edible part is not cross contaminated by the residues being on the peel. Large fruits (e.g., watermelon, pumpkin, jackfruit) should be cut into wedge-shaped sections and the flesh part removed with proper spoons as shown in Figure 2. It should be noted that for checking compliance with MRLs, the whole fruit shall be comminuted and further processed. It is recommended that one section from each of the five large crops making up one composite sample according to the Codex sampling standard [70] is used for determining the residues in edible portions and a second set of five sections is comminuted for determination of residues in/on whole fruits.

Figure 2. Processing of jackfruit: top left: cutting wedge-shaped section from jackfruit; top right: comminution of the whole fruit; bottom: peel remaining after removal of edible part.

2.3. Subsampling and Comminution of Selected Sample Portions

Because of the usually very large difference in the concentration of residues in individual crop units [78], the whole laboratory sample or representative part of each primary sample (crop unit) must be processed to obtain accurate information on the average residue in the laboratory sample. Omeroglu [82] and Ambrus [83] provided detailed graphical illustrations for obtaining representative subsamples and calculation of CV_L.

The distribution of residues within the natural crop units is also uneven. For instance, the residues concentrate on the lower part of fruits hanging on the trees or vines due to the runoff of the sprays. Therefore, slices should never be cut from crop units (Figure 3).

Figure 3. Distribution of residues in/on crop units: Left: Residues concentrate on the bottom part of fruits hanging on the trees; Right: Cutting the middle of cucumber leads to biased result. Slices should never be cut for subsampling as emphasized by the crossing red lines.

Obtaining representative portion of the large crop units (e.g., cabbage, watermelon, papaya, etc.) requires special attention making sure that each crop unit is proportionally represented in the subsample to be comminuted. Figure 4 illustrates the subsampling of large fruits.

The efficiency of cutting, blending of the sample materials may vary from day-to-day and sample-to-sample because of the changing physical properties and textural composition of crops depending on the variety and maturity. Moreover, it is strongly influenced by the sharpness of cutting blades. The fundamental sampling error defined by Gy [84] can be applied for characterizing the relative variance of the residues in comminuted materials [85].

$$CV_{Sp}^2 = \frac{C \times d^3}{w} \quad (5)$$

In Equation (5), the C is the sampling constant depending on the nature of the homogenized material, d is the diameter of the 95th percentile of the comminuted particles, and w is the mass of the test portion. Though CV_{Sp} cannot be calculated for plant materials applying Gy's theory, Equation (5) clearly indicates the importance of particle size (d^3) distribution. Reducing the particle size in a comminuted laboratory sample considerably reduces CV_{Sp} and consequently CV_L (Equation (2)). Therefore, the proper homogeneity of the comminuted materials should be checked for each sample.

A very quick and convenient method for this purpose is the 'Petri dish' test, in which a small portion of the comminuted material is spread on the glass surface and the particle size distribution is visually checked. If the particles are smaller than 2 mm, the homogeneity would be generally sufficient to keep CV_{Sp} smaller than 10–12% if 10–15 g test portions are taken for extraction [73,86]. Otherwise, the comminutions should be continued preferably by adding a further portion of dry ice [75,87]. Figure 5 provides some examples. Much smaller particles can be obtained, and considerably reduced test portions can be used when liquid nitrogen is used for cryogenic processing [88–90]. The two-stage sample processing can also be used in the combination of pre-homogenization of a large sample with proper choppers (C_L), then transferring its representative 100–150 g portion into a Waring laboratory blender (or a baker if Ultra Turrax is used for fine cutting), and adding about 10% known amount of distilled water for fine comminution (C_F) [83].

Figure 4. Cutting wedge-shaped portions of large fruits. Top left: papaya fruit; Top right: papaya subsamples; Bottom left: winter squash; Bottom right: sections of winter squash. Note: that the cut portions from each fruit should be kept in separate groups until taking the required number of pieces from each group for comminution.

Figure 5. Top row left Blending winter squash; top center and right: tomato homogenates on filter paper and Petri dish; second row: cabbage leaves homogenized to different particle sizes. The star marks the acceptable particle size distribution.

It is generally recommended to add a small portion of water to dry materials to improve the efficiency of comminution [62,63]. The exact amount of added water shall be accounted for in reporting the residue concentrations. The portions for further processing should be taken without delay in small increments (preferably > 10) of the test material from various positions of the blender to obtain representative test portion and avoid segregation.

The CV_{Sp} will be determined by the combined effects of the two comminution steps.

$$CV_{Sp} = \sqrt{\frac{C_L}{w_L} + \frac{C_F}{w_F}} \tag{6}$$

Equation (6) should also be applied for estimation of sample processing uncertainty in case of two-stage processing with liquid nitrogen [88,89]. The CV_{Sp} will depend on the C_L/w_L ratio. It is misleading to report the repeatability/reproducibility based on the analyses of spiked portions taken after fine comminution with liquid nitrogen.

The size of test portion significantly affects the reproducibility of the measurements. Based on Gy's sampling theory the relationship between the mass of the comminuted laboratory sample (m_L), the test portion (m_{TP}), and the CV_{Sp} can be described as [85]:

$$CV_{Sp}^2 = Cd^3\left(\frac{1}{m_{Tp}} - \frac{1}{m_L}\right) \tag{7}$$

Table 2 shows the change of CV_{Sp} depending on the test portion size taken from the same comminuted material.

Table 2. Change of CV_{Sp} as a function of the test portion mass.

	CV_{Sp}	
m_L [g]	>5000	1000
T_p [g]		
1	0.387	0.387
2	0.274	0.274
5	0.173	0.173
10	0.122	0.122
15	0.100	0.099
25	0.077	0.076

Table 2 indicates that reducing the test portion size from 15 g to 1 g will increase the CV_{Sp} by about 3.2 times. For instance, if the CV_{Sp} is 12.2% when a 10 g test portion containing incurred residues is extracted, and then one gramme portions are also taken from the same comminuted matrix, the theoretically expected CV_{Sp} would be about 38.7%. Naturally, the measurable CV_L will depend on the combined contribution of CV_{Sp} and CV_A according to Equation (2) (CV_{SS} is zero in this case). Provided that the CV_A from recovery tests is 10%, and the CV_{Sp}-s from Table 2 are 12.2% and 38.7%, the corresponding CV_L would be 15.8% and 40%, respectively, if 10 g and 1 g test portions were extracted from the same comminuted material. This significant effect remains unnoticed when the recoveries are determined with spiking the test portions. Therefore, making use of the high sensitivity of the recent MS systems and extracting 1–2 g test portions should only be done after careful checking of the reproducibility of the method with incurred residues, otherwise the real variability of the results may not be reflected [91–93]. A practical solution is to extract 5–10 g portions and dilute the extracts to utilize the sensitive detection and reducing the matrix effect [45,49].

2.4. *Definition of Residues*

Where the toxic metabolites or degradation products are present in a treated commodity in toxicologically significant proportion, they should be considered for the determination of the dietary exposure of consumers to pesticide residues. The principles are explained, for instance, in the FAO/WHO JMPR Manual [93]. The analyses of polar metabolites that are often present in conjugated form requires specific procedures and cannot be determined with the usual multi-residue methods. To facilitate testing the compliance with MRLs carried out in large number of samples, the regulatory authorities often establish different definitions of residues for monitoring and risk assessment purposes. The JMPR emphasized that the definition of residues for enforcement purposes should be as practical as possible and preferably based on a single residue component (the parent compound, a metabolite, or a derivative produced in an analytical procedure) as an indicator of the total significant residue, and it should be determinable with a multi-residue procedure whenever possible [93]. Some examples for the different residue definitions are highlighted in Tables 3 and 4 [94,95].

Table 3. Different residue definitions for flupyradifurone.

Flupyradifurone [4-[(6-chloro-3-pyridylmethyl)(2,2-difluoroethyl)amino]furan2(5H)-one	
definition of the residue (for compliance with MRLs) for plant commodities	flupyradifurone
definition of the residue (for dietary risk assessment) for plant commodities	sum of flupyradifurone, difluoroacetic acid and 6-chloronicotinic acid, expressed as parent equivalents

Table 4. Different residue definitions for fluxapyroxad.

Fluxapyroxad [3-(difluoromethyl)-1-methyl-N-(3′,4′,5′-trifluoro [1,1′-biphenyl]-2-yl)-1H-pyrazole-4-carboxamide]	
definition of the residue (for compliance with the MRL for plant and animal commodities)	fluxapyroxad
definition of the residue for estimation of dietary intake for plant commodities	sum of fluxapyroxad and 3-(difluoromethyl)-N-(3′,4′,5′-trifluoro[1,1′-biphenyl]-2-yl)-1H-pyrazole-4-carboxamide (M700F008) and 3-(difluoromethyl)- 1-(ß-D glucopyranosyl)-N-(3′,4′,5′-triflurobipheny-2-yl)-1Hpyrzaole-4-carboxamide (M700F048) and expressed as parent equivalents
for estimation of dietary intake for animal commodities	sum of fluxapyroxad and 3-(difluoromethyl)-N-(3′,4′,5′-trifluoro[1,1′-biphenyl]-2-yl)-1H-pyrazole-4-carboxamide (M700F008) expressed as parent equivalents; the residue is fat soluble

The definition of residues in commodities of animal origin is often much more complex. The list of Codex MRLs indicates the residues to be tested for checking compliance with MRLs and for risk assessment purposes [21]. The latest recommendations of the JMPR can be found in the JMPR reports [96]. Alternately, the proper composition of residues can be accessed from the websites of the national registration authorities [10,97].

The examples above underline the importance of adhering to the residue definition that fits for the objectives of the study in order to obtain accurate results. Due to the inclusion of metabolites the total residue for risk assessment purposes can be much higher than that for monitoring purposes. In such cases, the calculation of estimated daily intake (EDI) based on the residues defined for monitoring purposes will underestimate the real exposure of consumers and result in wrong conclusions.

2.5. Extraction of Residues and Cleanup of Extracts

The selection of solvents and adjusting the pH to obtain acceptable recoveries have been extensively studied, providing sufficient information for the optimization of the procedures for various matrix-analyte combinations. A detailed guidance document for testing the efficiency of extraction [98] provides the basis, if followed, for obtaining accurate results. The efficiency of extraction should always be tested with incurred residues in all kinds of samples.

2.6. Accuracy of Standard Solutions

It is evident for every analyst that the accuracy of standard solutions is one of the very basic pre-conditions for the correct quantification of the residues. We cannot assume that the analytical standard prepared in our laboratory is accurate unless it is verified. To assist the laboratories participating in EU proficiency tests to find out the reasons for unsatisfactory results, the EU Reference Laboratory for Pesticide Residues in Fruits and Vegetables organized a ring test for the determination of the concentrations of certified pesticide analytical standards provided in a mixture. Forty official and national reference laboratories from 20 countries took part in the tests [99]. The summary of results is given in Table 5. The accuracy and uncertainty of the analytical standards may be affected by their storage and handling conditions.

Table 5. Summary of results of EU-RT-FV-17 [a].

	Bupirimate	**Carbendazim**	**Diazinon**	**Difenoconazole**	**Diflubenzuron**	**Methoxyfenozide**	**Pendimethalin**	**Permethrin**	**Spinosad**	**Thiabendazole**	**Trifloxystrobin**
Certified conc. mg/L	5.00	5.00	5.04	18.99	18.96	14.95	4.97	15.05	15.03	19.04	19.00
No. Lab	33	31	36	34	25	30	35	32	30	32	33
Accurate	2	1	0	0	0	0	0	0	0	0	0
Rel dif.% Min	−74.2	−86.6	−41.5	−43.7	−40.9	−51.5	−59.8	−54.1	−36.1	−32.5	−36.0
Rel dif.% Max [b]	40.0	164	202	36.9	129	107	28.0	73.4	91.0	116	118
No ≥ 10%	17	23	18	19	16	18	19	23	19	16	19

[a]: Courtesy of Carmen Ferrer Amate; [b]: rounded to 3 digits; No. Number of laboratories: reported result; Accurate: Certified = reported; No ≥ 10%: number of laboratories reported >10% rel. difference.

All laboratory equipment used for the preparation of analytical standards have their own inherent uncertainty of the nominal volume that is combined with the variability of filling them to mark depending on the daily performance of the analysts. Various manufacturers provide volumetric glassware of different grades. The relative uncertainty of the measured volume can be calculated from their specified tolerance (e.g., 50 ± 0.05 mL) assuming triangular distribution [100],

$$u = 0.05/\sqrt{6} = 0.02 \text{ mL},\ CV = 0.02/50 = 4.08 \times 10^{-4}$$

The combined uncertainty (CV_{exp}) of volumetric measurements can be calculated from the tolerance of the glassware (CV_T) and the variability of filling them to mark (CV_{fil}):

$$CV_{exp} = \sqrt{CV_{fil}^2 + CV_T^2} \quad (8)$$

Involving our technicians making most accurately the volumetric measurements based on prior tests, we determined the relative uncertainties of filling in the volumetric glassware [100]. An example of the results is given in Table 6.

Table 6. Example of reproducibility of filling A-grade volumetric flasks.

Vol. Flasks	Specification	CV_T	CV_{Rfil}	CV_{Rexp}
25 mL	±0.03 mL	4.899×10^{-4}	7.30×10^{-3}	7.32×10^{-3}
50 mL	±0.05 mL	4.082×10^{-4}	7.59×10^{-4}	8.61×10^{-4}

Using our five-digit analytical balances, the weighing relative uncertainty of 25 mL water is 1.6×10^{-6}. It is three magnitudes lower than the volumetric measurement (Table 6). Therefore, the diluted standard solutions should be prepared based on weighing except the last step where an A-grade $\geq$ 25 mL volumetric flask should be used to obtain the standard concentration in mass/volume (e.g., µg/mL) [65].

We tested the reproducibility of the preparation of diluted standard solutions with the combination of weighing and volumetric measurements according to the regular practice in two of our laboratories [101]. The relative differences were calculated for the nominal concentrations. Some of the results are summarized in Table 7.

Table 7. Example for the reproducibility of preparation of analytical standard solutions.

C_0 mg/mL	CV_R	Effective Concentration			Deviation [1] from C_0 [%]		
		C_{min}	C_{ave}	C_{max}	Min	Average	Max
1	0.0079	0.9840	1.0044	1.0084	−1.6%	0.44	0.84
0.005	0.0086	0.0050	0.0052	0.0053	0%	4.00	6.00
0.001	0.0103	0.0010	0.0011	0.0011	0%	10.00	10.0

[1]: Deviation of the effective concentration from the nominal concentration [C_0].

The results indicated that the analytical standards can occasionally deviate by 10% from the nominal concentration even with the most careful and precise preparation. The relative uncertainty of nominal concentration (CV_{Rep}) increases with the increasing dilution of the standard solutions. These findings underline the importance of verifying the accuracy of analytical standard solutions. As a minimum, two new solutions should be prepared independently and their average chromatographic responses from minimum 5 replicate injections should be compared. If their relative difference is less than 10%, the two solutions can be combined for use. If the difference is larger, then a 3rd solution should be prepared, and the two closest ones can be combined.

The same procedure can be used for comparing the old standard solution with the new one. The SANTE Guidance document suggests accepting the two solutions (old and new or two new ones) if their relative difference is less than 10% [62]. It is pointed out that the *t*-test comparing the mean values cannot be used for this purpose, because it is designed to prove that the two mean values are not significantly different. Instead, the two-sample *t*-test (TOST) should be used to correctly verify that the relative difference between the two mean values is $\leq$10% [102]. As an alternative to the relatively complicated calculations, Figure 6 can be used for visually testing that the two standard solutions are within the targeted range ($\Delta_{rd} \leq 10\%$). Based on a minimum of 5 replicate injections of both standard solutions the relative difference Δ_{rd} is calculated as

$$\Delta_{rd} = \overline{C}_{diff}\% = 100 \times \frac{\overline{C}_{new} - \overline{C}_{old}}{C_{new}} \quad (9)$$

If the pooled relative standard deviation of the responses is above the critical decision line we cannot state with 95% probability that the relative difference is within the acceptance criterion (10%). Further on, the figure indicates that $|\Delta_{rd}|$ is inversely proportional to CV_p. The closer the $|\Delta_{rd}|$ to 10% the smaller the CV_p must be to verify compliance with the $\leq$10% criterion.

Since the typical repeatability CV of replicate injections into LC-MS/MS is about 2–2.5%, the maximum difference between the two standard solutions which can be stated 'equivalent' is about 7.67% and 7.09%, respectively. The calculation with Equation (10) is shown in Table 8.

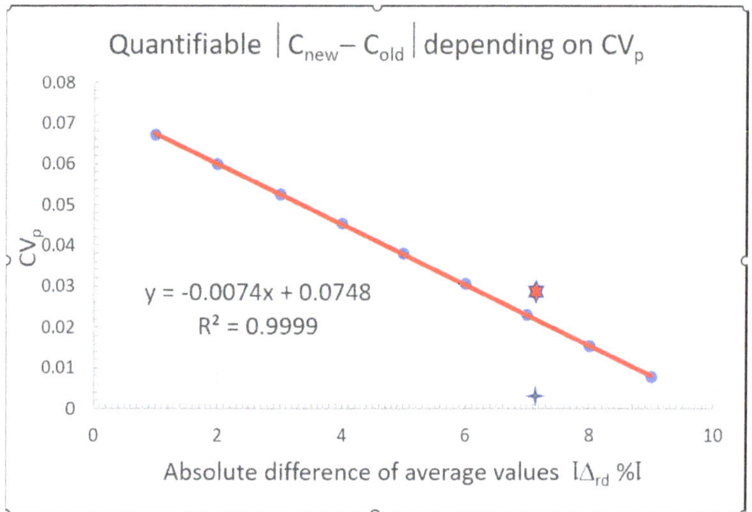

Figure 6. Relation of the absolute value of Δrd and maximum CV_p that can be used to verify that $\Delta rd \leq \Theta = 10\%$ at 95% probability level if $n_1 = n_2 = 5$. The symbols ✦ and ✶ indicate that the acceptance criterion ($\Delta rd \leq 10\%$) can or cannot be confirmed, respectively. Taken with permission from [103].

Table 8. Testing the difference in nominal concentrations of analytical standard solution.

	Standard A1	Standard A2	Standard B1	Standard B2
	121315	112823	123453	114811.3
	121525	112813	131282	122092.3
	121310	113000	123456	114814.1
	121401	113121	124356	115651.1
	121392	112802	123451	114809.4
Ave	121388.6	112911.8	125199.6	116435.6
Δrd	7.2%		7.3%	
CV	0.000718	0.001262	0.027337	0.027337
CVp	0.001027		0.027337	
	✦		✶	

The symbols ✦ and ✶ indicate that the acceptance criterion ($\Delta_{rd} \leq 10\%$) can or cannot be confirmed, respectively.

The pooled CV_P is calculated as:

$$uCV_P = \sqrt{\left(CV_{A1}^2 + CV_{A2}^2\right)/2} \quad (10)$$

The corresponding Δ_{rd} and CV_P values are plotted on Figure 6. It can be seen that for the B standard solutions the CV_p is above the critical line. Consequently, according to the TOST calculation, we cannot state with 95% probability that the difference between B1 and B2 standard solution is $\leq$10%.

2.7. Stability of Analytes

The pH of the plant fluids, enzymes released during the cutting, chopping of sample material can decompose sensitive analytes [87]. The analytes remaining in the final extract are also influenced by their physical−chemical properties, the temperature of comminution and mass of the laboratory sample. The disappearance of captan and dithiocarbamate residues during sample comminution was already observed in the middle of the 70s [104]. Hill reported the decomposition of chlorothalonil and phthalimide type of compounds especially in lettuce and onion [86]. The procedure for the determination of the stability of analytes was reported in the case of tomato, lettuce and maize [105]. It was found that buprofezin and chlorpyrifos did not decompose in the tested matrices at ambient temperature either. Their recoveries were well reproducible and close to 100%, therefore they can be used as reference compounds for assessing the stability of other analytes by comparing their residue concentrations surviving after comminution. Since the recovery tests performed with spiked test portions before extraction do not reveal any information on the stability of analytes, the stability tests should be executed as part of the method validation or performance verification for the new analyte matrix combinations with surface-treated sample material.

The stability test is practically the same as the procedural recovery. Its performance briefly described hereunder:

(a) Take about two or more kg of the crop in which the stability of analytes will be tested.
(b) Prepare the portion of the commodity to be analyzed according to Codex CAC/GL 41-1993 [81] from the whole laboratory sample.
(c) Use approximately half of the sample matrix for the stability test and the remaining part for the recovery tests performed with spiked test portions as usual.
(d) Prepare analytical standard mixture of exactly known concentration of compounds to be tested together with buprofezin (Bu) and chlorpyrifos (Ch) at well detectable concentrations keeping in mind the total mass of the sub-sample to be processed. The number of pesticides or metabolites included in the mixture is limited only by the capability of the chromatographic separation and detection system.
(e) Take about 1/3 of the part of the laboratory sample (e.g., 3–4 units out of 10 fruits) obtained in step 3;
(f) Treat the surface of the selected portion applying either Hamilton syringe for carefully spreading the standard mixture (step 4) on the surface of the crops or injecting the standard solution into the flesh of the fruit [106]. Use liquid dispenser to treat leafy vegetables or small-size crops. Perform the treatment in a fume cupboard over a tray with filter paper which can absorb the runoff. The exact amount of standard mixture that remains on the crop surface need not be known as the concentration ratios of the reference and test compounds will be calculated.
(g) Steps for the treatment of the surface of tomatoes:
    i. Place the surface-treated portions into the chopper together with the remaining 2/3 part of the sample and comminute the whole matrix. By this way you represent a potentially worst-case scenario for testing the efficiency of sample processing and determination of CV_L at the same time as testing the stability of analytes. The test may be performed both at ambient temperature and under cryogenic conditions applying dry ice or liquid nitrogen following the normal procedure applied in the laboratory.
    ii. Verify the efficiency of comminution with a Petri dish test (See Section 2.3). Continue the process until an acceptable particle size distribution is obtained. Note that a lengthy process may increase the decomposition.
    iii. Remove test portions from the comminuted matrix according to the normal procedure of the laboratory, but preferably from $\geq$ 10 different positions.
(h) Using the remaining part of the test material, perform the recovery test as usual by spiking the selected test portion with the standard mixture. Calculate the recovery for each compound.
(i) Determine the concentration of survived residues from the surface-treated material and their recovery from spike test portions with the method to be applied.
(j) Perform the test in $\geq$5 replicates.

Calculation of the Stability of Test Compounds

The measured concentrations of survived reference compounds are C_{ch} and C_{Bu}. Average recoveries of chlorpyrifos and buprofezin from spiked test portion are denoted as $\overline{\sigma}_{Ch}$ and: $\overline{\sigma}_{Bu}$, respectively. The measured concentrations of the 'ith' test compound from surface treatment is C_i. Average recovery of compound 'i' from the spiked test portion: $\overline{\sigma}_i$. Surviving residues are calculated with the average recoveries for each replicate test portion separately because the concentrations present in the test portions are different due to the inhomogeneity of the comminuted material (the efficiency of comminution). The survived portion of compound 'i' is calculated from the first test portion as:

$$\varphi_{iCh1} = \frac{C_{i1} \times \overline{\rho}_{Ch}}{C_{Ch1} \times \overline{\rho}_i};\ \varphi_{iBu1} = \frac{C_{i1} \times \overline{\rho}_{Bu}}{C_{Ch1} \times \overline{\rho}_i} \tag{11}$$

The first estimate of the survived portion ($\overline{\varphi}_{i1}$) is calculated as the average obtained from the comparison with chlorpyrifos (Φ_{iCh1}) and buprofezin (Φ_{iBu1}). The stability of an analyte is characterized with the estimated grand average of survived portions of the 'ith' compound obtained from the n replicate measurements:

$$\overline{\overline{\varphi}}_i = \frac{\sum_1^n \overline{\varphi}_i}{n} \tag{12}$$

The numerical calculations are illustrated in Tables 9–11. The surviving proportions were calculated with Equations (11) and (12). It is pointed out that each test portion was analyzed by different analysts taking part in one of our international training workshops. The test mixture used contained 17 pesticide active substances. The participants got acquainted with the QuEChERS method during the workshop and they had not used it before. Consequently, better reproducibility can be expected with the staff having experience with the method and working in their own laboratory. The results indicated the within-laboratory reproducibility of the analyses phase (CV_A) and whole process of determination of pesticide residues (CV_L). The summary of results is given in Table 9.

Table 9. Results of stability test performed with 1 g test portions at ambient temperature.

	Recovery Tests with 0.2 mg/kg Spike Residues Measured [mg/kg]					Survived Residues [mg/kg]			
	Bu	Ch	Etri	Etox		Bu	Ch	Etri	Etox
	0.177	0.165	0.137	0.176		0.161	0.157	0.122	0.139
	0.186	0.182	0.156	0.176		0.173	0.180	0.117	0.165
	0.204	0.183	0.151	0.153		0.129	0.116	0.100	0.133
	0.178	0.152	0.164	0.170		0.142	0.128	0.108	0.132
	0.169	0.169	0.150	0.164		0.135	0.137	0.106	0.125
$\overline{\rho}$	0.913	0.852	0.758	0.840	$\overline{\rho}$	0.207	0.231	0.176	0.217
CV_A	0.015	0.015	0.013	0.011	CV_L	0.698	0.937	0.908	0.885

Notes: Ch: chlorpyrifos, Etri: etridiazole; Etox: etoxazole; $\overline{\rho}$: average recovery. The Table shows rounded values, but the calculations were performed with four-digit numbers.

Comparing the CV_L values obtained with the analyses of 1 g and 10 g test portions clearly indicates the effect of test portion size on the reproducibility of the results which is about two times higher for the 1 g portion than the 10 g portion. The corresponding CV_A values (0.015 and 0.014) for buprofezin and chlorpyrifos do not show any dependence from the test portion size. It is not surprising because they reflect the repeatability of the procedure from the point of spiking of the test portions. The CV_{Sp} values, calculated with the rearranged Equation (2):

$$CV_{Sp} = \sqrt{CV_L^2 - CV_A^2} \tag{13}$$

are practically the same as the CV_L because the CV_A is much smaller. The average CV_{Sp1}:CV_{Sp10} ratio is about two which is smaller than that predicted with Equation (7) indicating that the latter provides only an approximate tendency. These results underline the importance of regular testing the reproducibility of the whole pesticide residue determination process that can be done most conveniently with the reanalysis of retained test portions described hereunder.

Table 10. Proportion of survived residues based on 1 g test portion [1].

	Etridiazole with Bu	Etridiazole with Ch	Etoxazole with Bu	Etoxazole with Ch
	0.523	0.502	0.662	0.636
	0.469	0.420	0.731	0.655
	0.533	0.555	0.788	0.819
	0.528	0.547	0.711	0.737
	0.542	0.500	0.709	0.654
$\overline{\varphi}$	0.519	0.505	0.720	0.700
$\overline{\overline{\varphi}}$	0.512		0.710	

Note: [1]: The proportions of survived residues were calculated applying both buprofezin (Bu) and chlorpyrifos (Ch) as stable reference compounds.

Table 11. Summary of recoveries, survived residues, CV_A and CV_L values obtained with the tests performed at ambient temperature during the training workshop.

Parameter	Bu	Ch	Etri	Etox
Spiking 1 g test portion (5 replicates)				
Average recovery	0.913	0.852	0.758	0.840
CV_A	0.015	0.015	0.013	0.011
Spiking 10 g test portions (5 replicates)				
Average recovery	0.970	0.949	0.815	0.957
CV_A	0.014	0.014	0.027	0.017
Extracting 1 g portion from surface-treated tomato				
Average survived [mg/kg]	0.148	0.143	0.111	0.139
CV_L	0.123	0.175	0.080	0.110
CV_{Sp}	0.122	0.175		
Extracting 10 g portion from surface-treated tomato				
Average survived [mg/kg]	0.146	0.146	0.135	0.145
CV_L	0.055	0.098	0.082	0.125
CV_{Sp}	0.053	0.097		

2.8. Determination and Demonstration of within-Laboratory Reproducibility

According to various guidance documents [62–64] the precision of the analysis steps (CV_A) should be determined with recovery tests. The individual recoveries are affected by the random and systematic errors. The sum of systematic errors is indicated by the average recovery, and the standard deviation of individual recovery values reflects the uncertainty (precision) of the measurements. Where the individual recoveries are within the 60–140% default range [62] and the average recoveries obtained for individual analyte sample matrix combinations are statistically not different (e.g., based on Grubb's outlier test [102] the average recovery and the pooled CV_A (CV_{AP}) can be calculated [107] and used for describing initially the performance of the method.

$$CV_{AP} = \sqrt{\frac{\sum df_i CV_{Ai}^2}{\sum df_i}} \quad (14)$$

Point to note: when the Grubb's test is applied: there are several websites offering the critical values for the test. To obtain a correct outcome the critical values should be selected for a two-sided test, as given by ISO 5725 [102]. The initial estimate of method performance, based on a limited number of tests, should be verified, or refined, if necessary, during the ongoing performance verification that requires testing the recovery in each analytical batch. Keeping in mind the wide scope of the methods, covering often over 300 residues, it would not be practical to include all of them in every batch. It is recommended to test at least 10% of analytes (minimum five) included in the scope of the method in the rolling program together with various representative commodities from different commodity groups [62]. Consequently, hundreds of recovery values are generally generated in a laboratory monitoring pesticide residues. Each recovery value obtained on different days provides one estimate for the precision (relative uncertainty) of the results under within-laboratory reproducibility conditions. It is usually assumed that the random error of analytical results conforms to normal distribution because the total error is made up of the combination of small independent random errors arising at the various stages of an analytical procedure [108].

Assuming normal distribution, we can expect that the individual recovery values vary around the average (μ). Provided that the determination process is under "statistical control" 95% and 99.7% of the recovery values should be within the average (μ) $\pm$ 1.96sd and $\mu \pm$ 3sd intervals (sd = standard deviation). Consequently, the control chart for individual recoveries is constructed based on the initial method validation data. The upper (UWL) and lower (LWL) warning limits encompass the $\mu \pm$ 2sd range, whereas the corresponding action limits (UAL, LAL) are at $\mu \pm$ 3sd. Since the probability for falling outside one of the action limits is very small (0.15%) such a situation would require immediate action by the operators. ISO 17025-2017 recommends preparing control charts to record the results in such a way that trends are detectable [109].

The original QuEChERS method has been used [52] with no or minor modifications in combination with GG-MS/MS and LC-MS/MS detection in one of our laboratories. The initial in-house validation of the method with representative analytes and sample materials resulted in an average recovery of 91.7% with a 'within-laboratory reproducibility' CV_{AR} = 9.6%. As part of the regular internal quality control 2354 recovery tests were performed at 0.01 mg/kg and 0.05 mg/kg levels altogether with 302 pesticide residues during the previous four months. The sample matrices included fruits and vegetables of high-water content such as apple, carrot, cucumber, eggplant, dragon fruit, grape, longan, mango, onion, orange, and sweet and chili pepper. Control charts were constructed for the selected groups of pesticides that were tested together. One example is shown in Figure 7, indicating only the results of the first 15 testing days with a limited number of pesticide residues to enable the graphical presentation and visual evaluation of the data.

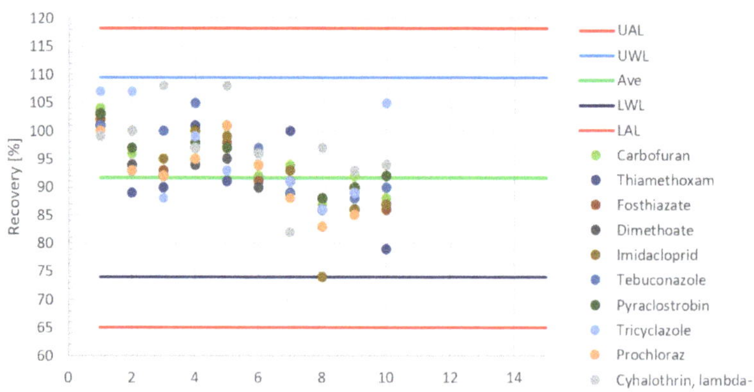

Figure 7. Example for the control chart demonstrating the within-laboratory reproducibility of the analyses phase of the determination of ten pesticide residues. Action (AL) and warning limits (WL) are indicated with red and blue lines, and the green line shows the average recovery.

Figure 8 shows that the recoveries were within the warning limits (74 and 109) of the randomly selected pesticides, though their distribution is not symmetrical, without displaying any clear tendency. In view of the size limitations of control charts, the periodic evaluation of a great number of recovery data (e.g., 2354 recoveries for the tested 302 compounds) can be better done based on their relative frequency diagram shown in Figure 9. The calculation can be easily done with Excel and has no size limitations.

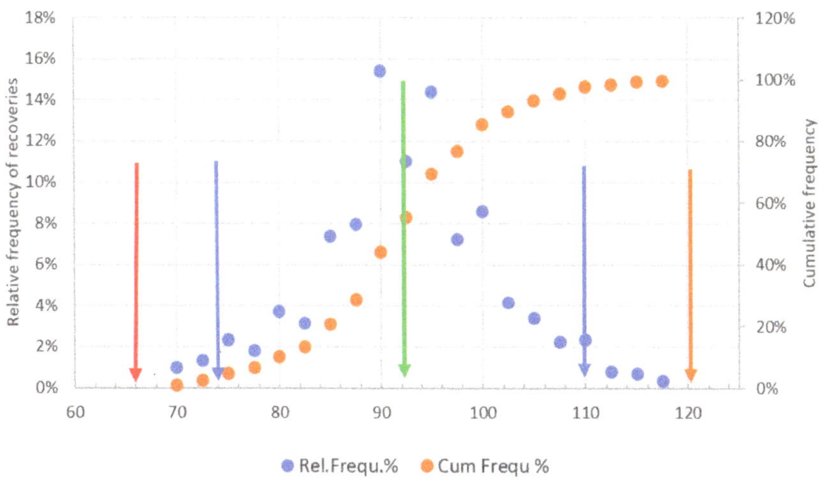

Figure 8. Relative and cumulative frequency distribution of 2354 recovery data obtained with 302 pesticide residues in commodities of high-water content during a four-month period. Red and blue arrows indicate the action and warning limits, respectively. Green arrow indicates the average recovery.

Figure 8 indicates that the highest recovery (117.5%) is within the upper action limit (UAL) (120%) and the lowest recovery is 70% well above the 65% lower action limit. Moreover, the mean recovery (91.7%) determined during the validation of the method is encompassed by the most frequently occurring 90% and 95% recoveries within the warning limits. The results of the 2354 recovery tests confirm that the tested 302 substances can be determined in fruits and vegetable samples with the typical performance parameters established during the validation of the method.

The long-term within-laboratory reproducibility (CV_L) of the residue determination process, which incorporates the contribution of subsampling, sample homogenization and analyses (Table 2), can be most conveniently determined [65] with the reanalyzes of the retained test portion that is also recommended by ISO17025:2017 as an internal quality control action [109]. The retained test portions must be obtained from samples containing incurred residues to demonstrate the efficiency of comminution. Analysts should be aware that only the CV_L can indicate the performance of the whole determination process and not the CV_A. Therefore, CV_L should be determined regularly for each type of commodity as part of the internal quality control plan of the laboratory. For performing the reanalyzes of retained test portions, prepare 10–15 test portions from each sample. If residues are detected, keep the test portions for further analyses. If no residue is detected a few test portions may be kept for preparing matrix-matched calibration solutions. The remaining test portions can be discarded. In due course of the regular analyses of various samples, a retained test portion should be included in the analytical batch and blindly reanalyzed. The results should be recorded in the format shown as an example in Table 12.

Table 12. Results of the analyses of retained test portions (example).

Test No [1]	Original Test Portion				Retained Test Portion		
	Sample code	Date of anal. [2]	Residue/commodity Name	[mg/kg]	Test portion Code [3]	Date of anal.	Residue [mg/kg]
1	M261	22 August 2022	Bupirimate/	0.205	M261/1	29 August 2022	0.216
2			orange		M261/2	6 September 2022	0.210
3					M261/3	14 September 2022	0.195
4	M283	15 September 2022	Lufenuron/	0.52	M283/1	22 September 2022	0.75
5			pepper		M283/2	26 September 2022	0.45
6					M283/3	3 October 2022	0.50
7					M283/4	10 October 2022	0.68

Notes: [1]: The repeated tests can be performed at various time intervals after the first analysis. [2]: Date of the first analysis of the sample. [3]: Test portions retained form the sample at the time of the first analyses.

The results of the reanalyzes of retained test portions may be evaluated based on the standard deviation of the difference of the two measurements made on "closely similar materials containing residues fairly close in amount present" [110]. The number of sample portion pairs analyzed should be ≥5 to obtain a realistic estimate for the CV_L. Since the average residues of the original and retained test portions are different, their relative difference should be used for the estimation of CV_L.

$$CV_L = \sqrt{\frac{\sum R_{\Delta i}^2}{2n}} \tag{15}$$

where $R_{\Delta I} = 2(R_{i1} - R_{i2})/(R_{i1} + R_{i2})$, R_{i1} and R_{i2} are the residues obtained from the analyses of the ith test portions and n is the number of test portion pairs. Assuming that only random error affects the duplicate measurements, their average must be zero, thus the degree of freedom is equal to 'n', the number of measurement pairs. Alternately, the range statistics [111] can be used for the estimation of CV_L that does not assume the above-mentioned preconditions specified by Youden. For the ith measurement pairs the CV_{Ri} is calculated with Equation (16). The d_2 for two replicate measurements is 1.128.

$$CV_{Ri} = \frac{R_{max} - R_{min}}{\overline{R} \times d_2} \tag{16}$$

The CV_L is calculated from 'n' test option pairs with pooling the CV_{Ri} values [112]):

$$CV_L = \sqrt{\frac{\sum CV_{Ri}^2}{n}} \tag{17}$$

The degree of freedom for the corresponding standard deviations [sd = $CV_L \times R$] of the measured residues (R) is equal to 'n'. The two estimates of CV_L with Equations (15) (0.1283) and (17) (0.1608) are slightly, but statistically not significantly, different. We recommend using the larger CV_L to avoid underestimating the long term within-laboratory reproducibility of the residue determination process.

2.9. Chromatographic Determination of Residues

The gas and liquid chromatographic separation and MS detection conditions are generally well described in the publications often following the guidance given by SANTE/11312/2021, SANTE/2020/12830, USFDA documents [62–64]. However, there are a few points that should be considered when the chromatographic conditions are characterized.

The reported LOD values or reporting limits should always be checked at the beginning and at the end of the analytical batch of sample extracts for all targeted analytes preferably in blank sample extract, because loading the column with coextracted materials may change the resolution of the column and or shift the retention times as illustrated in

Figures 9 and 10. This is especially important in the case of screening methods for unknown pesticide residues in monitoring programs.

The inertness and satisfactory operating conditions of gas chromatographic columns can be improved by applying the so-called analyte protectants [112,113] A critical review and re-assessment of analyte protectants in gas chromatography was published by Rodríguez-Ramos [114].

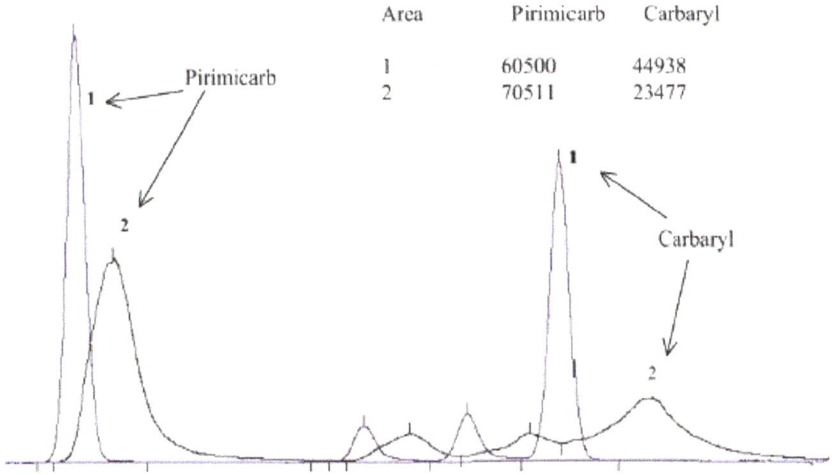

Figure 9. Deterioration of chromatographic peak shape after injecting 40 sample extracts (5 g sample/mL). Taken with permission form [115].

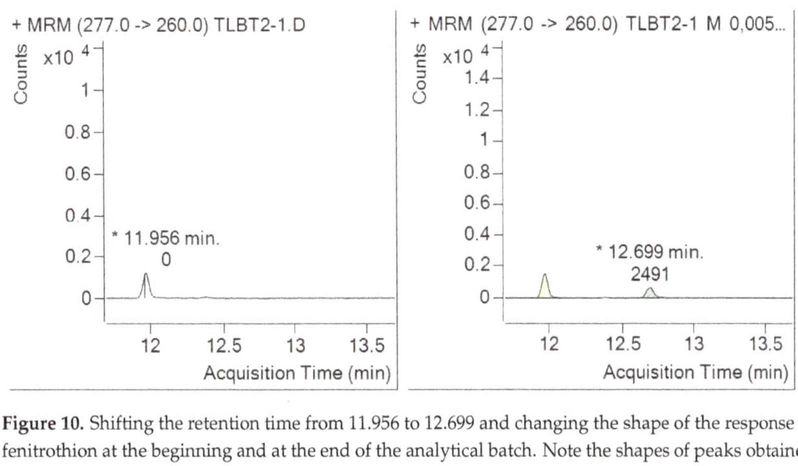

Figure 10. Shifting the retention time from 11.956 to 12.699 and changing the shape of the response of fenitrothion at the beginning and at the end of the analytical batch. Note the shapes of peaks obtained after injection of 0.005 µg/mL considered to be the LOD. MRM: multi reaction monitoring mode, TLBT2-1: sample identifier.

The data analyses reports provided by the software should not be viewed as a 'black box' and accept it without verification of its correctness. The modern data analyzers (e.g., Aglient Mass Hunter) usually offer six different curve fit types (linear, quadratic, power, first order ln, second order ln, and average of response factors), four possible choices for the origin (ignore, include, force, blank offset), and seven for weighing (none, 1/x, 1/x, 1/y, $1/y^2$, Log, $1/sd^2$). The reported results can be quite different depending on which integration options are selected. Attention is also required to assess the number of disabled points and the reported confidence limits of the slope and intercept of the regression equations. For instance, where three out of six calibration points are disabled the predicted analyte concentration should be critically considered, and possibly additional calibration injections should be made.

Chromatograms must be inspected by the analyst and the actual baseline fit examined and adjusted, if necessary. The response of the suspected peaks should always be checked to verify that the ion(s) acquisition includes the whole peek(s) and their integration is correct (Figure 11).

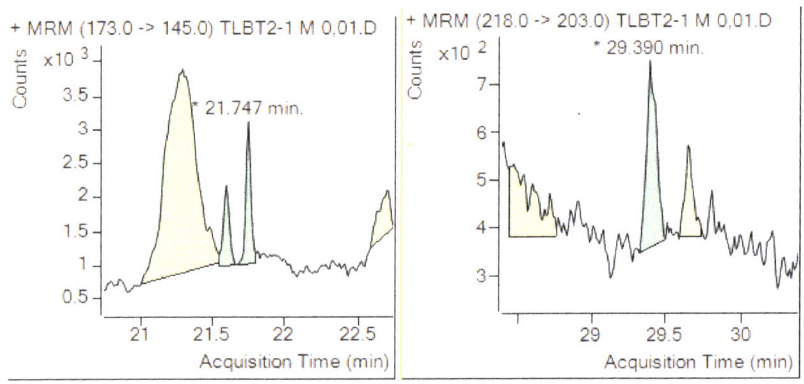

Figure 11. Examples for verification of correct identification of suspected peaks of propiconazole and indoxacarb in left and right pictures. MRM: multi reaction monitoring; TLBT2-1: sample identifier.

For multi-level calibration the standard concentrations should be equidistantly distributed over the calibrated range. Figure 12 illustrates a frequently applied questionable practice where four calibration points [ng/mL] were in the first 1/10 part of the calibrated range (0.05, 0.1, 0.5, 1, 5, 10 μg/kg).

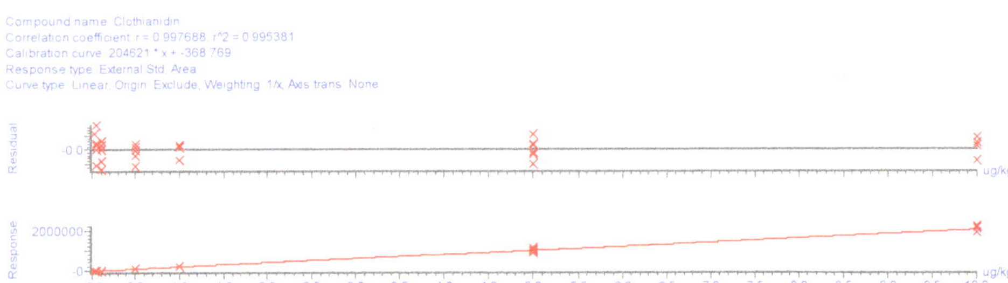

Figure 12. Improper selection of calibration points that should be equidistantly distributed. * indicates the position of the response obtained with the injection of standard solutions.

Such a calibration program type is only justified where analytes potentially present at low concentrations are looked for in screening analyses.

It should be recognized that the correlation coefficient (r) or the coefficient of determination (R^2) provides information only on the linearity of the calibration but does not characterize the quality of the calibration. It can be assessed based on confidence intervals, calculated by those of the data processing software for the slope and intercept of the regression line or from the standard deviation of the relative residuals. The latter parameters should also be reported together with 'r' or R^2.

Figures 13 and 14 show calibration charts with confidence and tolerance intervals around the linear regression line obtained with 1/x weighting [116]. Note that the R^2 values indicating the linear fit are practically the same, but the standard deviation of relative residuals (Sd_{rr}) indicating the scatter of the responses around the regression line as well as the width of the confidence and tolerance intervals are substantially different. The confidence intervals around the regression line are strongly influenced by the number of standard injections (not shown in the figure).

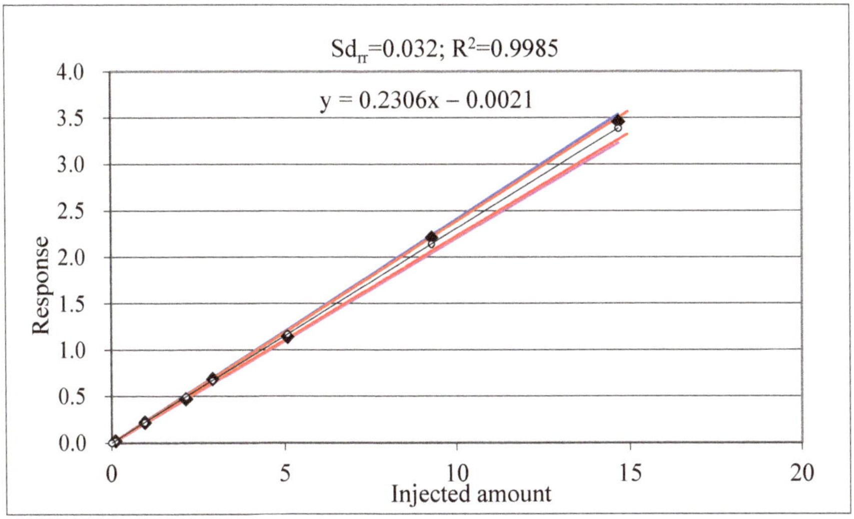

Figure 13. Terbuthylazine calibration charts. Blue lines indicate the confidence intervals, the red lines the tolerance intervals around the regression line that were calculated applying the approximation recommended by Miller and Miller [116].

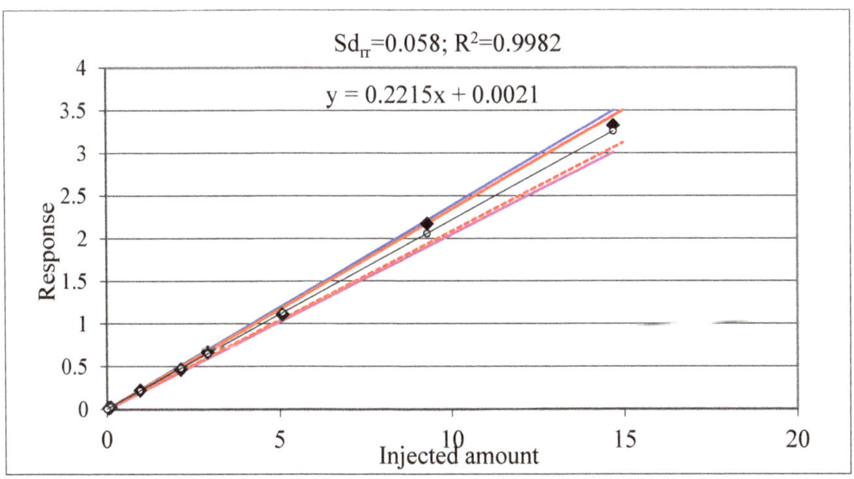

Figure 14. Terbuthylazine calibration charts. Blue lines indicate the confidence intervals, the red lines the tolerance intervals around the regression line that were calculated applying the approximation recommended by Miller and Miller [116].

The regression residual Δy_i describes the vertical distance of measured responses from the regression curve according to:

$$\Delta_{yi} = y_i - \hat{y}_i;\ \Delta_{yirel} = \frac{\Delta_{yi}}{\hat{y}_i} \tag{18}$$

The standard deviation of the relative residuals is calculated as:

$$Sd_{\Delta y/\hat{y}} = \sqrt{\frac{\left\langle \Delta_{yirel} - \overline{\Delta}_{rel}{}^2 \right\rangle}{n-2}} = Sd_{rr} \tag{19}$$

When each reference material is measured *k* times, the number of degrees of freedom is (nk–2).

Nonetheless the R^2 values are practically the same, the Sd_{rr} values indicating the large difference in the confidence/tolerance intervals in Figures 13 and 14. Table 13 shows further examples from our practice underlying the fact that the R^2 is not a proper indicator of the accuracy of the calibration [65]. Our experience suggests that for accurate calibration the Sd_{rr} should be <0.1 (10%). The Codex quality control guidelines suggest accepting a maximum of 20% relative residuals (30% near the instrument LOQ) [71].

Table 13. Examples for the corresponding Sd_{rr} and R^2 values.

Sd_{rr}	R^2
0.042	0.9937
0.061	0.9976
0.085	0.9988

3. Discussion

The monitoring programs are conducted around the world including large number of samples to provide data for carrying out:

- dietary exposure assessment of consumers;
- evaluating the residue levels and their compliance with national or international maximum residue limits or guidance values;
- assessing the contamination of the environment;
- providing the basis for the necessary corrective actions if the residues exceed the reasonably expectable levels in the treated crops.

Each analysis may have significant consequences. Therefore, the results should be representative and defendable even in legal proceedings. Analysts must be aware of their responsibilities and the fact that their credibility could be at stake. They should be able to verify the correctness of their measurements with documented evidence.

The international standards and guidelines provide the frame and acceptable performance criteria for performing the pesticide residue analytical measurements. They would facilitate obtaining accurate, defendable results only if the laboratory operations are performed by staff members (from the top manager, who has the key role, to each member) who are aware of their own responsibility and are working in coordination with each other.

It is not sufficient to validate our methods or test the performance of already validated methods once. The laboratories should establish their own internal quality control programs to be used daily for ensuring that their methods satisfy the specified performance characteristics when applied for instance to screen over several hundreds of analytes in samples of unknown origin or to test the residues in commodities before export.

The provisions of guidance documents should be fulfilled bearing in mind that the priorities of internal quality control are in order: (1) good analytical practice; (2) good science; (3) minimum bureaucracy; (4) facilitating reliability and (5) efficiency. The quality assurance/quality control (QA/QC)should only be an appropriate proportion of the activities related to the analyses of samples and reporting of the results.

Keeping in mind the above priorities, we emphasize that it is not sufficient to report the recoveries obtained with spiked test portions, the linearity of calibration, detection conditions, and confirmation of the identity of substances. In addition, we propose checking and preferably briefly reporting, for instance, the validity of samples considering the parameters that can be verified in the laboratory, accuracy of analytical standards, stability of analytes during the laboratory operations, quality of calibration characterized with the relative residuals or their standard deviation, and the reproducibility relative standard deviation of the measured residues.

Moreover, the selection of the parts of samples and the composition of the residues to be determined should always be matched with the objectives of the work.

It is advisable to take part regularly in proficiency tests that provide a means of objectively evaluating and demonstrating the accuracy and reliability of our measurements. Critical review of the Z-scores and identification of the sources of the potential errors can help to improve the technical operation standard of the laboratory. However, participating in proficiency tests does not replace the regular and rigorous internal quality control actions.

Finally, reliable results on which regulatory decisions are based can be expected only from well-trained analysts whose knowledge should be regularly updated to fully utilize the advantages of the high-performance instruments and benefit from the rapidly expanding methodical experience gained by other laboratories through the analyses of a great variety of samples.

Author Contributions: Conceptualization, draft preparation, manuscript finalization Á.A.; methodology Á.A., V.V.N.D., H.S.-D. and A.V.; review Á.A., J.S.-C. and A.V.; editing J.S.-C. All authors have read and agreed to the published version of the manuscript.

Funding: This research received no external funding.

Institutional Review Board Statement: This study did not require ethical approval as no human or animal tests were involved.

Informed Consent Statement: Not applicable.

Data Availability Statement: All data used are included in the manuscript.

Acknowledgments: The authors are indebted to Carmen Ferrer Amate, European Union Reference Laboratory for Pesticide Residues in Fruit and Vegetables, Universidad de Almería, Edificio de Química for providing the unpublished results of EU ring test. Moreover, the authors acknowledge the cooperation of Zs. Domak and T. Gönczöl Food Chain Safety Laboratory Directorate Plant Protection and Soils Conservation National Reference Laboratory, Á. H. Kamirán Centre for Agricultural Research, Agriculture Institute Martonvásár, Hungary who shared some of their experimental data.

Conflicts of Interest: The authors declare no conflict of interest.

Abbreviations

APVMA	Australian Pesticides and Veterinary Medicines Authority
EPA	US Environmental Protection Agency
ADI	Acceptable Daily Intake
ANVISA	Brazil National Health Surveillance Agency
CAC	Codex Alimentarius Commission
CCPR	Codex Committee on Pesticide Residues
DAFF	Australian Government Department of Agriculture, Fisheries and Forestry
EC	European Commission
EFSA	European Food Safety Authority
EPC	European Parliament and Council
EU	European Union
FAO	Food and Agriculture Organization of the United Nations
FIFRA	Federal Insecticide, Fungicide and Rodenticide Act
GAP	Good Agricultural Practice
JMH	Ministry of Health of Japan
JMPR	FAO/WHO Joint Meeting on Pesticide Residues
MRL	Maximum Residue Limit [mg/kg]

OECD	Organisation for Economic Cooperation and Development
USA	United States of America
US EPA	US Environmental Protection Agency
US FDA	US Food and Drug Administration

References

1. FAO/WHO. Synthesis Report on the Environmental and Health Impacts of Pesticides and Fertilizers and Ways to Minimize Them. 2022. Available online: www.fao.org/pest-and-pesticide-management/resources/en/ (accessed on 9 November 2022).
2. *National Research Council, 2020 Future Role of Pesticides in US Agriculture*; National Academies Press: Washington, DC, USA, 2020; p. 332. ISBN 0-309-50383-3.
3. Turner, A. Pesticide Manual, 19th ed.British Crop Production Council: Alton, UK, 2021; p. 1400. ISBN 978-1-9998966-5-2.
4. University of Hertfordshire (Hatfield, UK). Pesticide Property Database. Available online: http://sitem.herts.ac.uk/aeru/ppdb/en/Reports/3478.htm (accessed on 20 October 2022).
5. EFSA. Eurobarometer. 2022. Available online: https://www.efsa.europa.eu/sites/default/files/2022-09/EB97.2-food-safety-in-the-EU_report.pdf (accessed on 24 November 2022).
6. Australian Government. Agricultural and veterinary chemicals code, standards. 2022. Available online: https://www.legislation.gov.au/Details/F2022L00137 (accessed on 20 October 2022).
7. European Parliament and Council. Regulation (EC) no. 396/2005 of the European Parliament and of the Council of 23 February 2005 on Maximum Residue Levels of Pesticides in or on Food and Feed of Plant and Animal Origin and Amending Council Directive 91/414/EEC. Available online: https://eur-lex.europa.eu/legal-content/EN/TXT/?uri=celex%3A32005R0396 (accessed on 20 October 2022).
8. European Parliament and Council. Regulation (EC) no. 1107/2009 of the European Parliament and of the Council of 21 October 2009 Concerning the Placing of Plant Protection Products on the Market and Repealing Council Directives 79/117/EEC and 91/414/EEC. Available online: https://eur-lex.europa.eu/legal-content/HU/ALL/?uri=CELEX:32009R1107 (accessed on 20 October 2022).
9. EPA. Summary of the Federal Insecticide, Fungicide and Rodenticide Act. Available online: https://www.epa.gov/laws-regulations/summary-federal-insecticide-fungicide-and-rodenticide-act (accessed on 20 October 2022).
10. EPA. Regulation of Pesticide Residues on Food. 2022. Available online: https://www.epa.gov/pesticide-tolerances (accessed on 5 November 2022).
11. Japanese Ministry of Health, Labour and Welfare. Agricultural Chemical Residues in Foods. Available online: https://www.mhlw.go.jp/stf/seisakunitsuite/bunya/kenkou_iryou/shokuhin/zanryu/index_00016.htm (accessed on 5 November 2022).
12. Solecki, R.; Schumacher, D.M.; Pfeil, R.; Bhula, R. OECD Guidance documents and test guidelines. In *Food Safety Assessment of Pesticide Residues*; Ambrus, Á., Hamilton, D., Eds.; World Scientific: Hackensack, NJ, USA, 2017; pp. 507–510.
13. OECD. Test Guidelines for Chemicals. Available online: https://www.oecd.org/chemicalsafety/testing/oecdguidelinesforthetestingofchemicals.htm (accessed on 5 November 2022).
14. OECD. Work on Pesticides and Sustainable Pest Management. Vision for the Future. Available online: https://inergency.com/oecd-work-on-pesticides-and-sustainable-pest-management-vision-for-the-future (accessed on 5 November 2022).
15. OECD. Data Requirements for Pesticide Registration in OECD Member Countries: Survey Results. OECD/GD (94)47 Environment Monographs No. 77. Available online: https://www.oecd.org/env/ehs/pesticides-biocides/oecdguidancedocumentsforpesticideregistration.htm (accessed on 5 November 2022).
16. ASEAN. Maximum Residue Levels. Available online: https://asean.org/wp-content/uploads/images/2015/October/outreach-document/Edited%20MRLs-2.pdf (accessed on 5 November 2022).
17. Brazil Government. Biblioteca de Temas de Agrotóxicos. 2022. Available online: https://www.gov.br/anvisa/pt-br/assuntos/regulamentacao/legislacao/bibliotecas-tematicas/arquivos/agrotoxicos.pdf (accessed on 5 November 2022).
18. Vietnam. Ministry of Healthy Circular 50/2016 Regulations on MRL of Pesticide in Food. Available online: https://thuvienphapluat.vn/van-ban/The-thao-Y-te/Thong-tu-50-2016-TT-BYT-gioi-han-toi-da-du-luong-thuoc-bao-ve-thuc-vat-trong-thuc-pham-337490.aspx (accessed on 25 November 2022).
19. Hamilton, D.; Yoshida, M.; Wolterink, G.; Solecki, R. Evaluation of pesticide residues by FAO/WHO JMPR. In *Food Safety Assessment of Pesticide Residues*, 1st ed.; Ambrus, Á., Hamilton, D., Eds.; World Scientific Publishing: London, UK, 2017; pp. 113–196.
20. Yamada, Y. Importance of codex maximum residue limits for pesticides for the health of consumers and international trade. In *Food Safety Assessment of Pesticide Residues*, 1st ed.; Ambrus, Á., Hamilton, D., Eds.; World Scientific: Hackensack, NJ, USA, 2017; pp. 269–282.
21. Codex Alimentarius. Pesticide Index. Available online: http://www.fao.org/fao-who-codexalimentarius/codex-texts/dbs/pestres/pesticides/en/ (accessed on 5 November 2022).
22. Australian National Residue Survey. Available online: https://www.awe.gov.au/agriculture-land/farm-food-drought/food/nrs (accessed on 5 November 2022).
23. Austrian Agency for Health and Food Safety. National Pesticide Residue Control Programmes. Available online: https://www.ages.at/en/plant/pesticides/pesticide-residues#c5252 (accessed on 30 October 2022).

24. European Commission. Commission Implementing Regulation (EU) 2021/601 of 13 April 2021 Concerning a Coordinated Multiannual Control Programme of the Union for 2022, 2023 and 2024 to Ensure Compliance with Maximum Residue Levels of Pesticides and to Assess the Consumer Exposure to Pesticide Residues in and on Food of Plant and Animal Origin. Available online: http://data.europa.eu/eli/reg_impl/2021/601/oj (accessed on 22 October 2022).
25. EFSA. The 2020 European Union Report on Pesticide Residues. Available online: https://multimedia.efsa.europa.eu/pesticides-report-2020/ (accessed on 5 November 2022).
26. Federal Office of Consumer Protection, Tables for the National Reporting of Pesticide Residues in Food 2020. Available online: https://www.bvl.bund.de/DE/Arbeitsbereiche/01_Lebensmittel/01_Aufgaben/02_AmtlicheLebensmittelueberwachung/07_PSMRueckstaende/01_nb_psm_2020_tabellen/nbpsm_2020_tabellen_node.html (accessed on 20 October 2022).
27. EPA 2015 Compliance Monitoring Strategy for Federal Insecticide, Fungicide, and Rodenticide Act (FIFRA)EC 2022a EU Pesticide Database. Available online: https://ec.europa.eu/food/plant/pesticides/eu-pesticides-database/active-substances/?event=search.as (accessed on 20 September 2022).
28. European Commission. EU Pesticide Database. Available online: https://ec.europa.eu/food/plant/pesticides/eu-pesticides-database/active-substances/?event=search.as (accessed on 20 October 2022).
29. USFDA. Pesticide Residue Monitoring Program Reports and Data. Available online: https://www.fda.gov/food/pesticides/pesticide-residue-monitoring-program-reports-and-data (accessed on 5 November 2022).
30. Zhang, M.; Zeiss, M.R.; Geng, S. Agricultural pesticide use and food safety: California's model. *J. Integr. Agric.* **2015**, *14*, 2340–2357. [CrossRef]
31. Mac Loughlin, T.M.; Peluso, L.; Etchegoyen, A.; Alonso, L.L.; de Castro, C.; Percudani, C.; Marino, D.J.G. Pesticide residues in fruits and vegetables of the Argentine domestic market: Occurrence and quality. *Food Control* **2018**, *93*, 129–138. [CrossRef]
32. Rahman, M.; Hoque, M.S.; Bhowmik, S.; Ferdousi, S.; Kabiraz, M.P.; van Brakel, M.L. Monitoring of pesticide residues from fish feed, fish and vegetables in Bangladesh by GC-MS using the QuEChERS method. *Heliyon* **2021**, *7*, e06390. [CrossRef] [PubMed]
33. Jardim, A.N.O.; Caldas, E.D. Brazilian monitoring programs for pesticide residues in food-results from 2001 to 2010. *Food Control* **2012**, *25*, 607–616. [CrossRef]
34. Galani, Y.J.H.; Houbraken, M.; Wumbei, A.; Djeugap, J.F.; Fotio, D.; Gong, Y.Y.; Spanoghe, P. Monitoring and dietary risk assessment of 81 pesticide residues in 11 local agricultural products from the 3 largest cities of Cameroon. *Food Control* **2020**, *118*, 107416. [CrossRef]
35. Calderon, R.; García-Hernández, J.; Palma, P.; Leyva-Morales, J.B.; Zambrano-Soria, M.; Bastidas-Bastidas, P.J.; Godoy, M. Assessment of pesticide residues in vegetables commonly consumed in Chile and Mexico: Potential impacts for public health. *J. Food Compos. Anal.* **2022**, *108*, 104420. [CrossRef]
36. Zhang, Y.; Wang, M.; Silipunyo, T.; Huang, H.; Yin, Q.; Han, B.; Wang, M. Risk Assessment of Triflumezopyrim and Imidacloprid in rice through an evaluation of residual data. *Molecules* **2022**, *27*, 5685. [CrossRef]
37. Gebremichael, S.; Birhanu, T.; Tessema, D.A. Analysis of organochlorine pesticide residues in human and cow's milk in the towns of Asendabo, Serbo and Jimma in South-Western Ethiopia. *Chemosphere* **2013**, *90*, 1652–1657. [CrossRef]
38. El Hawari, K.; Mokh, S.; Al Iskandarani, M.; Halloum, W.; Jaber, F. Pesticide residues in Lebanese apples and health risk assessment. *Food Addit. Contam. B* **2019**, *12*, 81–89. [CrossRef]
39. Kottadiyil, D.; Mehta, T.; Thasale, R.; Sivaperumal, P. Determination and dietary risk assessment of 52 pesticide residues in vegetable and fruit samples by GC-MS/MS and UHPLC-QTOF/MS from Gujarat, India. *J. Food Compos. Anal.* **2023**, *115*, 104957. [CrossRef]
40. Tripathy, V.; Sharma, K.K.; Sharma, K.; Gupta, R.; Yadav, R.; Singh, G.; Aggarwal, A.; Walia, S. Monitoring and dietary risk assessment of pesticide residues in brinjal, capsicum, tomato, and cucurbits grown in Northern and Western regions of India. *J. Food Compos. Anal.* **2022**, *110*, 104543. [CrossRef]
41. Philippe, V.; Neveen, A.; Marwa, A.; Ahmad Basel, A.-Y. Occurrence of pesticide residues in fruits and vegetables for the Eastern Mediterranean Region and potential impact on public health. *Food Control* **2021**, *119*, 107457. [CrossRef]
42. Ngabirano, H.; Birungi, G. Pesticide residues in vegetables produced in rural south-western Uganda. *Food Chem.* **2022**, *370*, 130972. [CrossRef] [PubMed]
43. Osaili, T.M.; Al Sallagi, M.S.; Dhanasekaran, D.K.; Bani Odeh, W.A.M.; Al Ali, H.J.; Al Ali, A.A.S.A.; Radwan, H.; Obaid, R.S.; Holley, R. Pesticide residues in fresh vegetables imported into the United Arab Emirates. *Food Control* **2022**, *133*, 108663. [CrossRef]
44. Ramadan, M.F.A.; Abdel-Hamid, M.M.A.; Altorgoman, M.M.F.; AlGaramah, H.A.; Alawi, M.A.; Shati, A.A.; Shweeta, H.A.; Awwad, N.S. Evaluation of pesticide residues in vegetables from the Asir Region Saudi Arabia. *Molecules* **2020**, *25*, 205. [CrossRef]
45. Vu-Duc, N.; Nguyen-Quang, T.; Le-Minh, T.; Nguyen-Thi, X.; Tran, T.M.; Vu, H.A.; Chu, D.B. Multiresidue pesticides analysis of vegetables in Vietnam by ultrahigh-performance liquid chromatography in combination with high-resolution mass spectrometry (UPLC-Orbitrap MS). *J. Anal. Methods Chem.* **2019**, *2019*, 3489634. [CrossRef]
46. Nguyen, N.T.; Huynh Nguyen, K.; Duong, T.; Tran, V.D.; Kim, K.H.; Pham, L.T. Evaluation of pesticide residues in vegetables from Mekong, Delta, Vietnam using LC-MS/MS. *Trop. J. Pharm. Res.* **2021**, *20*, 1497–1504. [CrossRef]
47. Carrasco Cabrera, L.; Medina Pastor, P. The 2020 European Union report on pesticide residues in food. *EFSA J.* **2022**, *20*, e07215. [CrossRef]

48. Wu, P.; Wang, P.; Gu, M.; Xue, J.; Wu, X. Human health risk assessment of pesticide residues in honeysuckle samples from different planting bases in China. *Sci. Total Environ.* **2021**, *759*, 142747. [CrossRef]
49. Pang, G.-F.; Fan, C.-L.; Liu, Y.-M.; Cao, Y.-Z.; Zhang, J.-J.; Fu, B.-L.; Li, X.-M.; Li, Z.Y.; Wu, Y.P. Multi-residue method for the determination of 450 pesticide residues in honey, fruit juice and wine by double-cartridge solid-phase extraction/gas chromatography-mass spectrometry and liquid chromatography-tandem mass spectrometry. *Food Addit. Contam.* **2006**, *23*, 777–810. [CrossRef]
50. Pang, G.F.; Fan, C.L.; Chang, Q.Y.; Li, J.X.; Lu, J.K.L.M. Screening of 485 pesticide residues in fruits and vegetables by liquid chromatography-quadrupole-time-of-flight mass spectrometry based on TOF Accurate Mass Database and QTOF Spectrum Library. *J. AOAC Int.* **2018**, *101*, 1156–1182. [CrossRef] [PubMed]
51. Tran-Lam, T.-T.; Bui, M.Q.; Nguyen, H.Q.; Dao, Y.H.; Le, G.T. A combination of chromatography with tandem mass spectrometry systems (UPLC-MS/MS and GC-MS/MS), modified QuEChERS extraction and mixed-mode SPE clean-up method for the analysis of 656 pesticide residues in rice. *Foods* **2021**, *10*, 2455. [CrossRef]
52. Anastassiades, M.; Lehotay, S.J.; Štajnbaher, D.; Schenck, F.J. Fast and Easy Multiresidue Method Employing Acetonitrile Extraction/Partitioning and 'dispersive solid-phase extraction' for the determination of pesticide residues in produce. *J. AOAC Int.* **2003**, *86*, 412–431. [CrossRef] [PubMed]
53. Song, X.; Li, F.; Yan, T.; Tian, F.; Ren, L.; Jiang, C.; Wang, Q.; Zhang, S. Research progress in the sample pretreatment techniques and advanced quick detection methods of pesticide residues. *Process Saf. Environ. Prot.* **2022**, *165*, 610–622. [CrossRef]
54. Calza, P.; Guarino, B.; Bello, F.D.; Dioni, A.; Bergero, M.; Medana, C. Integrated approach for the analysis of neonicotinoids in fruits and food matrices. *Food Chem.* **2022**, *372*, 131153. [CrossRef]
55. Watanabe, E. Review of sample preparation methods for chromatographic analysis of neonicotinoids in agricultural and environmental matrices: From classical to state-of-the-art methods. *J. Chromatogr. A* **2021**, *1643*, 462042. [CrossRef]
56. Rutkowska, E.; Łozowicka, B.; Kaczyński, P. Three approaches to minimize matrix effects in residue analysis of multiclass pesticides in dried complex matrices using gas chromatography tandem mass spectrometry. *Food Chem.* **2018**, *279*, 20–29. [CrossRef]
57. Rutkowska, E.; Łozowicka, B.; Kaczyński, P. Compensation of matrix effects in seed matrices followed by gas chromatography-tandem mass spectrometry analysis of pesticide residues. *J. Chromatogr. A* **2019**, *1614*, 460738. [CrossRef]
58. Musarurwa, H.; Chimuka, L.; Pakade, V.E.; Tavengwa, N.T. Recent developments and applications of QuEChERS based techniques on food samples during pesticide analysis. *J. Food Compos. Anal.* **2019**, *84*, 103314. [CrossRef]
59. Xu, M.L.; Gao, Y.; Wang, X.; Han, X.X.; Zhao, B. Comprehensive strategy for sample preparation for the analysis of food contaminants and residues by GC–MS/MS: A review of recent research trends. *Foods* **2021**, *10*, 2473. [CrossRef]
60. García-Vara, M.; Postigo, C.; Palma, P.; José Bleda, M.; López de Alda, M. QuEChERS-based analytical methods developed for LC-MS/MS multiresidue determination of pesticides in representative crop fatty matrices: Olives and sunflower seeds. *Food Chem.* **2022**, *386*, 132558. [CrossRef] [PubMed]
61. Wang, F.; Li, S.; Feng, H.; Yang, Y.; Xiao, B.; Chen, D. An enhanced sensitivity and cleanup strategy for the nontargeted screening and targeted determination of pesticides in tea using modified dispersive solid-phase extraction and cold-induced acetonitrile aqueous two-phase systems coupled with liquid chromatography-high resolution mass spectrometry. *Food Chem.* **2019**, *275*, 530–538. [CrossRef] [PubMed]
62. European Commission. Analytical Quality Control and Method Validation Procedures for Pesticide Residues Analysis in Food and Feed 11312/2021. Available online: https://food.ec.europa.eu/system/files/2022-02/pesticides_mrl_guidelines_wrkdoc_2021-11312.pdf (accessed on 5 November 2022).
63. European Commission. Guidance Document on Pesticide Analytical Methods for Risk Assessment and Post-Approval Control and Monitoring Purposes SANTE/2020/12830, Rev. 1. Available online: https://food.ec.europa.eu/system/files/2021-02/pesticides_mrl_guidelines_2020-12830.pdf (accessed on 5 November 2022).
64. US FDA. *Guidelines for the Validation of Chemical Methods in Food, Feed, Cosmetics, and Veterinary Products*, 3rd ed.; Food and Drug Administration: Silver Spring, MD, USA, 2020. Available online: https://www.fda.gov/media/121751/download (accessed on 5 November 2022).
65. Ambrus, A. Reliability of measurements of pesticide residues in food. *Accredit. Qual. Assur.* **2004**, *9*, 288–304. [CrossRef]
66. Ambrus, Á.; Zentai, A.; Sali, J.; Ficzere, I. Hidden contributors to uncertainty and accuracy of results of residue analysis. *Accredit. Qual. Assur.* **2011**, *16*, 3–11. [CrossRef]
67. Lehotay, S.J.; Han, L.; Sapozhnikova, Y. Use of a quality control approach to assess measurement uncertainty in the comparison of sample processing techniques in the analysis of pesticide residues in fruits and vegetables. *Anal. Bioanal. Chem.* **2018**, *410*, 5465–5479. [CrossRef]
68. Lehotay, S.J.; Michlig, N.; Lightfield, A.R. Assessment of test portion sizes after sample comminution with liquid nitrogen in an improved high-throughput method for analysis of pesticide residues in fruits and vegetables. *J. Agric. Food Chem.* **2020**, *68*, 1468–1479. [CrossRef]
69. Ramsey, C.A.; Wagner, C. Sample quality criteria. *J. AOAC Int.* **2015**, *98*, 265–268. [CrossRef]
70. FAO. Recommended method of sampling for the determination of pesticide residues for compliance with mrls CXG33-1999. Available online: https://www.fao.org/fao-who-codexalimentarius/sh-proxy/en/?lnk=1&url=https%253A%252F%252Fworkspace.fao.org%252Fsites%252Fcodex%252FStandards%252FCXG%2B33-1999%252FCXG_033e.pdf (accessed on 5 November 2022).

71. FAO. CAC guidelines on performance criteria for methods of analysis for the determination of pesticide residues in food and feed CXG90-2017. Available online: https://www.fao.org/fao-who-codexalimentarius/sh-proxy/en/?lnk=1&url=https%253A%252F%252Fworkspace.fao.org%252Fsites%252Fcodex%252FStandards%252FCXG%2B90-2017%252FCXG_090e.pdf (accessed on 5 November 2022).
72. Bettencourt da Silva, R.J.N.; Santos, J.R.; Filomena, M.; Camoes, G.F.C. Evaluation of the analytical method performance for incurred samples. *Anal. Chim. Acta* **2003**, *485*, 241–252. [CrossRef]
73. Fussel, R.J.; Hetmanski, M.T.; MacArthur, R.; Findlay, D.; Smith, F.; Ambrus, Á.; Brodesser, P.J. Measurement uncertainty associated with sample processing of oranges and tomatoes for pesticide residue analysis. *J. Agric. Food Chem.* **2007**, *55*, 1062–1070. [CrossRef]
74. Suszter, K.G.; Ambrus, Á. Testing the efficiency of extraction of incurred residues from soil with optimized multi-residue method. *J. Env. Sci. Health B* **2017**, *52*, 547–556. [CrossRef]
75. Han, L.; Lehotay, S.J.; Sapozhnikova, Y. Use of an efficient measurement uncertainty approach to compare room temperature and cryogenic sample processing in the analysis of chemical contaminants in foods. *J. Agric. Food Chem.* **2017**, *66*, 4986–4996. [CrossRef] [PubMed]
76. Wang, W.; Cho, Y.-J.; Song, J.-W.; Kim, Y.-J.; Seo, J.-S.; Kim, J.-H. Residue behavior of methoxyfenozide and pymetrozine in Chinese cabbage and their health risk assessment. *Foods* **2022**, *11*, 2995. [CrossRef] [PubMed]
77. Ambrus, Á. Variability of pesticide residues in crop units. *Pest Manag. Sci.* **2006**, *62*, 693–714. [CrossRef] [PubMed]
78. Horváth, Z.; Ambrus, Á.; Mészáros, L.; Braun, S. Characterization of distribution of pesticide residues in crop units. *J. Environ. Sci. Health B* **2013**, *48*, 615–625. [CrossRef]
79. Ambrus, Á. Estimation of sampling uncertainty for determination of pesticide residues in plant commodities. *J. Env. Sci. Health B* **2009**, *44*, 627–639. [CrossRef]
80. Farkas, Z.; Cook, J.M.; Ambrus, Á. Estimation of uncertainty of measured residues and testing compliance with MRLs. In *Food Safety Assessment of Pesticide Residues*, 1st ed.; Ambrus, Á., Hamilton, D., Eds.; World Scientific: Hackensack, NJ, USA, 2017; pp. 404–466.
81. FAO. Portion of Commodities to Which Maximum Residue Limits Apply and Which is Analyzed CAC/GL 41-1993. Available online: www.fao.org/fao-who-codexalimentarius/sh-proxy/en/?lnk=1&url=https%253A%252F%252Fworkspace.fao.org%252Fsites%252Fcodex%252FStandards%252FCXG%2B41-1993%252FCXG_041e.pdf (accessed on 5 November 2022).
82. Omeroglu, P.Y.; Ambrus, Á.; Boyacioglu, D.; Majzik, E.S. Uncertainty of the sample size reduction step in pesticide residue analysis of large-sized crops. *Food Addit. Contam. Part A* **2013**, *30*, 116–126. [CrossRef]
83. Ambrus, Á.; Szemánné-Dobrik, H.; Domák, Z.; Kerekes, K. Sources of random variation of pesticide residue analytical results. *J. AOAC Int.* **2021**, *104*, 526–538. [CrossRef]
84. Gy, P. *Sampling for Analytical Purposes*; John Wiley & Sons Ltd.: Chichester, UK, 1999; p. 71.
85. Minkkinen, P. Practical applications of sampling theory. *Chemometr. Intell. Lab. Syst.* **2004**, *74*, 85–94. [CrossRef]
86. Hill, A.R.C.; Harris, C.A.; Warburton, A.G. Effect of sample processing on pesticide residues in fruit and vegetables. In *Principles of Method Validation*; Fajgelj, A., Ambrus, Á., Eds.; Royal Society of Chemistry: Cambridge, UK, 2000; pp. 41–48.
87. Ambrus, Á.; Suszter, G. Quality Control and Quality Assurance. In *Analysis of Pesticides in Food and Environmental Samples*, 2nd ed.; Toledo, J., Ed.; CRC Press: Roca Baton, FL, USA, 2019; pp. 135–173.
88. Ritter, L.S.; Lynn, K.J.; Wujcik, C.E.; Buchholz, L.M. Interlaboratory assessment of cryomilling sample preparation for residue analysis. *J. Agric. Food Chem.* **2015**, *63*, 4405–4408. [CrossRef]
89. Ritter, L.S.; Wujcik, C.E. Novel two-stage fine milling enables high-throughput determination of glyphosate residues in raw agricultural commodities. *J. AOAC Internat.* **2018**, *101*, 867–875. [CrossRef]
90. Roussev, M.; Lehotay, S.J.; Pollaehne, J. Cryogenic sample processing with liquid nitrogen for effective and efficient monitoring of pesticide residues in foods and feeds. *J. Agric. Food Chem.* **2019**, *67*, 9203–9209. [CrossRef]
91. Thi Le, L.H.; Tran-Lam, T.T.; Cam, T.Q.; Nguyen, T.N.; Dao, Y.N. Pesticides in edible mushrooms in Vietnam. *Food Addit. Contam. B* **2021**, *14*, 139–148. [CrossRef]
92. Ly, T.-K.; Behra, P.; Nhu-Trang, T.-T. Quantification of 397 pesticide residues in different types of commercial teas: Validation of high accuracy methods and quality assessment. *Food Chem.* **2022**, *370*, 130986. [CrossRef] [PubMed]
93. FAO. Submission and evaluation of pesticide residues data for the estimation of maximum residue levels in food and feed. In *FAO Plant Production and Protection paper 225*; FAO: Rome, Italy, 2016; p. 298.
94. FAO. Acibenzolar-S-methyl, benzovindiflupyr, flupyradifurone, penconazole, sulfoxaflor. In *Pesticide Residues in Food–2016 Evaluations*; FAO Plant Production and Protection Paper 231; FAO: Rome, Italy, 2017; p. 2403.
95. FAO. Ethephon, flutriafol, fluxapyroxad. In *Pesticide Residues in Food–2015 Evaluations*; FAO Plant Production and Protection Paper 226; FAO: Rome, Italy, 2016; p. 1629.
96. FAO. Pesticide Residues in Food—Reports. FAO Plant Production and Protection Papers Series. Available online: http://www.fao.org/agriculture/crops/core-themes/theme/pests/jmpr/jmpr-rep/en/ (accessed on 7 November 2022).
97. European Commission. Download MRLs Data. Available online: https://ec.europa.eu/food/plant/pesticides/eu-pesticides-database/start/screen/mrls/download (accessed on 5 November 2022).
98. European Commission. Technical Guideline on the Evaluation of Extraction Efficiency of Residue Analytical Methods, SANTE/2017/10632 Rev. 4. 2022. Available online: https://food.ec.europa.eu/system/files/2022-02/pesticides_mrl_guidelines_wrkdoc_2017-10632.pdf (accessed on 10 November 2022).

99. Ferrer, A.C. Personal communication 2022, EU-RT-FV17 Preliminary Results. 2022; *unpublished.*
100. Ellison, S.L.R.; Williams, A. (Eds.) *EURACHEM/CITAC Guide: Quantifying Uncertainty in Analytical Measurement*, 3rd ed.; CITAC: Kowloon Bay, Hong Kong, 2012; Available online: https://www.eurachem.org/index.php/publications/guides/quam (accessed on 5 November 2022).
101. Ambrus, Á.; Kamirán, Á.H.; Kötelesné-Suszter, G.; Németh, A.; Solymosné, M.E. Accuracy of analytical standard solutions and uncertainty in their nominal concentrations. *J. Food Investig.* **2017**, *63*, 1398–1421.
102. *ISO 5725-2:2019*; Accuracy (Trueness and Precision) of Measurement Methods and Results Part 2: Basic Method for the Determination of Repeatability and Reproducibility of a Standard Measurement Method. 2019. Available online: www.iso.org/standard/69419.html (accessed on 5 November 2022).
103. Ambrus, Á; Noonan, G.; Németh, A.; Kötelesné-Suszter, G.; Anastassiades, M.; Kamirán, Á.H.; Solymosné, M.E. Testing the accuracy of analytical standard solutions used for quantitative determination of pesticide residues. *J. AOAC Internat.* **2017**, *100*, 1–4. [CrossRef] [PubMed]
104. Ambrus, A.; Lantos, J.; Visi, E.; Csatlós, I.; Sárvári, L. General method for the determination of pesticide residues in samples of plant origin, soil and water, i extraction and cleanup. *J. Assoc. Off. Anal. Chem.* **1981**, *64*, 733–742. [CrossRef]
105. Ambrus, Á.; Buczkó, J.; Hamow, K.Á.; Juhász, V.; Solymosné Majzik, E.; Szemánné Dobrik, H.; Szitás, R. Contribution of sample processing to variability and accuracy of the results of pesticide residue analysis in plant commodities. *J. Agric. Food Chem.* **2016**, *64*, 6071–6081. [CrossRef] [PubMed]
106. Fussell, R.J.; Jackson Addie, K.; Reynolds, S.L.; Wilson, M.F. Assessment of the stability of pesticides during cryogenic sample processing. 1. Apples. *J. Agric. Food Chem.* **2002**, *50*, 441–448. [CrossRef]
107. Magnusson, B.; Örnemark, U. *Eurachem Guide: The Fitness for Purpose of Analytical Methods–a Laboratory Guide to Method Validation and Related Topics*, 2nd ed. EURACHEM. 2014. Available online: https://www.eurachem.org/images/stories/Guides/pdf/MV_guide_2nd_ed_EN.pdf (accessed on 5 November 2022). ISBN 978-91-87461-59-0.
108. Thompson, M.; Howarth, R.J. The frequency distribution of analytical error. *Analyst* **1980**, *105*, 1188–1195. [CrossRef]
109. *ISO/IEC 17025:2017*; General Requirements for the Competence of Testing and Calibration Laboratories. International Organization for Standardization: Geneva, Switzerland, 2018.
110. Youden, W.J. *Statistical Techniques for Collaborative Tests*; AOAC: Washington, DC, USA, 1967.
111. Anderson, L.R. *Practical Statistics for Analytical Chemists. Appendix D12*; Van Nostrand Reinhold Company: New York, NY, USA, 1987; ISBN 0442209738.
112. Anastassiades, M.; Maštovská, K.; Lehotay, S.J. Evaluation of analyte protectants to improve gas chromatographic analysis of pesticides. *J. Chromatogr. A* **2003**, *1015*, 163–184. [CrossRef]
113. Maštovská, K.; Lehotay, S.J.; Anastassiades, M. Combination of analyte protectants to overcome matrix effects in routine GC analysis of pesticide residues in food matrixes. *Anal. Chem.* **2005**, *77*, 8129–8137. [CrossRef] [PubMed]
114. Rodríguez-Ramos, R.; Lehotay, S.J.; Michlig, N.; Socas-Rodríguez, B.; Rodríguez-Delgado, M.Á. Critical review and re-assessment of analyte protectants in gas chromatography. *J. Chromatogr. A* **2020**, *1632*, 461596. [CrossRef] [PubMed]
115. Soboleva, E.; Ambrus, Á. Application of a system suitability test for quality assurance and performance optimisation of a gas chromatographic system for pesticide residue analysis. *J. Chromatogr. A* **2004**, *1027*, 55–65. [CrossRef] [PubMed]
116. Miller, J.C.; Miller J., N. *Statistics and Chemometrics for Analytical Chemistry*, 6th ed.; Ellis Horwood PTR, Prentice Hall Pearson: Harlow, UK, 2010; p. 297. ISBN 978-0-273-73042-2.

molecules

MDPI

Article

Ionic Liquid-Dispersive Micro-Extraction and Detection by High Performance Liquid Chromatography–Mass Spectrometry for Antifouling Biocides in Water

Li Zhou [1,2,†], Tong Wu [1,†], Chuanshan Yu [1], Shaowen Liu [1] and Canping Pan [1,*]

1 College of Science, China Agricultural University, Beijing 100193, China
2 Tea Research Institute, Chinese Academy of Agricultural Sciences, Hangzhou 310008, China
* Correspondence: canpingp@cau.edu.cn; Tel.: +86-10-62731978; Fax: +86-10-62733620
† These authors contributed equally to this work.

Abstract: A simple analytical method was developed and evaluated for the determination of two antifouling biocides using an ionic liquid-dispersive liquid–liquid micro-extraction (IL-DLLME) and a high-performance liquid chromatography–electrospray ionization mass spectrometry (LC-ESI-MS) analysis. Irgarol 1051 and Sea-Nine 211 were extracted from deionized water, lake water, and seawater using IL 1-hexyl-3-methylimidazolium hexafluorophosphate ([HMIm][PF6]) and ethyl acetate as the extraction solvent and the dispersion solvent. Several factors were considered, including the type and volume of extraction and dispersive solvent, IL amount, sample pH, salt effect, and cooling temperature. The developed method resulted in a recovery range of 78.7–90.3%, with a relative standard deviation (RSD, $n = 3$) less than 7.5%. The analytes were enriched greater than 40-fold, and the limits of detection (LOD) for two antifouling biocides were 0.01–0.1 μg L^{-1}. The method was effectively applied for the analysis of real samples of freshwater as well as samples of seawater.

Keywords: antifouling biocides; high-performance liquid chromatography–mass spectrometry (LC-MS); ionic liquid-dispersive liquid–liquid micro-extraction (IL-DLLME)

Citation: Zhou, L.; Wu, T.; Yu, C.; Liu, S.; Pan, C. Ionic Liquid-Dispersive Micro-Extraction and Detection by High Performance Liquid Chromatography–Mass Spectrometry for Antifouling Biocides in Water. *Molecules* **2023**, *28*, 1263. https://doi.org/10.3390/molecules28031263

Academic Editor: Miguel Ángel González-Curbelo

Received: 13 December 2022
Revised: 13 January 2023
Accepted: 23 January 2023
Published: 28 January 2023

1. Introduction

The marine industry is hindered by marine biofouling, which damages submerged equipment and raises production costs. Antifouling biocides are commonly employed to prevent the attachment of fouling organisms to ships and other equipment [1]. Since harmful antifouling paints containing tributyltin (TBT) have been banned [2], new organic booster biocides have become the main constituents of antifouling paints to enhance their efficacy. These biocides include metal-based compounds such as zinc pyrithione and zineb, as well as non-metallic compounds such as Irgarol 1051, Sea-nine 211, Kathon 5287, chlorothalonil, dichlofluanid, and thiram [3]. However, the use of these compounds appears to be hazardous due to their residues, toxicity, and resultant contamination of the aquatic environment, as well as the potential impact on public health. Irgarol 1051 is highly toxic to non-target marine algae [4], as it destabilizes aquatic herbivorous mammal populations [5] and causes coral bleaching [6]. The use of Irgarol 1051 in antifouling paints is restricted in the European Union and the United States [7,8]. Despite having a significantly better environmental profile, Sea-Nine 211 is still hazardous to fish [9], sea urchins, and embryos [10].

Recently, these antifouling biocides have been widely identified in marinas and harbors throughout the world [11]. In aquatic environments, concentrations of Irgarol 1051 ranged from 0.12–4800 ng L^{-1} [12,13], whereas concentrations of Sea-Nine 211 ranged from 0.1–3300 ng L^{-1} [14,15]. Due to their prevalence at low concentrations, pre-concentration techniques and sensitivity detection are generally highlighted. Most analysis strategies in recent years have been based on liquid–liquid extraction (LLE) [16], solid-phase extraction (SPE) [17,18], and a few others, including solid-phase microextraction (SPME) [19], stir bar sorptive extraction and thermal desorption (SBSE-TD) [20], and microfunnel-supported liquid-phase microextraction (MF-LPME) [21]. In terms of the examination of antifouling biocides, the aforementioned approaches have various drawbacks, such as being tedious, time-consuming, expensive, complex, and harmful to the environment. Therefore, the development of less complex, more effective, and safe extraction approaches for the identification of antifouling biocides is receiving a lot of focus.

Dispersive liquid–liquid micro-extraction (DLLME) is a simple, rapid, inexpensive, sensitive, and effective technique for the extraction of target analytes. According to the DLLME principle, a water sample containing target analytes is quickly mixed with a mixture of an extraction solvent and a dispersion solvent to produce a ternary component solvent system, in which the target analytes are enriched into the micro-extraction solvent [22]. Organic solvents with a high density, incompatibility in water, and high solubility for target analytes are commonly utilized as extraction solvents. Organic solvents that are miscible with the extraction solvent and water are employed as disperser solvents to assist the extraction solvent in forming dispersed micro-droplets in the water sample, hence increasing the contact area between the extraction solvent and the target analytes [23]. Traditional DLLME employs highly toxic extraction solvents such as chlorobenzene, carbon tetrachloride, chloroform, dichloro-methane, and tetrachloroethylene [24,25]. The DLLME technique is currently being improved by employing low toxicity and new extraction solvents [25]. Since the majority of target analytes are polar compounds, the ideal DLLME extraction solvents must be liquid under standard conditions, have a low vapor pressure, be incompatible with water, have a high polarity, and have a high density.

Ionic liquids (ILs) are organic salts with melting points lower than 100 °C, composed of organic cations and organic or inorganic anions [22]. ILs have distinctive characteristics, including high thermal stability, low vapor pressure, high viscosity, and low toxicity [26]. Particularly, their physicochemical properties can be modified by selecting a particular combination of anions and cations to enhance the solubility of specific analytes [27]. Therefore, various ILs have been used as extraction solvents of DLLME, such as 1-hexyl-3-methylimidazolium hexafluorophosphate ([HMIM][PF6]), 1-butyl-3-methylimidazolium hexafluorophosphate, tetradecyl (trihexyl) phosphonium chloride, and 1-butyl-3-methylimidazolium bis (trifluoromethylsulfonyl) imide [23,28], which are typically recognized as green solvents in analytical chemistry [29], thereby deriving the IL-DLLME approach. Neurotransmitters [30], anthraquinones [31], phthalate esters [32], organic dyes [33], metal ions [34,35], pesticides [36,37], antibiotics [38,39], and other biological compounds, as well as food and environmental pollutants, have all been focused using IL-DLLME. However, the IL has not yet been used to extract antifouling biocides from water samples.

In this study, the potential application of IL-DLLME and LC–MS for the identification of two kinds of antifouling biocides in water samples was investigated. An IL ([HMIm][PF6]) was used as the extraction solvent, whereas ethyl acetate was selected as the dispersion solvent. The effects of various experimental factors on the extraction were examined, and the process was validated via linearity, precision, and accuracy investigations. The developed method can be used to analyze real lake water and seawater samples.

2. Results and Discussion

2.1. Optimization of IL-DLLME Procedure

All parameters of IL-DLLME conditions were optimized using deionized water (5.0 mL) spiked with antifouling biocides. Each data point was determined using the mean of three separate extractions.

2.1.1. Effect of Amount of IL

Because of its low water solubility, low volatility, and higher density than water, [HMIm][PF6] has been widely used as an extraction solvent for pesticides [40], metal ions [41], mycotoxins [24], and polycyclic aromatic hydrocarbons [42]. Quantities of 30 mg, 40 mg, 50 mg, 60 mg, 70 mg, 80 mg, 90 mg, and 100 mg of [HMIm][PF6] were analyzed in deionized water that was spiked with 2 μg L^{-1} of Irgarol 1051 and 10 μg L^{-1} of Sea-Nine 211 at a constant volume of disperser solvent (0.4 mL) (Figure 1). As the amount of IL increased from 30 to 60 mg, the recoveries exhibited a similar linear sign increase. However, when the amount of IL exceeded 60 mg, the recoveries declined or remained nearly constant. Wang et al. discovered that when [HMIm][PF6] exceeded 60 μL in their study on the analysis of fungicides in fruit juice, the recoveries decreased [43]. The distribution coefficient and recovery of analytes in IL may have been reduced as a result of the larger amounts of IL being dissolved, which could have decreased the polarity of the aqueous phase [44]. The optimal amount of IL was therefore determined to be 60 mg.

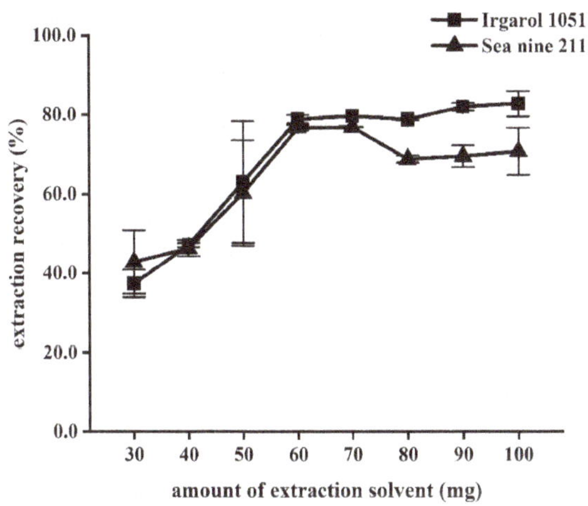

Figure 1. Effect of amount of [HMIm][PF6] on extraction recovery. Extraction conditions: water sample, 5.0 mL; disperser solvent, ethyl acetate 0.4 mL; NaCl 0% (*w*/*v*).; pH 6; and cooling temperature 20 °C.

2.1.2. Selection of Disperser Solvent and Effect of Volume

The disperser solvent must be miscible with the extraction solvent and the water sample, thereby increasing the contact area and interaction between the two phases to enhance the extraction efficiency. The selection of a disperser is crucial for achieving excellent preconcentration and extraction effects. Consequently, four potential disperser solvents, acetone, methanol, acetonitrile, and ethyl acetate, were tested. The sample solutions for this, and the subsequent tests used 5 mL of deionized water spiked with 1 μg L^{-1} of Irgarol 1051 and 5 μg L^{-1} of Sea-Nine 211. A series of sample solutions were analyzed using 0.5 mL of each disperser solvent containing 60 mg of [HMIm][PF6]. The results showed that Irgarol 1051 (85.0%) and Sea-Nine 211 (86.0%) had higher recoveries when ethyl acetate was used as the dispersant than those of acetone (Irgarol 1051 33.6%, Sea-Nine 211 50.1%), methanol (Irgarol 1051 36.4%, Sea-Nine 211 45.8%), and acetonitrile (Irgarol 1051 61.9%, Sea-Nine 211 62.8%). Kong et al. also examined vitamins and carotenoids in human serum using ethyl acetate as the disperser solvent [45]. The use of ethyl acetate as the disperser solvent resulted in good media miscibility and the best recoveries. As a result, ethyl acetate was selected for further investigation.

The volume of the disperser affects the dispersion degree of the extraction phase in the aqueous phase, thereby influencing the extraction efficiency. When the disperser volume is small, the extraction solvent cannot be completely dispersed in the aqueous phase, preventing the formation of a good ternary cloudy solution of water/disperser/extraction solvent, and lowering the extraction efficiency. In contrast, when the volume of the disperser is increased, the distribution coefficient of analytes in the water rises, and the extraction efficiency decreases. To assess the impact of the organic solvent on the yield of the IL-DLLME process, various ethyl acetate volumes were tested. To determine the optimal volume, experiments were conducted with varying volumes of ethyl acetate (0.30 mL, 0.40 mL, 0.50 mL, 0.55 mL, and 0.60 mL) mixed with 60 mg [HMIm][PF6]. Figure 2 shows that, in contrast to the enrichment factor (EF), the recoveries increased initially and then decreased as the volume of ethyl acetate increased. A total of 0.4 mL of ethyl acetate yielded the highest recoveries for all analytes. Similar behavior was observed when parabens were analyzed using IL-DLLME [46]. This can be explained by the possibility that if there is insufficient dispersion solvent, the extraction solvent may not make good contact with the analytes in the sample solution, which could lower the recovery. On the other hand, more disperser solution resulted in a more settled phase, which decreased the EF. The results showed that 0.4 mL was selected to achieve a high EF and a good extraction recovery (ER).

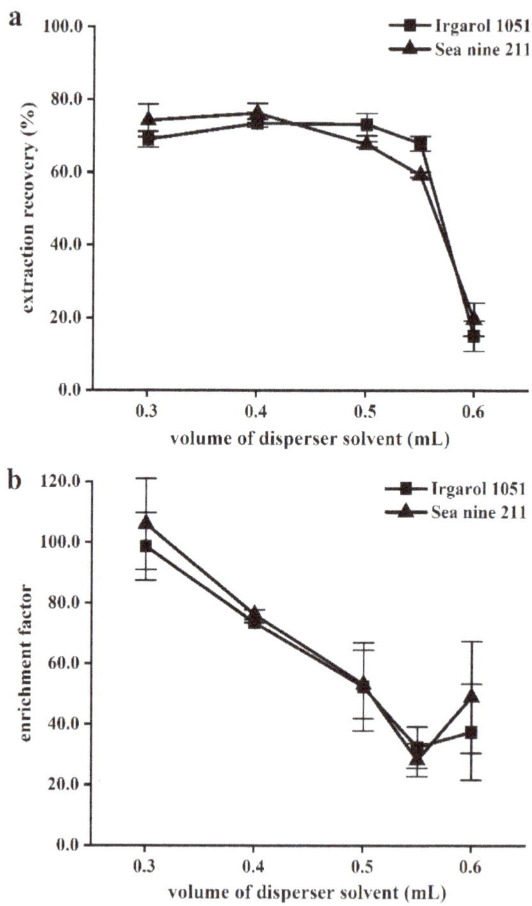

Figure 2. Effect of ethyl acetate volume on (**a**) extraction recovery and (**b**) enrichment factor. Extraction conditions: water sample, 5.0 mL; extraction solvent [HMIm][PF6] 60 mg; NaCl 0% (*w*/*v*).; pH 6; and cooling temperature 20 °C.

2.1.3. Salt Effect

In general, an increase in ionic strength frequently results in better extraction performance with salting out, which has an impact on the analyte partitioning coefficients between the aqueous and organic phases. In contrast, the addition of salt increases the ionic liquid's solubility in water, resulting in low recovery [47]. Different NaCl concentrations (0%, 2%, 4%, 8%, and 12%, *w*/*v*) were added to deionized water to assess the impact of the ionic strength on the effectiveness of extraction and enrichment. As depicted in Figure 3, the addition of salt had no discernible effect on either the EF or ER at concentrations of NaCl less than 8%. With a higher concentration and an increase in ILs solubility in the aqueous phase, the sediment volume decreased, resulting in a low ER and a high EF. In the study that used 1-octyl-3-methylimidazolium hexafluorophosphate ([C8MIM][PF6]) to extract pyrethroid pesticides, Zhang et al. also discovered that a high salt concentration increased the viscosity of the water phase and improved the solubility of IL in water, thereby reducing the extraction efficiency [48]. As a result, no NaCl was added to the water samples, allowing the proposed method to be used for the preconcentration of Irgarol 1051 and Sea-Nine 211 in both fresh and salty water.

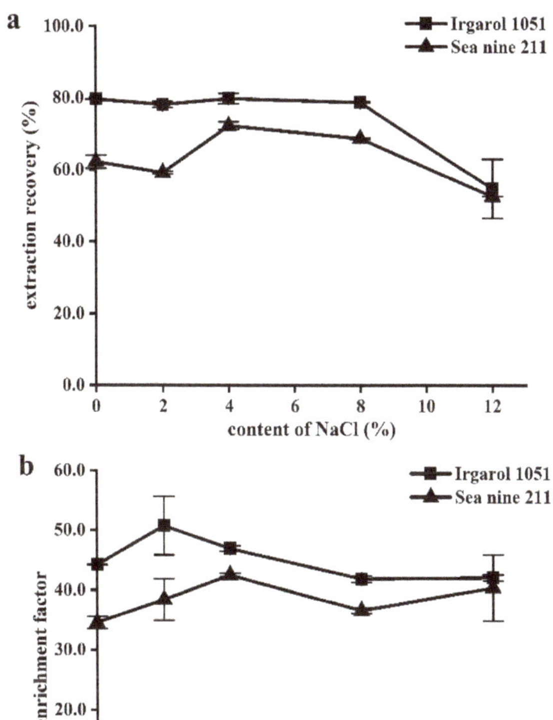

Figure 3. Influence of content of NaCl on (**a**) extraction recovery and (**b**) enrichment factor. Extraction conditions: water sample, 5.0 mL; disperser solvent, ethyl acetate 0.4 mL; extraction solvent [HMIm][PF6] 60 mg; pH 6; and cooling temperature 20 °C.

2.1.4. Sample pH

The effect of various pH levels (4, 5, 6, 7, and 9) on IL-DLLME ER and EF was examined by adding the appropriate amount of hydrochloric acid or sodium hydroxide solution to water samples. The results are displayed in Figure 4, which shows that pH 5 or pH 6 provided the best analyte recovery. Similar behavior was observed in a prior study that used IL-DLLME to identify organophosphorus pesticides [40]. The results indicated that Irgarol 1051 (pK_a 4.13 ± 0.10) and Sea-Nine 211 (pK_a −6.09 ± 0.60) were relatively stable and had a high IL distribution coefficient in neutral and weakly acidic media, and that they could be decomposed in strong bases. A pH of 6 was selected due to the ease of operation. Since the pH of the utilized deionized water was approximately 6, pH adjustments were avoided throughout the entire optimization procedure. After being diluted with deionized water, the real water samples were examined.

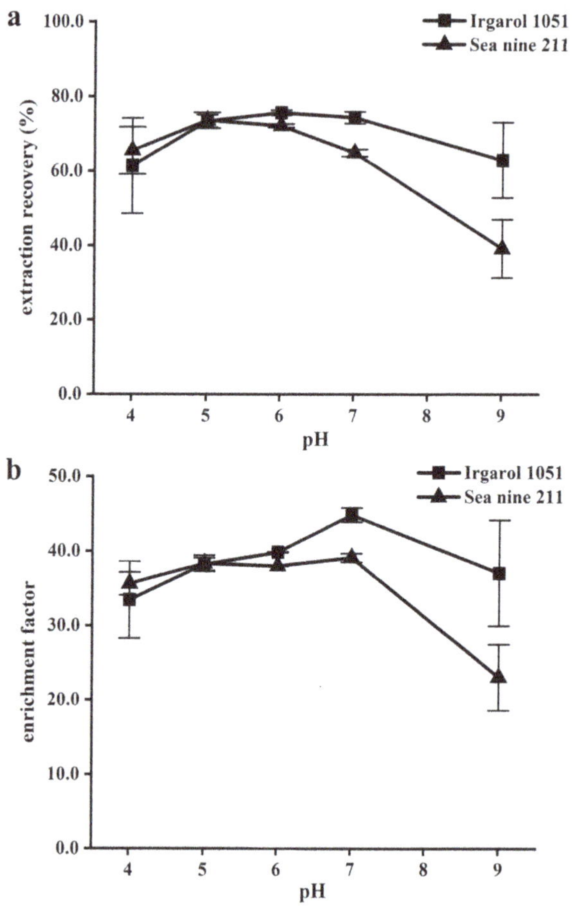

Figure 4. Influence of pH on (**a**) extraction recovery and (**b**) enrichment factor. Extraction conditions: water sample, 5.0 mL; disperser solvent, ethyl acetate 0.4 mL; extraction solvent [HMIm][PF6] 60 mg; NaCl 0% (*w*/*v*); and cooling temperature 20 °C.

2.1.5. Effect of Cooling Temperature

Temperature can influence analyte partition coefficients, IL solubility in water, and phase separation [49]. The different cooling temperatures (10 °C, 15 °C, 20 °C, 25 °C, and 30 °C) in the water bath (defined as the temperature before centrifugation and after extraction) were investigated at 30 °C of the extraction temperature. As shown in Figure 5, as the temperature decreased from 30 °C, the recovery initially increased and then reduced. In varying temperatures, the EF exhibited the same characteristics as the ER. Therefore, it can be concluded that the partition coefficient of analytes between IL and water had a significant impact on recovery and enrichment. The cooling temperature was found to have the greatest contribution of all the optimized factors. The recovery of Irgarol 1051 increased from 75.8% to 94.9%, while the recovery of Sea-Nine 211 increased from 57.2% to 96.4%, with a decrease in temperature from 30 °C to 20 °C. In the following method validation studies, 20 °C was used.

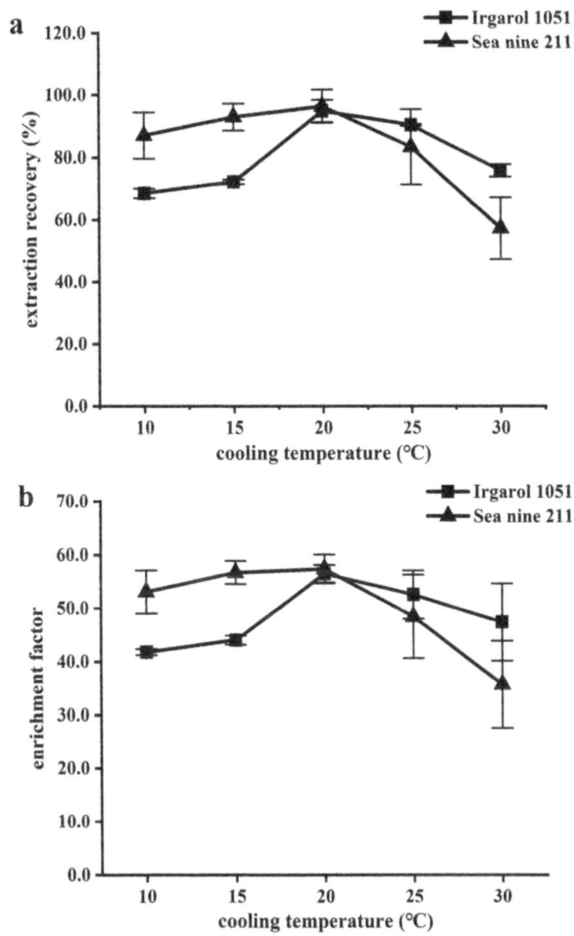

Figure 5. Influence of cooling temperature on (**a**) extraction recovery and (**b**) enrichment factor. Extraction conditions: water sample, 5.0 mL; disperser solvent, ethyl acetate 0.4 mL; extraction solvent [HMIm][PF6] 60 mg; NaCl 0% (*w*/*v*); and pH 6.

2.2. Method Validation

To validate the analytical approach, the series levels of spiked samples in deionized water, lake water, and seawater were examined (Table 1). Linearities were determined using deionized water spiked with five different concentrations of Irgarol 1051 (0.02 $\mu g\ L^{-1}$, 0.2 $\mu g\ L^{-1}$, 2 $\mu g\ L^{-1}$, 20 $\mu g\ L^{-1}$, and 100 $\mu g\ L^{-1}$) and Sea-Nine 211 (0.1 $\mu g\ L^{-1}$, 1 $\mu g\ L^{-1}$,

10 μg L^{-1}, 100 μg L^{-1}, and 500 μg L^{-1}). Calibration curves exhibited the linear relationships between analyte peak regions and concentrations. The equations for the calibration curves of Irgarol 1051 and Sea-Nine 211 were y = 70,515,778x − 32,368 and y = 11,067,977x + 53,400, respectively, and their respective correlation coefficients (R^2) were 0.9995 and 0.9993. The accuracy and precision of this method were validated using a recovery experiment. Analytes were spiked at three concentration levels in deionized water, lake water, and seawater samples, respectively, and each concentration level was repeated in triplicate. The mean recoveries ranged from 78.7% to 90.3%, and all relative standard deviations (RSDs) were less than 7.5%. The accuracy and precision of this method met the requirements for reliable analyte detection (recoveries were 70–120%, RSD < 20%) [50]. The limits of detection (LOD) and quantification (LOQ) were determined as the analyte concentrations corresponding to the instrument responses of 3 and 10 signal/noise, respectively, by injecting spiked samples of deionized water, lake water, and seawater. This method had LODs and LOQs of 0.01–0.1 μg L^{-1} and 0.02–0.5 μg L^{-1}, respectively, with the EF ranging from 22 to 45. Figure 6 depicts a typical chromatogram of antifouling biocides in a spiked water sample.

Table 1. Recoveries and RSDs of Irgarol 1051 and Sea-Nine 211 spiked in water samples (*n* = 3).

Sample		Deionized Water			Lake Water			Seawater		
Irgarol 1051	Spiked level (μg L^{-1})	0.02	0.1	1.0	0.02	0.1	1.0	0.1	1.0	5.0
	Recovery (%)	87.0	80.3	85.7	85.7	90.2	81.0	79.1	82.4	85.2
	RSD (%)	5.4	1.2	2.5	3.9	4.2	5.6	5.6	1.2	3.1
	LOQ, LOD (μg L^{-1})		0.02, 0.01			0.02, 0.01			0.1, 0.05	
Sea-Nine 211	Spiked level (μg L^{-1})	0.1	1.0	5.0	0.1	0.5	5.0	0.5	5.0	10.0
	Recovery (%)	86.9	90.3	84.7	84.4	83.3	80.6	78.7	86.1	89.3
	RSD (%)	7.5	4.5	2.7	5.0	4.1	3.1	1.7	6.7	4.6
	LOQ, LOD (μg L^{-1})		0.06, 0.02			0.06, 0.02			0.5, 0.1	

2.3. *Real Water Samples Analysis*

Finally, the developed analytical methodology was evaluated for its practical application in extracting antifouling biocides from freshwater and seawater. The environmental risk limit (ERL) is the concentration level at which pollutants pose a possible threat to the environment. The previous literature revealed the 0.024 μg L^{-1} ERL for Irgarol 1051 in water [51]. According to the European Union directive, the maximum allowable concentration of environmental quality standards (EQS) for Irgarol was 0.016 μg L^{-1} in water [52]. The limit standard for Sea-Nine 211 is still undefined. The suggested IL-DLLME technique has LODs 0.01–0.1μg L^{-1}. Therefore, this method typically achieved the criteria for detecting antifouling biocides from real water samples. The freshwater was collected from the North Sea Lake and Xiaoqing River in the city of Beijing, China, while the seawater was collected from Qing Dao, China. The outcomes revealed that the examined water samples were well below the LODs of the proposed method. Therefore, the antifouling biocides did not represent a significant threat to the aquatic ecosystem described above.

2.4. *Comparison of IL-DLLME with Other Sample Preparation Techniques*

Table 2 represents the performance of the proposed IL-DLLME approach in comparison to existing reported extraction procedures for the determination of antifouling biocides in water samples, such as LLE, SPE, SPME, SBSE, and LPME. Large sample volumes and substantial enrichment are responsible for the drastically reduced LOD obtained using SPE and LLE techniques. However, the enormous number of samples results in a prolonged extraction time and considerable consumption of organic solvent. IL-DLLME only requires a small amount of sample and organic solvent for extraction, and its recovery and RSD values are comparable to those of SPE and LLE. The extraction solvent is not necessary for LPME, SBSE, or SPME; however, these processes require a long time and requirements for specialized equipment. With a lower LOD than the SPME approach, the simple operation of the IL-DLLME procedure facilitates the whole sample treatment; just a few minutes are required before instrument analysis. All of these results indicate that the optimized IL-DLLME procedure appears to be a reproducible, rapid, simple, and low-cost alternative that can be used for the preconcentration of antifouling biocides such as Irgarol 1051 and Sea-Nine 211 from water samples.

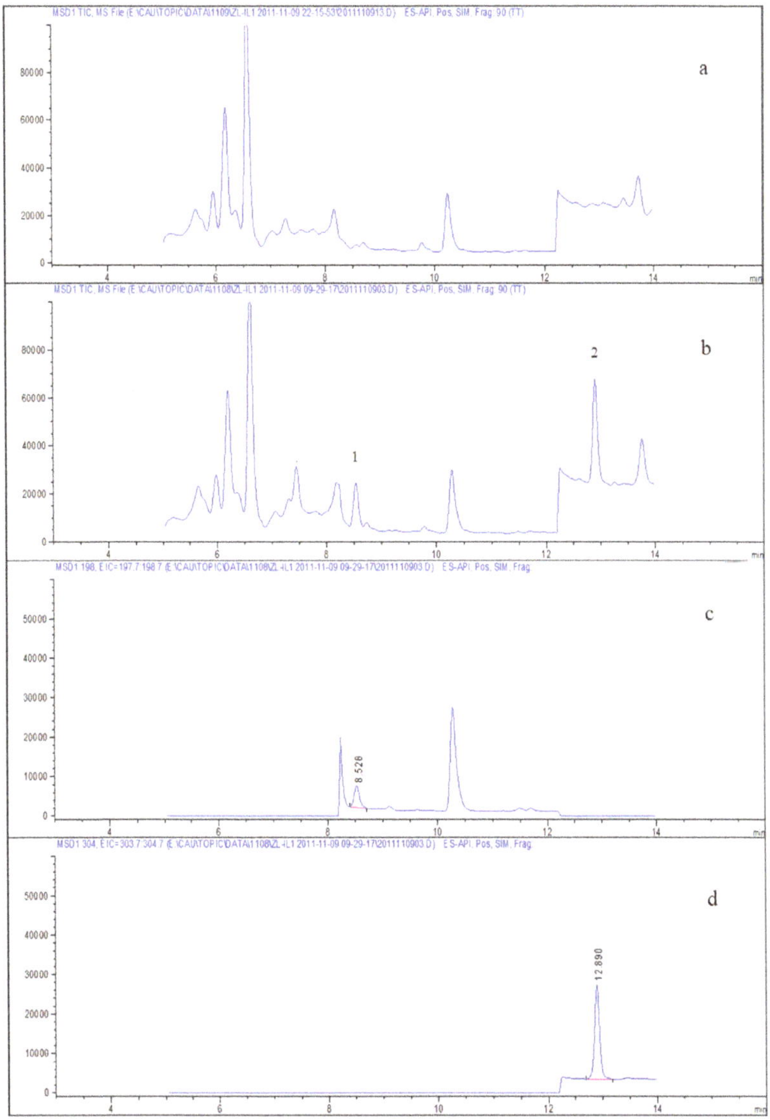

Figure 6. Chromatograms of analytes spiked in the water sample. (**a**) Sum chromatogram of blank, (**b**) sum chromatogram of spiked water, (**c**) extracted ion chromatogram (EIC) of ion 198, and (**d**) EIC chromatogram of ion 304. 1-Irgarol 1051, spiked 0.02 μg L^{-1}; 2-Sea-Nine 211, spiked 0.12 μg L^{-1}.

Table 2. Comparison of the IL-DLLME method with other procedures for the determination of antifouling biocides in water samples.

Method	Sample Amount (mL)	Extraction Solvent	Solvent Volume [a] (mL)	Extraction Time [b] (min)	Extraction Recovery (%)	LOD (μg L^{-1})	RSD%
SPE-GC-MS [53]	200	EA	15	46	42–95	0.0012–0.0015	<10
SPE-LC-MS/MS [54]	100	ACN	12	Not given	77–93	0.002	<8
SPME-GC-MS [55]	3	—	—	60	Not given	0.05–0.2	<20
SPE-LC-MS/MS [56]	1000	MeOH, DCM	9	200	80–120	0.001	<18
SPE-LC-MS/MS [57]	250	MeOH, DCM	8	25	78–120	0.0003–0.0027	<13
SPE-LC-QTOF/MS [58]	200	MeOH, DCM	8	60	79.7–119.2	Not given	17.7–27.7
LLE-GC-MS [59]	2000	DCM	50	Not given	70–120	0.001	30
SPE-GC-MS [60]	2000	EA, AC	15	145	>90	0.001	<10
SPE-LC-MS [61]	500	10 mM HAc MeOH	15	65	82.5–111	0.0002–0.001	3–5
SBSE-TD-GC-MS [20]	10	—	—	90	72–125	0.005–0.9	7–15
LLE-GC-MS [62]	1000	Toluene	1	60	73.55–120.28	0.00177–0.01242	1.64–4.87
MF-LPME-HPLC-UV [21]	300	Toluene	0.4	90	Not given	0.001–0.0048	<12
IL-DLLME method	5	[HMIm][PF6]	0.046	1	80–90	0.01–0.1	<8

[a] Solvent consumption only in the extraction stage; solvent consumption in solvent exchanges not included. [b] Time employed in the extraction stage; any other operations were not included. GC-MS, gas chromatography-mass spectrometry; LC-MS/MS, liquid chromatography-tandem mass spectrometry; LC-QTOF/MS, liquid chromatography-quadrupole time-of-flight mass spectrometry; EA, ethyl acetate; ACN, acetonitrile; MeOH, methanol; DCM, dichloromethane; AC, acetone; and HAc, acetic acid.

3. Materials and Methods

3.1. Reagents and Chemicals

Analytical standards for Irgarol 1051 were supplied by Dr. Ehrenstorfer (Augsburg, Germany), and Sea-Nine 211 was supplied by Pure Chemistry Scientific Inc. (Newton, MA, USA). The basic information about analytes is detailed in Table 3. The standard stock solution of 1 mg mL^{-1} was prepared in acetonitrile. The stock solution was diluted with acetonitrile to provide a working standard solution of 10 μg mL^{-1}. Both standard stock solutions and working solutions were stored at −20 °C. HPLC-grade acetonitrile, methanol, and ethyl acetate (Fisher Scientific, Waltham, MA, USA) were used. IL [HMIm][PF6] was acquired from the Lanzhou Institute of Chemical Physics of Chinese Academy of Sciences (Lanzhou, China). Sodium chloride (NaCl, AR) was purchased from Sinopharm Beijing Chemical and Reagent Ltd. (Beijing, China). Deionized water (18 M/cm) was prepared by a MILI-Q Pure treatment system (Millipore, St. Louis, MO, USA). The freshwater was collected from the North Sea Lake, an artificial lake in the city of Beijing. The seawater was collected from the Yellow Sea.

Table 3. Basic information and chromatographic parameters of the analytes.

Analyte	Chemical Structure	Molecular Weight	Retention Time (min)	Mass Ions (*m/z*)	Fragmentor Voltage (V)
Irgarol 1051	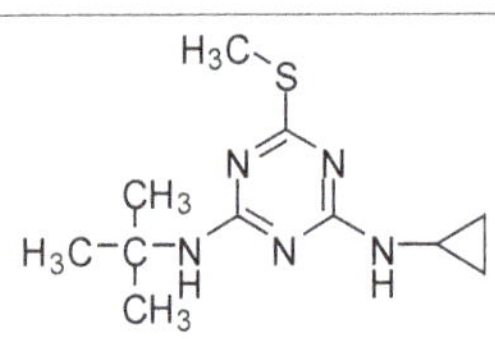	253.1	8.53	254.0 $[M + H]^+$ 198.0 *	120 230
Sea-Nine 211	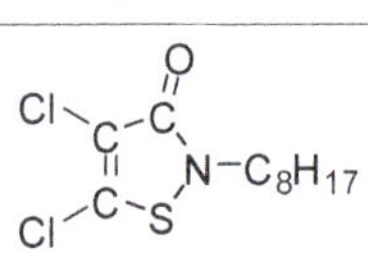	281.0	12.89	282.0 $[M + H]^+$ 304.0 * $[M + Na]^+$	90 100

* Quantitative ion.

3.2. *Apparatus*

The analytes were separated from the extracts using the Agilent 1260 series HPLC (Agilent Technologies, Palo Alto, CA, USA). A ZORBAX SB-C18 column (150 mm × 4.6 mm i.d., 3.5 μm; Agilent Technologies, Palo Alto, CA, USA) was employed. The mobile phase was comprised of methanol (A) and 0.1% formic acid in water (B). The gradient program was as follows: 0–5 min, 55–85% A; 5–7 min, 85% A; 7–10 min, 85–95% A; 10–13 min, 95% A; 13–15 min, 95–55% A; and 15–19 min, 55% A. The flow rate was 0.6 mL min^{-1} and the injection volume was 10 μL. The column temperature was maintained at 30 °C.

The HPLC system was coupled to an Agilent 6130 Single Quadrupole mass spectrometer equipped with an electrospray source in positive ionization mode. The operational parameters were as follows: drying gas flow 10.0 L min^{-1}, drying gas temperature 350 °C, nebulizer gas pressure 35 psi., and capillary voltage 3000 V. Flow injection analysis (FIA) was used to optimize the fragmentor, and analytes were quantified in the selected ion monitoring mode (SIM). The chromatographic parameters of the analytes are presented in Table 3.

3.3. *IL-DLLME Procedure*

The environmental samples, including lake water and seawater, were filtered with 0.45 μm water phase membrane prior to analysis. After that, the seawater had to be diluted fourfold with deionized water. In a 15 mL conical-bottomed centrifuge tube, 5.0 mL of water samples were placed. The aqueous phase was then rapidly injected with 60 mg of the [HMIm][PF6] and 0.4 mL of ethyl acetate as extraction and disperser solvents, followed by 1 min of manual shaking. After cooling the cloudy solution in a 20 °C water bath and centrifuging at 3800× *g* rpm for 5 min, the IL phase settled at the bottom of the tube. The IL phase was collected and diluted with acetonitrile to a final volume of 150 μL after the upper aqueous phase had been removed using a syringe.

4. Conclusions

This research used an IL-DLLME methodology coupled with LC-MS to identify two types of commonly used booster biocides in water samples. The quantity of the IL ([HMIm][PF6]) utilized as an extraction solvent for Irgarol 1051 and Sea-Nine 211 was first optimized. Furthermore, the type and volume of the disperser solvent, the amount of salt, the pH, and the cooling temperature were studied to determine the optimal extraction conditions. A systematic validation demonstrated that the proposed method has acceptable linearity ($R^2 > 0.999$), recovery (78.7–90.3%), and repeatability (RSD ≤ 7.5%). The LOD and LOQ of this method were found to be 0.01–0.1 μg L^{-1} and 0.02–0.5 μg L^{-1}, respectively. The successful utilization of lake water and seawater samples revealed that the method is acceptable for determining antifouling biocides in real water samples. Furthermore, the use of IL provides a simple, quick, less toxic, and ecologically favorable technique for determining the booster biocides in water samples.

Author Contributions: L.Z. performed experiments, investigation, formal analysis, and writing—original draft. T.W. performed investigation, formal analysis, and writing—review and editing. C.Y. performed validation and investigation. S.L. performed visualization and writing—review and editing. C.P. provided conceptualization, experimental mentorship, supervision, and resources. All authors have read and agreed to the published version of the manuscript.

Funding: This study was supported the by Guangxi Key Research and Development Program, NO: AB21196039 and the 2115 Talent Development Program of China Agricultural University.

Institutional Review Board Statement: Not applicable.

Informed Consent Statement: Not applicable.

Data Availability Statement: Data sharing is not applicable to this article.

Conflicts of Interest: The authors declare no conflict of interest.

Sample Availability: Samples of the compounds are not available from the authors.

References

1. Amara, I.; Miled, W.; Slama, R.B.; Ladhari, N. Antifouling processes and toxicity effects of antifouling paints on marine environment. A review. *Environ. Toxicol. Pharmacol.* **2018**, *57*, 115–130. [CrossRef] [PubMed]
2. Konstantinou, I.K.; Albanis, T.A. Worldwide occurrence and effects of antifouling paint booster biocides in the aquatic environment: A review. *Environ. Int.* **2004**, *30*, 235–248. [CrossRef] [PubMed]
3. Yebra, D.M.; Kiil, S.; Dam-Johansen, K. Antifouling technology—Past, present and future steps towards efficient and environmentally friendly antifouling coatings. *Prog. Org. Coat.* **2004**, *50*, 75–104. [CrossRef]
4. Liu, D.; Pacepavicius, G.J.; Maguire, R.J.; Lau, Y.L.; Okamura, H.; Aoyama, I. Survey for the occurrence of the new antifouling compound Irgarol 1051 in the aquatic environment. *Water Res.* **1999**, *33*, 2833–2843. [CrossRef]
5. Scarlett, A.; Donkin, P.; Fileman, T.W.; Morris, R.J. Occurrence of the antifouling herbicide, irgarol 1051, within coastal-water seagrasses from Queensland, Australia. *Mar. Pollut. Bull.* **1999**, *38*, 687–691. [CrossRef]
6. Ishibashi, H.; Takaichi, D.; Takeuchi, I. Effects of the herbicide Irgarol 1051 on the transcriptome of hermatypic coral Acropora tenuis and its symbiotic dinoflagellates. *Sci. Total Environ.* **2021**, *780*, 146542. [CrossRef]
7. US EPA. Irgarol. Interim Registration Review Decision. Case Number 5031. Docket Number EPA-HQ-OPP-2010-0003. 2021. Available online: https://www.regulations.gov/document/EPA-HQ-OPP-2010-0003-0017 (accessed on 9 December 2022).
8. EU Commission. Commission Implementing Decision (EU) 2016/107 of 27 January 2016 not Approving Cybutryne as an Existing Active Substance for Use in Biocidal Products for Product-Type 21. 2016. Available online: https://www.legislation.gov.uk/eudn/2016/107/2020-12-31 (accessed on 9 December 2022).
9. Okamura, H.; Watanabe, T.; Aoyama, I.; Hasobe, M. Toxicity evaluation of new antifouling compounds using suspension-cultured fish. *Chemosphere* **2002**, *46*, 945–951. [CrossRef]
10. Kobayashi, N.; Okamura, H. Effects of new antifouling compounds on the development of sea urchin. *Mar. Pollut. Bull.* **2002**, *44*, 748–751. [CrossRef]
11. de Campos, B.G.; Figueiredo, J.; Perina, F.; Abessa, D.M.d.S.; Loureiro, S.; Martins, R. Occurrence, effects and environmental risk of antifouling biocides (EU PT21): Are marine ecosystems threatened? *Crit. Rev. Environ. Sci. Technol.* **2022**, *52*, 3179–3210. [CrossRef]
12. Lam, N.H.; Jeong, H.-h.; Kang, S.-d.; Kim, D.-J.; Ju, M.-J.; Horiguchi, T.; Cho, H.-S. Organotins and new antifouling biocides in water and sediments from three Korean Special Management Sea Areas following ten years of tributyltin regulation: Contamination profiles and risk assessment. *Mar. Pollut. Bull.* **2017**, *121*, 302–312. [CrossRef]
13. Diniz, L.G.R.; Jesus, M.S.; Dominguez, L.A.E.; Fillmann, G.; Vieira, E.M.; Franco, T. First appraisal of water contamination by antifouling booster biocide of 3rd generation at Itaqui Harbor (São Luiz-Maranhão-Brazil). *J. Braz. Chem. Soc.* **2014**, *25*, 380–388. [CrossRef]
14. Mochida, K.; Hano, T.; Onduka, T.; Ichihashi, H.; Amano, H.; Ito, M.; Ito, K.; Tanaka, H.; Fujii, K. Spatial analysis of 4,5-dichloro-2-n-octyl-4-isothiazolin-3-one (Sea-Nine 211) concentrations and probabilistic risk to marine organisms in Hiroshima Bay, Japan. *Environ. Pollut.* **2015**, *204*, 233–240. [CrossRef] [PubMed]
15. Martínez, K.; Ferrer, I.; Barceló, D. Part-per-trillion level determination of antifouling pesticides and their byproducts in seawater samples by off-line solid-phase extraction followed by high-performance liquid chromatography–atmospheric pressure chemical ionization mass spectrometry. *J. Chromatogr. A* **2000**, *879*, 27–37. [CrossRef] [PubMed]
16. Basheer, C.; Tan, K.S.; Lee, H.K. Organotin and Irgarol-1051 contamination in Singapore coastal waters. *Mar. Pollut. Bull.* **2002**, *44*, 697–703. [CrossRef] [PubMed]
17. Sánchez-Rodríguez, Á.; Sosa-Ferrera, Z.; Santana-Rodríguez, J.J. Applicability of microwave-assisted extraction combined with LC–MS/MS in the evaluation of booster biocide levels in harbour sediments. *Chemosphere* **2011**, *82*, 96–102. [CrossRef] [PubMed]
18. Bester, K.; Lamani, X. Determination of biocides as well as some biocide metabolites from facade run-off waters by solid phase extraction and high performance liquid chromatographic separation and tandem mass spectrometry detection. *J. Chromatogr. A* **2010**, *1217*, 5204–5214. [CrossRef]
19. Lambropoulou, D.A.; Sakkas, V.A.; Albanis, T.A. Analysis of antifouling biocides Irgarol 1051 and Sea Nine 211 in environmental water samples using solid-phase microextraction and gas chromatography. *J. Chromatogr. A* **2002**, *952*, 215–227. [CrossRef]
20. Giraldez, I.; Chaguaceda, E.; Bujalance, M.; Morales, E. Determination of five booster biocides in seawater by stir bar sorptive extraction-thermal desorption-gas chromatography-mass spectrometry. *J. Chromatogr. A* **2013**, *1271*, 17–26. [CrossRef]
21. Saleh, A.; Fumani, N.S.; Molaei, S. Microfunnel-supported liquid-phase microextraction: Application to extraction and determination of Irgarol 1051 and diuron in the Persian Gulf seawater samples. *J. Chromatogr. A* **2014**, *1356*, 32–37. [CrossRef]
22. Trujillo-Rodríguez, M.J.; Rocío-Bautista, P.; Pino, V.; Afonso, A.M. Ionic liquids in dispersive liquid-liquid microextraction. *TrAC Trends Anal. Chem.* **2013**, *51*, 87–106. [CrossRef]
23. Rykowska, I.; Ziemblińska, J.; Nowak, I. Modern approaches in dispersive liquid-liquid microextraction (DLLME) based on ionic liquids: A review. *J. Mol. Liq.* **2018**, *259*, 319–339. [CrossRef]

24. Lai, X.; Ruan, C.; Liu, R.; Liu, C. Application of ionic liquid-based dispersive liquid–liquid microextraction for the analysis of ochratoxin A in rice wines. *Food Chem.* **2014**, *161*, 317–322. [CrossRef] [PubMed]
25. Sajid, M. Dispersive liquid-liquid microextraction: Evolution in design, application areas, and green aspects. *TrAC Trends Anal. Chem.* **2022**, *152*, 116636. [CrossRef]
26. Zheng, Y.-Z.; Wang, K.; Liang, Q.; Xue, X.-F.; Zhao, L.-W.; Chen, D.-F.; Wu, L.-M.; Guo, R.; Xiong, C.-L. Ionic liquid dispersive liquid–liquid microextraction for pesticide residue analysis in honey. *J. Apic. Res.* **2020**, *59*, 458–467. [CrossRef]
27. Ullah, H.; Wilfred, C.D.; Shaharun, M.S. Ionic liquid-based extraction and separation trends of bioactive compounds from plant biomass. *Sep. Sci. Technol.* **2019**, *54*, 559–579. [CrossRef]
28. Feng, J.; Loussala, H.M.; Han, S.; Ji, X.; Li, C.; Sun, M. Recent advances of ionic liquids in sample preparation. *TrAC Trends Anal. Chem.* **2020**, *125*, 115833. [CrossRef]
29. Lim, J.R.; Chua, L.S.; Mustaffa, A.A. Ionic liquids as green solvent and their applications in bioactive compounds extraction from plants. *Process Biochem.* **2022**, *122*, 292–306. [CrossRef]
30. Zhou, G.-S.; Yuan, Y.-C.; Yin, Y.; Tang, Y.-P.; Xu, R.-J.; Liu, Y.; Chen, P.-D.; Yin, L.; Duan, J.-A. Hydrophilic interaction chromatography combined with ultrasound-assisted ionic liquid dispersive liquid–liquid microextraction for determination of underivatized neurotransmitters in dementia patients' urine samples. *Anal. Chim. Acta* **2020**, *1107*, 74–84. [CrossRef]
31. Zhang, H.-F.; Shi, Y.-P. Temperature-assisted ionic liquid dispersive liquid–liquid microextraction combined with high performance liquid chromatography for the determination of anthraquinones in Radix et Rhizoma Rhei samples. *Talanta* **2010**, *82*, 1010–1016. [CrossRef]
32. Zhang, H.; Chen, X.; Jiang, X. Determination of phthalate esters in water samples by ionic liquid cold-induced aggregation dispersive liquid–liquid microextraction coupled with high-performance liquid chromatography. *Anal. Chim. Acta* **2011**, *689*, 137–142. [CrossRef]
33. Unsal, Y.E.; Soylak, M.; Tuzen, M. Ultrasound-assisted ionic liquid-based dispersive liquid–liquid microextraction for preconcentration of patent blue V and its determination in food samples by UV–visible spectrophotometry. *Environ. Monit. Assess.* **2015**, *187*, 203. [CrossRef]
34. Elik, A.; Demirbas, A.; Altunay, N. Analysis of Zinc and Chromium in grain samples using ionic liquid-based ultrasound-assisted microextraction followed by flame-AAS after microwave digestion. *Biol. Trace Elem. Res.* **2020**, *198*, 697–706. [CrossRef] [PubMed]
35. Adhami, K.; Asadollahzadeh, H.; Ghazizadeh, M. Preconcentration and determination of nickel (II) and copper (II) ions, in vegetable oils by [TBP] [PO4] IL-based dispersive liquid–liquid microextraction technique, and flame atomic absorption spectrophotometry. *J. Food Compos. Anal.* **2020**, *89*, 103457. [CrossRef]
36. Yang, J.; Fan, C.; Tang, G.; Zhang, W.; Dong, H.; Liang, Y.; Wang, Y.; Zou, M.; Cao, Y. Relationship between the structure of ionic liquid and its enrichment ability to trace fungicides from an environmental water sample. *J. Agric. Food. Chem.* **2018**, *66*, 9418–9425. [CrossRef] [PubMed]
37. Zhang, R.; Tan, Z.; Zhao, J.; Wen, Y.; Fan, S.; Liu, C. Determination of pyrethroid residues in herbal tea using temperature-controlled ionic liquid dispersive liquid-liquid microextraction by high performance liquid chromatography. *Sci. Rep.* **2020**, *10*, 4709. [CrossRef] [PubMed]
38. Hong, J.; Liu, X.; Yang, X.; Wang, Y.; Zhao, L. Ionic liquid-based dispersive liquid–liquid microextraction followed by magnetic solid-phase extraction for determination of quinolones. *Microchim. Acta* **2021**, *189*, 8. [CrossRef]
39. Gao, S.; Jin, H.; You, J.; Ding, Y.; Zhang, N.; Wang, Y.; Ren, R.; Zhang, R.; Zhang, H. Ionic liquid-based homogeneous liquid–liquid microextraction for the determination of antibiotics in milk by high-performance liquid chromatography. *J. Chromatogr. A* **2011**, *1218*, 7254–7263. [CrossRef]
40. Zhang, L.; Yu, R.; Zhang, X.; Zhang, D. Ionic liquid–based dispersive liquid–liquid micro-extraction of five organophosphorus pesticides in coarse cereals. *Food Anal. Methods* **2021**, *14*, 10–17. [CrossRef]
41. Taziki, M.; Shemirani, F.; Majidi, B. Robust ionic liquid–based dispersive liquid–liquid microextraction method for determination of chromium(vi) in saline solutions. *Commun. Soil Sci. Plant Anal.* **2013**, *44*, 3400–3411. [CrossRef]
42. Liu, Q.; Tang, J.; Chen, D.; Zhou, Y.; Lin, Q.; Ma, X.; Zhang, M.; Hu, H. [Hmim]PF6 enhanced the extraction of polycyclic aromatic hydrocarbons from soil with the QuEChERS method. *Arabian J. Chem.* **2020**, *13*, 4102–4110. [CrossRef]
43. Wang, S.; Liu, C.; Yang, S.; Liu, F. Ionic liquid-based dispersive liquid–liquid microextraction following high-performance liquid chromatography for the determination of fungicides in fruit juices. *Food Anal. Methods* **2013**, *6*, 481–487. [CrossRef]
44. Marube, L.C.; Caldas, S.S.; dos Santos, E.O.; Michaelsen, A.; Primel, E.G. Multi-residue method for determination of thirty-five pesticides, pharmaceuticals and personal care products in water using ionic liquid-dispersive liquid-liquid microextraction combined with liquid chromatography-tandem mass spectrometry. *J. Braz. Chem. Soc.* **2018**, *29*, 1349–1359. [CrossRef]
45. Kong, L.; Wang, J.; Gao, Q.; Li, X.; Zhang, W.; Wang, P.; Ma, L.; He, L. Simultaneous determination of fat-soluble vitamins and carotenoids in human serum using a nanostructured ionic liquid based microextraction method. *J. Chromatogr. A* **2022**, *1666*, 462861. [CrossRef] [PubMed]
46. Yin, Q.; Zhu, Y.; Yang, Y. Dispersive liquid–liquid microextraction followed by magnetic solid-phase extraction for determination of four parabens in beverage samples by ultra-performance liquid chromatography tandem mass spectrometry. *Food Anal. Methods* **2018**, *11*, 797–807. [CrossRef]
47. Lubomirsky, E.; Padró, J.M.; Reta, M.R. Development of a dispersive liquid-liquid microextraction technique for the analysis of aryloxyphenoxy-propionate herbicides in soy-based foods. *Microchem. J.* **2016**, *129*, 63–70. [CrossRef]

48. Zhang, J.; Gao, H.; Peng, B.; Li, S.; Zhou, Z. Comparison of the performance of conventional, temperature-controlled, and ultrasound-assisted ionic liquid dispersive liquid–liquid microextraction combined with high-performance liquid chromatography in analyzing pyrethroid pesticides in honey samples. *J. Chromatogr. A* **2011**, *1218*, 6621–6629. [CrossRef] [PubMed]
49. Padilla-Alonso, D.J.; Garza-Tapia, M.; Chávez-Montes, A.; González-Horta, A.; Waksman de Torres, N.H.; Castro-Ríos, R. New temperature-assisted ionic liquid-based dispersive liquid–liquid microextraction method for the determination of glyphosate and aminomethylphosphonic acid in water samples. *J. Liq. Chromatogr. Relat. Technol.* **2017**, *40*, 147–155. [CrossRef]
50. European Commission. Analytical Quality Control and Method Validation Procedures for Pesticide Residues Analysis in Food and Feed SANTE/11312/2021. 2021. Available online: https://www.eurl-pesticides.eu/userfiles/file/EurlALL/SANTE_11312_2021.pdf (accessed on 10 January 2023).
51. van Wezel, A.P.; van Vlaardingen, P. Environmental risk limits for antifouling substances. *Aquat. Toxicol.* **2004**, *66*, 427–444. [CrossRef]
52. European Union. Directive 2013/39/EU of the European Parliament and of the Council of 12 August 2013 Amending Directives 2000/60/EC and 2008/105/EC as Regards Priority Substances in the Field of Water Policy. 2013. Available online: https://eur-lex.europa.eu/legal-content/EN/TXT/?uri=celex%3A32013L0039 (accessed on 7 January 2023).
53. Agüera, A.; Piedra, L.; Hernando, M.a.D.; Fernández-Alba, A.R. Multiresidue method for the analysis of five antifouling agents in marine and coastal waters by gas chromatography–mass spectrometry with large-volume injection. *J. Chromatogr. A* **2000**, *889*, 261–269. [CrossRef]
54. Lamoree, M.H.; Swart, C.P.; van der Horst, A.; van Hattum, B. Determination of diuron and the antifouling paint biocide Irgarol 1051 in Dutch marinas and coastal waters. *J. Chromatogr. A* **2002**, *970*, 183–190. [CrossRef]
55. Peñalver, A.; Pocurull, E.; Borrull, F.; Marcé, R.M. Solid-phase microextraction of the antifouling Irgarol 1051 and the fungicides dichlofluanid and 4-chloro-3-methylphenol in water samples. *J. Chromatogr. A* **1999**, *839*, 253–260. [CrossRef] [PubMed]
56. Kock-Schulmeyer, M.; Postigo, C.; Farre, M.; Barcelo, D.; de Alda, M.L. Medium to highly polar pesticides in seawater: Analysis and fate in coastal areas of Catalonia (NE Spain). *Chemosphere* **2019**, *215*, 515–523. [CrossRef] [PubMed]
57. Batista-Andrade, J.A.; Caldas, S.S.; Arias, J.L.D.; Castro, I.B.; Fillmann, G.; Primel, E.G. Antifouling booster biocides in coastal waters of Panama: First appraisal in one of the busiest shipping zones. *Mar. Pollut. Bull.* **2016**, *112*, 415–419. [CrossRef] [PubMed]
58. Zhang, R.H.; Du, J.; Dong, X.B.; Huang, Y.; Xie, H.J.; Chen, J.W.; Li, X.H.; Kadokami, K. Occurrence and ecological risks of 156 pharmaceuticals and 296 pesticides in seawater from mariculture areas of Northeast China. *Sci. Total Environ.* **2021**, *792*, 148375. [CrossRef]
59. Fernandez, M.V.; Gardinali, P.R. Risk assessment of triazine herbicides in surface waters and bioaccumulation of Irgarol and M1 by submerged aquatic vegetation in Southeast Florida. *Sci. Total Environ.* **2016**, *541*, 1556–1571. [CrossRef]
60. Ali, H.R.; Ariffin, M.M.; Omar, T.F.T.; Ghazali, A.; Sheikh, M.A.; Shazili, N.A.M.; Bachok, Z. Antifouling paint biocides (Irgarol 1051 and Diuron) in the selected ports of Peninsular Malaysia: Occurrence, seasonal variation, and ecological risk assessment. *Environ. Sci. Pollut. Res.* **2021**, *28*, 52247–52257. [CrossRef]
61. Ansanelli, G.; Manzo, S.; Parrella, L.; Massanisso, P.; Chiavarini, S.; Di Landa, G.; Ubaldi, C.; Cannarsa, S.; Cremisini, C. Antifouling biocides (Irgarol, Diuron and dichlofluanid) along the Italian Tyrrhenian coast: Temporal, seasonal and spatial threats. *Reg. Stud. Mar. Sci.* **2017**, *16*, 254–266. [CrossRef]
62. Lee, S.; Lee, D.; Lee, Y.W. Determination of five alternative antifouling agents found along the Korean coasts. *Water Environ. Res* **2017**, *89*, 622–628. [CrossRef]

molecules

Article

Risk Assessment of Pesticide Residues by GC-MSMS and UPLC-MSMS in Edible Vegetables

Mohamed T. Selim [1], Mohammad M. Almutari [2,3,4], Hassan I. Shehab [4] and Mohamed H. EL-Saeid [5,*]

1 Institute of Research and Consultancy, Ministry of Environment, Water and Agriculture, Riyadh 11442, Saudi Arabia
2 National Agriculture & Animal Resources Research Center, Riyadh 11442, Saudi Arabia
3 Plant Resource, Ministry of Environment, Water and Agriculture, Riyadh 11442, Saudi Arabia
4 Food and Environmental Research Center in Unayzah, Ministry of Municipal and Rural Affairs, Unayzah 51911, Al-Qassim, Saudi Arabia
5 Chromatographic Analysis Unit, Soil Science Department, College of Food & Agricultural Sciences, King Saud University, P.O. Box 2460, Riyadh 11451, Saudi Arabia
* Correspondence: elsaeidm@ksu.edu.sa

Abstract: In recent years, there has been a significant increase related to pesticide residues in foods, which may increase the risks to the consumer of these foods with the different quality and concentrations of pesticide residues. Pesticides are used for controlling pests that reduce yields. On the other hand, it has become a major public health concern due to its toxic properties. Thus, the objective of the current study employed the application of Quick Easy Cheap Effective Rugged Safe (QuEChERS) method, in combination with gas and liquid chromatography-tandem mass spectrometric detection (GCMSMS, LCMSMS) in order to determine 137 pesticide residues (63 insecticides, 41 acaricides, 40 herbicide, 55 fungicide, nematicide, growth regulator, Chitin synthesis inhibitors, and Juvenile hormone mimics), in 801 vegetables such as 139 tomatoes, 185 peppers, 217 squash, 94 eggplants, and 166 cucumbers from different locations in Hail and Riyadh cities. The results showed that the majority of pesticide residues were detected for each of the following pesticides: acetaimpride, metalaxyl, imidaclopride, bifenthrin, pyridaben, difenoconazole, and azoxystrobien, which were repeated in the samples studied 39, 21, 11, 10, 8, 7, and 5, respectively. In addition, results observed that the tomato was the most contaminated with pesticide residues; it was contaminated with 19 compounds and was followed by pepper, cucumber, and squash, and the last commodity in the contaminated ranking was eggplant. The highest calculated estimated daily intakes (EDIs) were recorded for tomatoes which were estimated between 0.013 to 0.516 mg/kg of body weight per day (bw/day) while the lowest EDIs value was between 0.000002 to 0.0005 mg/kg of bw/day for cucumber. Results indicated that the EDIs values were lower than the acceptable daily intake (ADI) values. Results observed that the most of pesticide residues exposure in food consumption in Saudi Arabia were lower than ADIs. In addition, the highest value for health risk index (HRI) was recorded with Ethion residue in tomato, but in sweet pepper, the highest value for HRI was 127.5 in the form of fipronil residue. On the other hand, results found that the highest values of HRI were 1.54, 1.61, and 0.047 for difenoconazole, bifenthrin, and pyridaben residues in squash, eggplant, and cucumber.

Keywords: pesticide residues; risk assessment; QuEChERS; EDIs; ADI; HRI; GC–MS/MS; LC-MSMS

Citation: Selim, M.T.; Almutari, M.M.; Shehab, H.I.; EL-Saeid, M.H. Risk Assessment of Pesticide Residues by GC-MSMS and UPLC-MSMS in Edible Vegetables. *Molecules* **2023**, *28*, 1343. https://doi.org/10.3390/molecules28031343

Academic Editor: Miguel Ángel González-Curbelo

Received: 11 December 2022
Revised: 23 January 2023
Accepted: 25 January 2023
Published: 31 January 2023

1. Introduction

In recent years, we have observed a substantial increase in the importance placed on aspects related to pesticide residues and a growing demand for better agricultural practices, transparency, and traceability in the production and marketing of conventional food. On one hand, pesticides make it possible to increase food production by destroying weeds and pests that attack cultivable plants and agricultural crops, and they also limit losses sustained during the transport and storage of food. On the other hand, pesticides are one of the most dangerous chemical compounds due to their toxic properties, environmental persistence, and bioaccumulation capability. Thus, the presence of pesticide residues in food commodities is a source of great worry; what makes it more complex is that some of these vegetables are consumed fresh or semi-processed, which may contain elevated levels of chemicals compared to other food crops of plant origin. Exposure to pesticides through diet is thought to be five orders of magnitude higher than other exposure routes, for example, air and drinking water [1–3]. The level of pesticide residues in foodstuffs is generally legislated so as to minimize consumers' exposure to harmful or unnecessary pesticide intakes, their maximum concentrations are controlled by the European Union Council Directive 91/414/EEC [4], and established maximum residue limits (MRLs) for pesticides in foodstuffs and animal feed in Directive No. 396/2005 (Regulation2005) [5].

LC and GC coupled to MS/MS detection provides accurate methods of identifying and quantifying numerous pesticides in food extracts. Several articles have recently been published where these techniques were successfully utilized for the analysis of pesticides in fruits and vegetables. Due to the high selectivity provided by MS/MS detection, simple extraction techniques with little cleanup are employed [6].

In the last few years, the so-called QuEChERS (Quick, easy, cheap, effective, rugged, and safe) sample preparation procedure has become a widely used technique because of its applicability on a wide range of pesticides [7–10]. It has several advantages over traditional methods of pesticide residue analysis, for example, high recoveries (>85%) are achieved for a wide polarity, very accurate (true and precise) results are achieved, solvent usage and waste are very small, and the MeCN is added by dispenser to an unbreakable vessel that is immediately sealed, thus, solvent exposure to the worker is minimal, the method is very inexpensive [11–15].

In this study, we aimed to apply the QuEChERS methodology in combination with gas and liquid chromatography-tandem mass spectrometric detection (GC-MS/MS, LC-MS/MS), for the analysis of 137 pesticides, to determine residues of chemical pesticides (Organophosphates, OPs; acaricides, ACs; fungicides, FUs and insecticides of biological origin, INsB) used in vegetable farming in Hail and Riyadh cities; and to assess the health risk of adults due to the ingestion of pesticides in and on their vegetables.

2. Results and Discussion

2.1. Pesticide Residues in Raw Foods

A wide range of pesticide residues (63 insecticides, 41 acaricides, 40 herbicide, 55 fungicide, nematicides growth regulators Chitin synthesis inhibitors and Juvenile hormone mimics) in 801 vegetables such as 139 tomatoes, 185 peppers, 217 squash, 94 eggplants, and 166 cucumbers from different locations in Hail and Riyadh cities were detected in the Kingdom of Saudi Arabia during 2020. Regarding the pesticides that were screened, results showed that most of the pesticide groups that were detected belonged to different groups as fungicides (10 compounds), insecticides (8 compounds), acaricides (2 compounds), and multifunctional groups, such as insecticides/acaricides (4 compounds), insecticides/IGR (1 compound), and insecticides/nematicides (one compound). These compounds belong to many chemicals groups, as we found that the most frequent chemical group was the Triazole chemical group which has three compounds (penconazole, propiconazole, and triadmenol) as a fungicide with a percentage to reach 38%. Following this, each of the other groups (carbamate, dicarboximide, neonicotinoid, organophosphate and pyrethroid) were repeated twice. On other hand, the other remaining groups (phenylpyrazoles, chlorophenyl, dioxolanes, Hydrazine carboxylate, hydroxyanilides, methoxyacrylates, oxadiazine, phenylamide, pyridazinone, quinazoline, tetronic acid, triazolinthione, and unclassified) were repeated one time. All of these chemical groups use 31% insecticides. Following in the most frequency is the mixed group of insecticides/acaricides with 15%. After that, there is the acaricides group with 8%, and the last groups both insecticides/acaricides and fungicides/nematicides have 4% for both. Figure 1. Data was mentioned previously partially in agreement with [1–3].

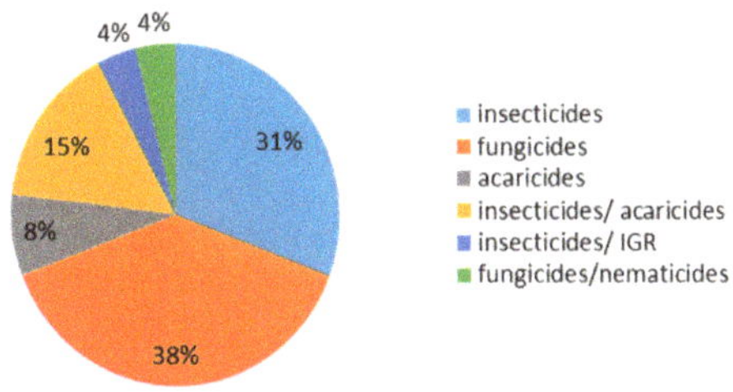

Figure 1. Illustration of the most important chemical groups and their percentages.

In our study, we observed the represented data in Table 1 and Figures 1–3 and the majority of residue compound was detected to be acetaimpride, followed with metalaxyl, imidaclopride, bifenthrin, pyridaben, difenoconazole, and azoxystrobien with a frequency of 39, 21, 11, 10, 8, 7, and 5, respectively. After that, triadmenol, ethion, deltamethrin, tolclofos-meth, spiromesifen, propiconazole, penconazole, fenitrothion, bifenazate, and buprofezine had frequencies of 4, 3, 3, 3, 2, 2, 2, 2, 2, 2, and 2. Finally, tebuconazole, procymidone, oxamyl, methomyl, indoxacarb, fipronil, fenhexamid, and fenazaquin had frequencies of 1 for all previous compounds, respectively.

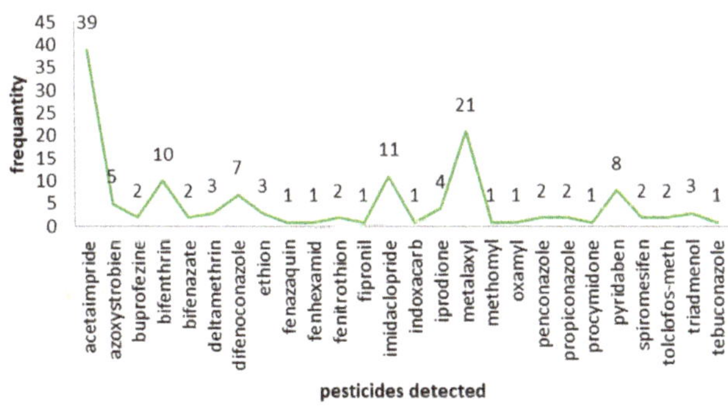

Figure 2. Frequency of occurrence of pesticides.

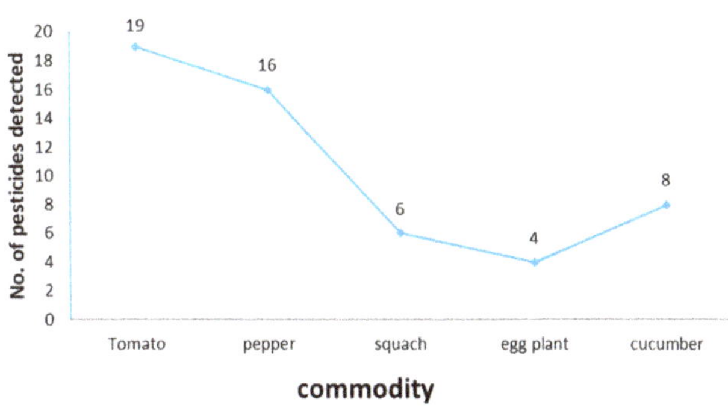

Figure 3. Demonstrates the contamination with pesticide residues.

Table 1. Demonstrates the frequency of occurrence of pesticides.

Pesticide	Tomato	Pepper	Squash	Eggplantt	Cucumber	Frq.
Acetaimprid	(0.017–0.347)	(0.011–0.358)	(0.011–0.118)	(0.008–0.085)	(0.018–0.209)	39
Azoxystrobien	(0.17 9–0.318)	0.216	0.39		0.054	5
Buprofezine		0.056		0.827		2
Bifenthrin	(0.03–0.362)	(0.064–0.145)	(0.01–0.125)	0.23		10
Bifenazate	0.1	0.052				2
Deltamethrin	(0.155–0.016)	0.16				3
Difenoconazole	(0.07–0.261)	(0.158–0.178)	(0.058–0.188)		0.012	7
Ethion	(0.125–0.137)	0.044				3
Fenazaquin	0.037					1
Fenhexamid	0.017					1
Fenitrothion	(0.159–0.161)					2
Fipronil		0.45				1
Imidaclopride	(0.076–0.38)	(0.018–0.721)				11
Indoxacarb	0.382					1
Iprodione	(0.178–0.0305)	0.086				4
Metalaxyl	(0.04–0.117)	(0.08–0.52)	(0.007–0.03)	0.005	(0.007–0.267)	21
Methomyl					0.026	1
Oxamyl	0.007					1
Penconazole			0.058		0.209	2
Propiconazole	(0.047–0.107)					2
Procymidone					0.053	1
Pyridaben	0.378	(0.018–0.36)			(0.08–0.328)	8
Spiromesifen		(0.06–0.196)				2
Tolclofos-meth	0.198	0.096				2
Triadmenol	(0.011–0.116)					3
Tebuconazole		0.309				1
Frq.	19	16	6	4	8	

On the other hand, we observed that the commodity in our study most contaminated with pesticide residues was tomato, as it is contaminated with acetaimpride, azoxystrobien, bifenthrin, bifenazate, deltamethrin, difenoconazole, ethion, fenazaquin, fenhexamid, fenitrothion, imidaclopride, indoxacarb, iprodione, metalaxyl, oxamyl, propiconazole, pyridaben, tolclofos-meth, and triadmenol with a frequency of 19 times followed by the frequency of 16 times for pepper, which was contaminated with acetaimpride, azoxystrobien, buprofezine, bifenthrin, bifenazate, deltamethrin, difenoconazole, ethion, fipronil, imidaclopride, iprodione, metalaxyl, pyridaben, spiromesifen, tolclofos-meth and tebuconazole after that in ranking cucumber was contaminated with acetaimpride, azoxystrobien, difenoconazole, metalaxyl, methomyl, penconazole, procymidone, and pyridaben. Following with a frequency of 8 times, squash was contaminated with acetaimpride, azoxystrobien, bifenthrin, metalaxyl and penconazole. With a frequency of 6 times, the last commodity in contaminated rankings was eggplant with a frequency of 4 times for acetaimpride, buprofezine, bifenthrin, and metalaxyl. Table 1, Figures 2 and 3. Overall, the pesticide residues which were found in this study were approximately similar to other studies [8,9,11].

2.2. Estimation of Dietary Intake

The objective of risk assessment from the point of view of food safety is, to ensure that in order to evaluate a dietary risk assessment, the ADI values were determined by summing the quotes of the pesticide ingested from various alimentary sources (i.e., vegetables and fruits). The Codex Alimentarius Commission of the FAO of the United Nations and WHO (FAO/WHO 2004) (17) recommended abiding with MRLs in fruits and vegetables. Monitoring of pesticide residues is a key tool for ensuring conformity with regulations and providing a check on compliance with good agricultural practice. The consideration of possible exposure to pesticide residues is an integral part of the risk assessment process to ensure that the ADI of the pesticides are not exceeded. As long as the residue of the pesticides ingested by consumers does not exceed the corresponding ADI, consumers are considered to be adequately protected. This is useful for assessing human exposure to pesticides through the food supply and for understanding the magnitude of health risks.

Additionally, the annual disappearance figures for a food commodity can be divided by the national population and by 365 days to obtain a "per capita" estimate of the food that is available for consumption per day expressed as grams per person per day (g/p/d). Disappearance data cannot be used to estimate intake for targeted sub-populations (e.g., young children, diabetics, or specific age-sex groups). The levels of contaminant pesticide residues used to estimate dietary intake of those substances can be obtained by combining the analytical results with amounts of food consumed reported in national food consumption surveys (Table 2).

Table 2. Estimated food consumption rate (g/day) in food basket: The Global Environment Monitoring System/Food Contamination Monitoring and Assessment Program (GEMS/Food).

Commodity	Consumption in the Middle East Grams per Person per Day
Tomato	81.5
Sweet Pepper	3.4
Squash	10.5
Eggplant	6.3
Cucumber	4.8

2.3. Estimation of Pesticide Exposure

The estimated daily intake for each monitoring pesticide residue was calculated with the next formula:

$$\text{EDI} = (\text{commodity consumption} \times \text{pesticide residue concentration})/\text{body weight}$$

2.4. Estimation of Health Risks from Pesticides

Estimation of the exposure risk to an adult person based on potential health risk by using the following formula:

$$\text{HRI} = \text{EDI}/\text{ADI}$$

In our study, the authors compiled the available data on pesticide residues in different plants that generate food commodities, such as vegetables. On the basis of previously conducted studies in different cities of Saudi Arabia, it was possible to conduct a human risk assessment using the hazard risk index (HRI). The results are summarized in Table 2, for HRI assessment, the estimated daily intake (EDI) (mg/kg/day) and acceptable daily intake (ADI) values (mg/kg/day) were taken and calculated by following international guidelines [16–20], where EDI is the estimated average daily intake (mg/kg/day), C is pesticide residue concentration (mg/kg) multiplying by the food consumed, and W is the average weight of an adult. Reference values for the food consumption rate of vegetables and fruits were taken from literature as 0.3 kg/person/day for vegetables and 0.4 kg/person/day of fruits, respectively, while 60 kg was considered an average adult weight [21–25]. The HRI value for the risk estimation of different toxic metals and pesticides via food consumption was calculated, and the general consumption rates were used (regardless of seasonal and generic wise consumption) due to data scarcity (Figures 4–8).

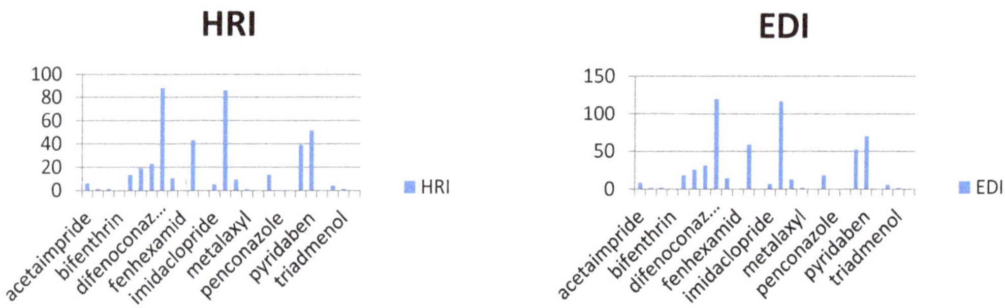

Figure 4. Estimation of EDI and HRI for pesticide residues detected in tomatoes.

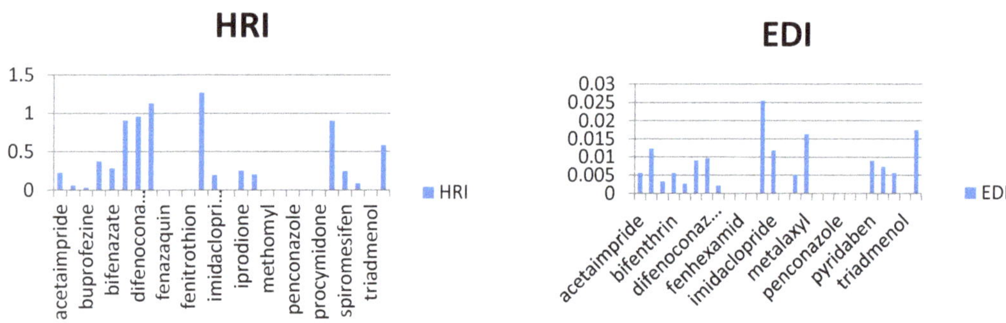

Figure 5. Estimation of EDI and HRI for pesticide residues detected in sweet peppers.

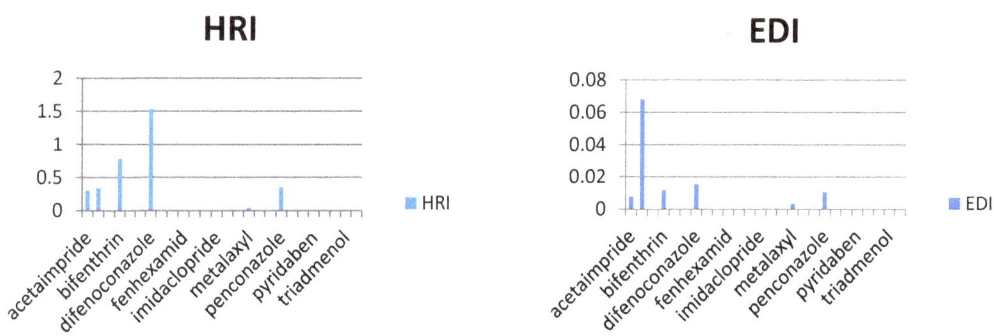

Figure 6. Estimation of EDI and HRI for pesticide residues detected in squash.

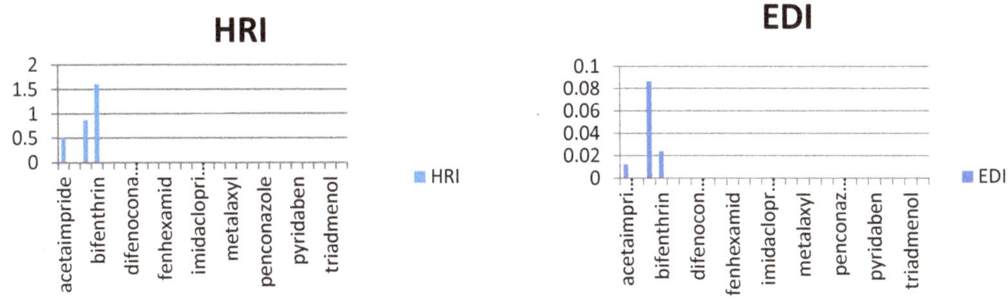

Figure 7. Estimation of EDI and HRI for pesticide residues detected in eggplant.

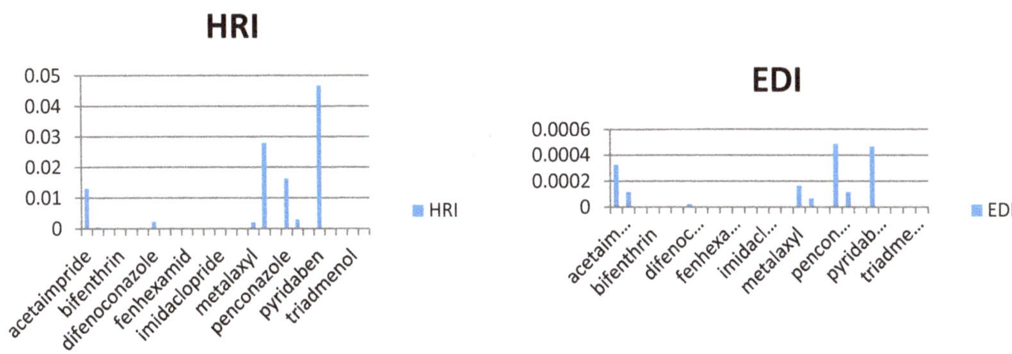

Figure 8. Estimation of EDI and HRI for pesticide residues detected in cucumber.

As we observed in Table 3, the calculated EDIs of tomatoes had been estimated between 0.013 to 0.516 mg/kg of bw/day. For sweet pepper, the EDIs value was between 0.0028 to 0.025 mg/kg of bw/day. However, in squash, the EDIs value was between 0.004 to 0.015 mg/kg of bw/day and in eggplant, the EDIs value was between 0.001 to 0.086 mg/kg of bw/day. Lastly, the EDIs value was between 0.000002 to 0.0005 mg/kg of bw/day in the cucumber. We observed that the EDIs values were lower than the ADI values. We reported that most pesticide residue exposure was lower than ADIs, and this depends on style of food consumption in Saudi Araba (Figures 4–8).

Furthermore, the EDIs values were used to estimate the hazard index (HRI) for each corps. We found a higher value for HRI for Ethion residue in tomato, but in sweet pepper, the higher value for HRI 127.5 was to fipronil residue. On the other hand, we found that the high value of HRI was 1.54 for difenoconazole residue in squash and 1.61 for bifenthrin residue in eggplant. Lastly, in cucumber, the high value was HRI 0.047 for pyridaben residue. We noticed all estimated data of the Hazard Index were exceeding the value of MRL, which may indicate a bad use of pesticides and failure to follow the application rates and the pre-harvest interval, which leads to exposure to health risks.

Table 3. Estimation of EDI and HRI for pesticide residues detected in Tomato, Sweet pepper, Squash and Cucumber samples.

Detected Pesticide	ADI	Tomato				Sweet Pepper				Squash				Eggplant				Cucumber			
		MRL	Mean	EDI	HRI	MRL	Mean	EDI	HRI	MRL	Mean	EDI	HRI	MRL	Mean	EDI	HRI	MRL	Mean	EDI	HRI
acetaimpride	0.025	0.5	0.12	0.163	6.52	0.1	0.1	0.0056	0.226	0.2	0.044	0.008	0.309	0.2	0.12	0.013	0.504	0.3	0.14	0.0003	0.013
azoxystrobien	0.2	3	0.25	0.34	1.697	3	0.22	0.0125	0.062	1	0.39	0.068	0.341	-	-	-	-	1	0.05	0.000117	0.001
buprofezine	0.1	0.3	0.12	0.163	1.63	0.5	0.06	0.0034	0.034	-	-	-	-	0.3	0.827	0.086	0.868	-	-	-	-
bifenthrin	0.015	-	-	-	-	0.5	0.1	0.0056	0.378	0.01	0.068	0.012	0.787	0.3	0.23	0.024	1.61	-	-	-	-
bifenazate	0.01	0.5	0.1	0.134	13.58	3	0.05	0.0028	0.283	-	-	-	-		-	-	-	-	-	-	-
deltamethrin	0.01	0.07	0.14	0.19	19.02	0.7	0.16	0.0091	0.907		-	-	-		-	-	-	-	-	-	-
difenoconazole	0.01	2	0.17	0.23	23.09	0.9	0.17	0.0096	0.963	0.2	0.088	0.015	1.54	-	-	-	-	0.3	0.01	2.33×10^{-5}	0.002
ethion	0.002	0.01	0.13	0.176	88.29	0.01	0.04	0.0023	1.13	-	-	-	-		-	-	-	-	-	-	-
fenazaquin	0.005	0.5	0.04	0.054	10.86	-	-	-	-	-	-	-	-	-	-	-	-	-	-	-	-
fenhexamid	0.2	2	0.02	0.027	0.135	-	-	-	-	-	-	-	-		-	-	-	-	-	-	-
fenitrothion	0.005	0.01	0.16	0.217	43.46	-	-	-	-	-	-	-	-	-	-	-	-	-	-	-	-
fipronil	0.0002	0.005	-	-	-	-	0.45	0.025	127.5	-	-	-	-	-	-	-	-	-	-	-	-
imidaclopride	0.06	0.5	0.23	0.312	5.206	1	0.21	0.0119	0.198	-	-	-	-	-	-	-	-	-	-	-	-
indoxacarb	0.006	0.5	0.38	0.516	86.02	-	-	-	-	-	-	-	-	-	-	-	-	-	-	-	-
iprodione	0.02	5	0.14	0.19	9.508	7	0.09	0.0051	0.255	-	-	-	-		-	-	-	-	-	-	-
metalaxyl	0.08	0.3	0.08	0.108	1.358	0.5	0.29	0.0164	0.205	0.01	0.02	0.004	0.044	0.01	0.005	0.001	0.007	0.01	0.071	0.0002	0.002
methomyl	0.0025			-	-	-	-	-	-	-	-	-	-	-	-	-	-	0.04	0.03	0.0001	0.028
oxamyl	0.001	0.01	0.01	0.013	13.58	-	-	-	-		-	-	-	-	-	-	-	-	-	-	-
penconazole	0.03	-	-	-	-	-	-	-	-	0.1	0.06	0.011	0.35	-	-	0	0	0.1	0.21	0.0005	0.016
propiconazole	0.04	-	-	-	-	-	-	-	-	-	-	-	-	-	-	-	-	0	0.05	0.000117	0.003
procymidone	0.0028	0.01	0.08	0.108	38.8	-	-	-	-	-	-	-	-	-	-	-	-	-	-	-	-
pyridaben	0.01	0.15	0.38	0.516	51.61	0.5	0.16	0.009	0.906	-	-	-	-	-	-	-	-	0.15	0.2	0.0005	0.047
spiromesifen	0.03			-	-		0.13	0.007	0.246		-	-	-		-	-	-		-	-	-
tolclofos-meth	0.064		0.2	0.271	4.244		0.1	0.006	0.088		-	-	-		-	-	-		-	-	-
triadmenol	0.05		0.05	0.067	1.358			-	-		-	-	-		-	-	-		-	-	-
tebuconazole	0.03			-	-		0.31	0.018	0.585		-	-	-		-	-	-		-	-	-

3. Materials and Methods

3.1. Chemicals and Reagents

Ultra-gradient HPLC-grade acetonitrile was purchased from J.T. Baker (Deventer, The Netherlands). Deionized water was obtained from a Milli-Q SP Reagent Water System (Millipore; Bedford, MA, USA). Formic acid (98% purity) and anhydrous magnesium sulfate were ordered from Fluka–Sigma–Aldrich (Steinheim, Germany). Each sample was filtered through a 13 mm × 0.45 um PTFE filter before injection, whilst PSA-bonded (primary secondary amine) silica was used as a sample clean-up step—both of them were from Supelco (Bellefonte, PA, USA). Acetic acid (Merck; Darmstadt, Germany) and sodium acetate-3-hydrate (Panreac; Castellonde Valles, Barcelona, Spain) were used for the sample preparation procedure. All certified pesticide standards obtained from Dr. Ehrenstorfer (Augsburg, Germany) were of 95 % or higher purity.

3.2. Study Area

This study was conducted in the Hail and Riyadh regions, which lie between longitude and latitude (43 N and 26 E and 34 N and 46 E), respectively. The city of Riyadh is characterized by a high population density, which is approximately six million people. On the contrary, the Hail region is characterized by a low population density, which reaches one and a half million people. These areas are dominated by a hot summer climate where temperatures reach 48 °C, and in winter, the average temperature drops to 9 °C.

3.3. Collection of Samples and Pretreatment

A total of 801 vegetable samples (139 tomatoes, 185 peppers, 217 squash, 94 eggplants, and 166 cucumber) from five local markets (three from Hail and two from Riyadh) were collected during the different seasons of 2019. Altogether, 2–3 units of fresh vegetables were collected from each local market (>1 kg) in accordance with the procedures described in the FAO, (1999). Samples were not rinsed. A portion of each sample, without tops such as the sepal and peduncle, was prepared according to annex I of European Commission regulation, 396/2005 EU (2010) using a knife and a chopping board and then thoroughly mixed. Two hundred gram of each sample were kept in a separate plastic bag at −20 °C until pesticide extraction and analysis could be carried out.

3.4. Extraction of Pesticide Residues by QuEChERS and Cleanup of Vegetables Samples

Vegetable samples were purchased from a local market and the preparation procedure was the same as the well-known and accepted QuEChERS (16), sample preparation procedure was applied to all the samples. After homogenization with the stainless-steel cutter (Sammic, Azpeitia, Spain), a 15 g portion of the homogenized sample was weighed in a 50 mL PTFE centrifuge tube. Then, 15 mL of acetonitrile were added with 6 g MgSO4and 2.5 g sodium acetate-3-hydrate and the samples were shaken vigorously by hand for 4 min. The extract was then centrifuged (3700 rpm) for 5 min. A 5 mL volume of the supernatant was removed to a 1-mL PTFE centrifuge tube containing 750 mg of MgSO4 and 250 mg of PSA. The extract was shaken in a vortex intensively for 20 s and centrifuged again (3700 rpm) for 5 min. Following this, an aliquot of the supernatant was evaporated under a nitrogen stream and reconstituted with acetonitrile/water (20/80) for LC analysis. Prior to injection into the LC–MS system, the sample was filtered through a 0.45-um PTFE filter. With this treatment, a 1 mL sample extract represents 1 g of sample.

3.5. Standard Preparation

A standard stock solution of each pesticide was prepared in acetonitrile at a concentration of 2000 µg/mL. A mixed standard solution was prepared at a concentration of 10 µg/mL from the individual stock solutions. The calibration curve for the LC measurements was prepared by diluting 10 µg/mL of the mixed standard solution to achieve final concentrations of 0.25, 0.5, 1, 2.5, 5, 10, 25, and 100 ng/mL in a mixture of acetonitrile and water (1:1, v/v). Stock and working solutions were stored at 4 °C until use. Pesticides were analyzed through Liquid Chromatography Triple Quadrupole Mass Spectrometry (LC-MSMS) and Gas Chromatography Triple Quadrupole Mass Spectrometry (GC-MSMS).

3.6. Analytical Techniques by Liquid Chromatography Triple Quadrupole Mass Spectrometry (LC-MSMS)

An ultra-high performance liquid chromatography (ACQUITY) coupled with a tandem quadrupole MS (XEVO TQD) was used with Mass Lynx 4.1 software (Waters Corporation, Milford, MA, USA). For the chromatographic separation, a reversed phase column, Atlantis T3 (100 × 3 mm, 5 µm), was used. The mobile phases (A and B) were water: methanol (98:2, v/v), and methanol, respectively, with 0.1% formic acid (FA) in each. The flow rate was maintained at 0.45 mL/min. The gradient program was initially set at 5% B (1 min), then linearly increased over the next 7.75 min to 100% and kept constant until 8.50 min. Thereafter, it was linearly decreased to 5%, and maintained for another 3.50 min (a total run time of 12 min). The MS was operated with Electrospray Ionization (ESI+). The optimized parameters included desolvation temperature (450 °C), desolvation gas flow rate (1000 L/hour), cone gas flow rate (50 L/hour), ion source temperature (120 °C), and capillary voltage (1 kV). The MS parameters are presented Table 4.

Table 4. LC-MS/MS retention times and multi reaction monitoring MRM transitions for the LC amenable pesticides.

Pesticide	Application	Parent	CV (V)	Product 1	CE (eV)	Product 2	CE (eV)	RT
3,4,5-Trimethacarb	Insecticide	194.1	22	137.1	12	122.1	26	5.41
Acephate	Insecticide	184.1	17	143.0	8	125.1	18	1.47
Acetamiprid	Insecticide	223.0	34	126.0	20	56.1	15	3.41
Alachlor	Herbicide	271.1	28	162.1	20	238.1	11	6.30
Aldicarb	Acaricide	213.1	30	89.1	16	116.1	11	3.98
Aldicarb sulfone	Metabolite	223.0	31	148.0	10	86.0	14	2.04
Aldicarb sulfoxide	Metabolite	207.0	22	89.0	14	132.0	10	1.91
Ametryn	Herbicide	228.1	38	186.1	18	68.1	36	4.88
Anilazine	Fungicide	274.9	46	153.0	26	178.0	24	5.98
Anilofos	Herbicide	367.9	30	124.9	34	198.9	15	6.57
Atraton	Herbicide	212.0	40	170.1	18	100.0	28	3.96
Atrazine	Herbicide	216.1	39	174.1	18	96.1	23	5.20
Atrazine-desethyl	Metabolite	188.0	34	146.0	16	78.9	26	3.69
Azaconazole	Fungicide	300.0	34	159.0	28	231.1	18	5.44
Azinphos-ethyl	Insecticide	346.0	16	132.0	16	77.1	36	6.20
Azinphos-methyl	Insecticide	318.0	20	160.0	8	261.0	8	5.56
Azoxystrobin	Fungicide	404.0	28	372.0	15	329.0	30	5.73
Benalaxyl	Fungicide	326.1	26	148.0	20	91.0	34	6.62
Bendiocarb	Insecticide	224.1	26	167.0	8	109.0	18	4.62
Benfluralin	Herbicide	336.0	34	57.0	18	236.0	15	no
Benfuracarb	Insecticide	411.1	23	195.0	23	190.0	13	7.07
Benomyl	Fungicide	291.0	22	160.0	28	192.0	16	5.50
Boscalid	Fungicide	342.9	41	307.0	20	139.9	20	5.90
Buprofezin	Insecticide	306.1	31	201.0	12	57.4	20	6.96
Butachlor	Herbicide	312.2	26	57.3	22	238.2	12	7.17

Table 4. *Cont.*

Pesticide	Application	Parent	CV (V)	Product 1	CE (eV)	Product 2	CE (eV)	RT
Cadusafos	Insecticide	271.1	28	159.0	16	131.0	22	6.88
Carbaryl	Insecticide	202.0	28	145.0	22	117.0	28	4.86
Carbendazim	Fungicide	192.1	33	160.1	18	132.1	28	2.20
Carbofuran	Insecticide	222.1	34	165.1	16	123.0	16	4.63
Carbosulfan	Insecticide	381.0	40	118.0	22	76.0	34	7.89
Carboxin	Fungicide	236.0	34	143.0	16	87.0	22	4.79
Chlorfenvinphos	Acaricide	358.9	28	155.0	12	99.0	30	6.65
Chlorpropham	Herbicide	214.1	18	172.0	8	154.0	18	6.01
Chlorpyrifos	Insecticide	349.9	36	97.0	32	198.0	20	7.35
Chlorpyriphos-methyl	Insecticide	321.8	34	125.0	20	289.9	16	6.87
Clethodim	Herbicide	360.0	32	164.0	18	268.1	12	7.02
Coumaphos	Insecticide	363.0	32	307.0	16	289.0	24	6.60
Cyanazine	Herbicide	241.0	41	214.0	17	96.0	25	4.39
Cyanofenphos	Insecticide	304.0	34	157.0	22	276.0	12	6.57
Cymoxanil	Fungicide	199.0	23	128.0	8	111.0	18	3.58
Deltamethrin	Insecticide	505.9	28	280.9	12	93.2	46	7.64
Desmetryn	Herbicide	214.1	38	172.1	20	82.1	30	4.26
Diazinon	Insecticide	305.1	31	169.0	22	96.9	35	2.55
Dichlorvos	Acaricide	221.0	34	109.0	22	79.0	34	4.53
Dicrotophos	Insecticide	238.0	28	112.0	10	193.0	10	2.97
Diethofencarb	Fungicide	268.0	28	226.0	10	124.0	40	5.71
Difenoconazole	Fungicide	406.0	46	251.1	25	111.1	60	6.90
Dimethoate	Acaricide	230.1	24	125.0	20	199.0	10	3.32
Diniconazole	Fungicide	326.1	46	70.2	25	159.0	34	6.87
Disulfoton	Acaricide	274.9	16	89.0	20	61.1	35	6.80
Disulfoton-sulfone	Metabolite	307.1	24	97.1	28	153.1	12	5.16
Disulfoton-sulfoxide	Metabolite	291.0	24	185.0	14	97.0	31	5.08
Diuron	Herbicide	233.0	34	72.1	18	46.3	14	5.37
Epoxiconazole	Fungicide	330.0	34	121.0	22	101.0	50	6.27
Ethion	Acaricide	284.9	25	199.1	10	97.0	46	5.22
Famphur	Insecticide	326.0	32	93.0	31	217.0	20	5.19
Fenamiphos	Nematicide	304.1	36	217.1	24	202.1	36	6.39
Fenarimol	Fungicide	331.0	46	268.0	22	81.0	34	6.26
Fenazaquin	Acaricide	307.2	36	57.2	25	161.0	19	7.70
Fenhexamid	Fungicide	302.1	41	97.2	22	55.3	38	6.22
Fenitrothion	Insecticide	278.0	38	109.1	20	79.1	34	6.06
Fenobucarb	Insecticide	208.0	22	94.9	14	152.0	8	5.72
Fenoxycarb	Insecticide	302.1	28	88.0	20	116.1	11	6.45
Fenpropathrin	Insecticide	350.1	24	125.0	14	97.0	34	7.51
Fenthion	Insecticide	279.1	36	169.1	16	247.1	13	6.57
Fonofos	Insecticide	247.1	24	109.0	20	137.0	10	6.60

Table 4. *Cont.*

Pesticide	Application	Parent	CV (V)	Product 1	CE (eV)	Product 2	CE (eV)	RT
Heptenophos	Insecticide	251.0	26	127.0	14	125.0	14	5.43
Hexaconazole	Fungicide	314.0	40	70.1	22	159.0	28	6.74
Imazalil	Fungicide	297.0	40	159.0	22	69.0	22	5.03
Imidacloprid	Insecticide	256.1	34	175.1	20	209.1	15	3.08
Indoxacarb	Insecticide	528.0	34	150.0	22	203.0	40	6.91
Iprobenphos	Fungicide	289.0	18	91.0	20	205.0	10	6.47
Iprodione	Fungicide	330.0	21	244.7	16	288.0	15	6.40
Isocarbofos	Insecticide	291.1	21	121.1	30	231.1	13	5.39
Kresoxim-methyl	Fungicide	314.1	24	116.0	12	206.0	7	6.50
Linuron	Herbicide	249.1	31	160.1	18	181.1	16	5.75
Malathion	Acaricide	331.0	20	127.0	12	99.0	24	5.95
Metalaxyl	Fungicide	280.1	26	220.1	13	192.1	17	6.27
Metamitron	Herbicide	203.1	34	175.1	16	104.0	22	3.25
Methacrifos	Acaricide	241.1	20	125.0	20	209.1	8	5.47
Methidathion	Insecticide	303.0	18	85.1	20	145.0	10	5.45
Methiocarb	Acaricide	226.0	28	121.0	22	169.0	10	5.83
Methomyl	Insecticide	163.0	26	88.0	10	106.0	10	2.34
Metolachlor	Herbicide	284.1	26	176.1	25	252.1	15	6.33
Metolcarb	Insecticide	166.0	20	109.0	12	94.1	27	4.29
Metribuzin	Herbicide	215.0	41	131.0	18	89.0	20	4.53
Mevinphos	Acaricide	225.1	24	127.1	15	193.1	8	3.37
Monocrotophos	Acaricide	224.1	26	127.1	16	98.1	12	2.71
Myclobutanil	Fungicide	289.1	34	70.2	18	125.1	32	6.08
Omethoate	Acaricide	214.1	26	125.1	22	183.1	11	1.76
Oxadixyl	Fungicide	279.0	40	219.0	10	132.0	34	4.32
Oxamyl	Insecticide	237.0	21	72.0	10	90.0	10	2.13
Paclobutrazol	Growth Regulator	294.1	36	125.1	38	70.2	20	5.95
Penconazole	Fungicide	284.0	34	70.1	16	159.0	34	7.35
Pendimethalin	Herbicide	282.2	21	212.2	10	194.1	17	8.04
Phenmedipham	Herbicide	301.0	34	168.0	10	136.0	22	5.57
Phenthoate	Insecticide	321.0	18	163.0	12	135.0	20	6.47
Phorate	Insecticide	261.0	17	75.0	12	97.0	32	6.74
Phorate sulfone	Metabolite	293.0	24	96.9	30	115.0	24	5.20
Phosmet	Insecticide	318.0	28	160.0	22	77.0	46	4.22
Phosphamidon	Insecticide	300.1	28	174.1	14	127.1	25	4.40
Phoxim	Insecticide	299.0	22	129.0	13	153.0	7	6.69
Pirimicarb	Insecticide	239.1	34	72.0	18	182.1	15	3.55
Pirimiphos-ethyl	Insecticide	334.1	42	198.1	23	182.1	25	7.09
Probenazole	Fungicide	224.0	22	41.5	10	196.1	13	4.38
Procloraz	Fungicide	376.0	22	307.1	16	70.1	34	6.53

Table 4. *Cont.*

Pesticide	Application	Parent	CV (V)	Product 1	CE (eV)	Product 2	CE (eV)	RT
Procymidone	Fungicide	284.1	42	67.1	28	256.1	17	8.13
Profenofos	Insecticide	372.9	36	302.6	20	127.9	40	7.12
Promecarb	Insecticide	208.1	26	151.0	9	109.0	15	5.94
Propachlor	Herbicide	212.1	31	170.1	14	94.1	25	5.31
Propetamphos	Insecticide	282.0	17	138.0	20	156.0	12	6.07
Propham	Herbicide	180.0	14	138.0	8	120.0	16	5.15
Propiconazole	Fungicide	342.0	46	69.0	22	159.0	34	6.65
Propoxur	Insecticide	210.0	21	111.0	16	168.0	10	4.58
Pyracarbolid	Fungicide	218.1	32	125.1	18	97.1	28	4.66
Pyraclostrobin	Fungicide	388.1	31	163.0	25	193.9	12	6.70
Pyrazophos	Fungicide	374.0	44	222.1	22	194.0	32	6.75
Pyroquilon	Fungicide	174.0	41	132.0	23	117.0	30	4.49
Quinalphos	Acaricide	299.0	24	162.9	24	96.9	30	6.47
Quinmerac	Herbicide	222.2	28	204.2	15	141.1	30	3.36
Rotenone	Insecticide	395.0	46	213.1	24	192.1	24	6.39
Simazine	Herbicide	202.0	40	124.0	16	96.0	22	4.57
Simetryn	Herbicide	214.0	41	124.0	20	95.9	25	4.27
Spiromesifen	Insecticide	371.1	16	273.1	10	255.1	24	7.43
Spiroxamine	Fungicide	298.0	38	144.0	20	100.0	32	5.44
Sulfotep	Insecticide	323.0	28	97.0	32	171.0	15	6.51
Terbutryn	Herbicide	242.1	40	186.1	20	91.0	28	5.49
Thiacloprid	Insecticide	253.0	41	126.0	20	90.1	40	3.76
Thiamethoxam	Insecticide	292.0	28	211.2	12	132.0	22	2.56
Thiophanate	Fungicide	371.0	28	151.0	22	93.1	50	5.37
Tolcofos methyl	Fungicide	301.1	41	125.0	17	174.9	29	6.8
Triadimefon	Fungicide	294.1	31	69.3	20	197.2	15	5.94
Triadimenol	Fungicide	296.1	21	70.2	10	99.1	15	6.15
Triazophos	Acaricide	314.1	31	161.9	18	118.9	35	6.12
Vamidothion	Acaricide	288.0	28	146.0	10	118.0	28	3.38
Vernolat	Herbicide	204.1	28	128.1	11	86.1	14	6.83

CV = cone voltage; CE = collision energy; Rt: Retention Time.

3.7. Compound Identification

Identification and confirmation of the target compounds on GC-MSMS, was performed by using the software (TraceFinder and Xcalibar) with an updated pesticides library consisting of a more than 900 pesticides and endocrine disruptors. The software incorporates the data such as retention time (with RT< ± 0.1 min), the parent/target ion (used for quantification), and 2 other ions (as qualifiers), for all the isomers, metabolites for almost all the included compounds (in the database).

MS analysis was carried out on a TSQ 8000 EVO GC triple stage quadrupole mass spectrometer. (Thermo Fisher Scientific, San Jose, CA). The MS conditions were as follows: Ionization mode: EI positive ion. Emission current: 50 µA. Ion source temperature: 220 °C. Scan type: SRM and Scan time: 0.02 s.

On the other hand, identification and confirmation of the target compounds on LC-MSMS, and two MRM transitions for each pesticide were generated using QUANPEDIA, ™. The data were acquired using MassLynx Software and processed using TargetLynx Application Manager. Peak shapes were adequate in most cases as shown in Figure 9.

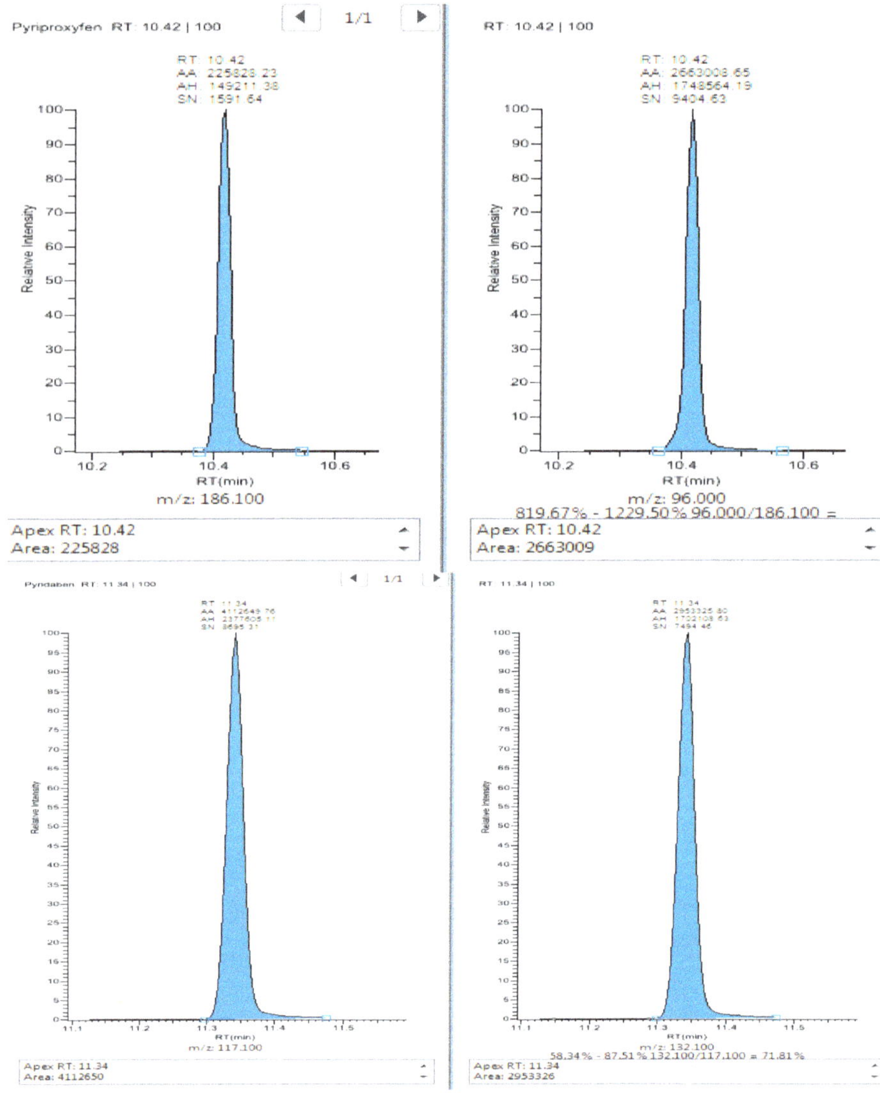

Figure 9. Representative multi-reaction monitoring SRM chromatograms for (1) Pyriprofexen and Pyridaben, spiked at a level of 0.100 µg/g in tomato and extracted using the sample preparation protocol reported (GC MSMS peaks).

3.8. Validation Design

The optimized analytical method was validated to ensure that it was fit for the intended purpose. The method was validated in terms of accuracy (mean recovery of the spiked samples at three different spiking levels), precision (intra-day and inter-day repeatability in terms of percent relative standard deviation, %RSD), selectivity, sensitivity (limit of quantitation (LOQ) and linearity (or linear range of measurement). The LOQ was calculated as the lowest concentration at which the recovery and precision was within the acceptable limits (recovery: 70–120%, precision: RSD < ±20%) (SANTE, 2019).

The calibration curve is determined by the analysis of each of the analysts at 6 calibration levels within the range of 0.25, 0.5, 1, 2.5, 5, 10, 25, and 100 ng/mL. The calibration curves were, in general, best fitted to a linear curve. The quantification was performed from the mean of three bracketing calibration curves. Most of the correlation coefficients (R2) were higher than or equal to 0.99.

For each level, three genuine replicates were performed. The method's acceptance criteria were accuracy, precision, sensitivity, and the qualifier/quantifier ion ratio of the detected pesticides in real samples (to be <30%). The ion ratio was calculated as: 'the mean ratio of the qualifier to quantifier ions for a pesticide calculated from an MMS batch' subtracted from 'the ion ratio for that pesticide in the positive sample', and then dividing the resultant by the mean ion ratio calculated for the MMS of the same batch, the value thus obtained was multiplied by 100 to get the percentage value (SANTE, 2019).

To ensure the quality of the analytical work, the analytical batch was designed every time in a way to include a solvent/reagent blank, one matrix blank, and three replicates for all the three spiking levels. The solvent/reagent blanks were processed according to the complete extraction procedure under investigation, to eliminate any chances of laboratory and glassware contamination. One sample as matrix blank (extract of the sample viz. free of the targeted pesticides and which was used in the validation of the method) was also analyzed, and three replicates for all the three, i.e., highest spiking level (HSL), medium spiking level (MSL), and lowest spiking level (LSL), were also run in the same batch. The instrumental samples' sequence was designed to be in the following order: reagent/solvent blank, then calibration standards in pure solvent, followed by matrix-matched standards (at the same concentration range as that of the standards-in-solvent) and then the real samples, bracketed by the standards-in-solvent, at the end. To eliminate the chances of carryover from previous samples' injections, the instrument was also configured at back-flush settings, supported by an additional post-run column flushing of one minute.

Uncertainty (U) of the proposed multi-residue method was calculated by bottom-up empirical model in accordance with the ISO 21748. Uncertainty of the method's repeatability, reproducibility, and trueness estimated was calculated as mentioned previously, partially in [5,10,13,15].

3.9. Pesticide Residue Analysis by Gas Chromatography Mass Spectrometry (GC–MSMS)

A Thermo Scientific TRACE 1310 Gas Chromatography coupled with TSQ 8000 Evo Triple Quadrupole Detector and AI 1310 Auto Sampler was used with Thermo Xcalibur 2.2 mass spectrometry data system (Software). For the chromatographic separation, a Thermo Scientific™ Trace GOLD™ TG-5SilMS 30 m × 0.25 mm I.D. × 0.25 µm film capillary column was used. The flow rate was maintained at constant flow 1.2 mL/min (He, inert carrier gas). The GC oven program was initially set at 70 °C (2 min), then increased 25 °C/min to 180 °C, 5 °C/min to 200 °C, and 10 °C/min to 280 °C, kept constant 5 min. The MS was operated with electrospray ionization (ESI+). The optimized parameters included transfer line (280 °C), electron energy(eV) 70, acquisition mode (SRM), and ion source temperature (320 °C). The Thermo ScientificTM TraceFinderTM software was used for method setup and data processing. For all pesticide compounds two SRM transitions were chosen for the overall MRM acquisition method. The first transition was used for quantitation, the second transition for confirmation. Table 5 and Figure 9, lists the SRM parameters for the compounds analyzed in this method.

Table 5. GC-MS/MS retention times and multi reaction monitoring SRM transitions for the LC amenable pesticides.

Pesticide	Application	Quantitation m/z	CE (eV)	Confirmation m/z	CE (eV)	RT (min)
Acephate	Insecticide	136.01 > 42.00	10	136.01 > 94.01	15	7.42
Alachlor	Herbicide	161.07 > 146.06	12	188.08 > 160.07	10	12.58
Atrazine	Herbicide	215.09 > 173.08	10	215.09 > 200.09	10	10.65
Azinphos-ethyl	Acaricide	132.01 > 77.01	20	160.02 > 132.01	5	19.18
Benfluralin	Herbicide	292.10 > 160.05	21	292.10 > 264.09	10	9.54
Bifenthrin	Acaricide	181.05 > 153.05	6	181.05 > 166.05	15	17.86
Boscalid	Fungicide	342.03 > 140.01	15	344.03 > 142.01	15	20.95
Bromophos-ethyl	Acaricide	358.89 > 302.91	20	358.89 > 330.90	10	14.58
Buprofezin	Chitin synthesis inhibitors	172.09 > 57.03	10	249.13 > 193.10	10	15.88
Butralin	Herbicide	266.14 > 190.10	15	266.14 > 220.11	15	13.56
Cafenstrole	Herbicide	100.04 > 72.03	15	188.08 > 119.05	15	20.21
Carbaryl	Acaricide	144.06 > 115.05	20	144.06 > 116.05	20	12.54
Chlordane	Insecticide	372.81 > 265.87	18	374.81 > 267.87	15	14.67
Chlorpropham	Herbicide	213.00 > 127.00	5	213.00 > 171.00	5	9.66
Cyfluthrin	Insecticide	163.02 > 91.01	12	163.02 > 127.02	10	20.08
Cypermethrin	Acaricide	163.03 > 127.02	10	181.03 > 152.03	25	20.66
Cyprodinil	Fungicide	224.13 > 208.12	20	225.13 > 210.12	18	14.08
Deltamethrin	Insecticide	252.99 > 93.00	18	252.99 > 173.99	18	22.19
Diazinon	Acaricide	137.05 > 84.03	10	304.10 > 179.06	15	10.09
Dimethachlor	Herbicide	197.08 > 148.06	10	199.08 > 148.06	10	12.06
Diniconazole	Fungicide	268.06 > 232.05	15	270.06 > 234.05	15	16.18
Dioxathion	Acaricide	125.00 > 97.00	15	125.00 > 141.00	15	10.78
Edifenphos	Fungicide	173.01 > 109.01	15	310.03 > 173.01	10	16.77
Ethion	Acaricide	230.99 > 202.99	15	383.99 > 230.99	10	16.18
Ethoprophos	Insecticide	158.00 > 80.90	15	158.00 > 114.00	5	9.58
Fenarimol	Fungicide	139.01 > 111.01	15	219.02 > 107.01	15	19.26
Fenobucarb	Insecticide	121.07 > 77.05	15	150.09 > 121.07	10	9.18
Fenpropathrin	Acaricide	181.09 > 152.07	23	265.13 > 210.10	15	18.06
Fipronil	Acaricide	212.97 > 177.98	16	366.95 > 212.97	25	13.94
Fluopicolide	Fungicide	208.80 > 182.00	20	261.00 > 175.00	24	16.94
Formothion	Acaricide	126.00 > 93.00	8	172.00 > 93.00	5	11.88
Imazalil	Fungicide	173.03 > 145.02	20	215.04 > 173.03	15	18.22
Iprodione	Fungicide	187.02 > 124.01	20	187.02 > 159.02	40	17.58
Isoprothiolane	Fungicide	290.06 > 118.03	15	290.06 > 204.05	15	15.28
Kresoxim-methyl	Fungicide	206.09 > 116.05	15	206.09 > 131.06	15	15.34
Lactofen	Herbicide	344.04 > 223.02	15	344.04 > 300.03	15	18.88
Malathion	Acaricide	127.01 > 99.01	10	173.02 > 127.01	10	13.05
Mecarbam	Acaricide	226.04 > 198.03	5	329.05 > 160.03	10	14.23
Mepanipyrim	Fungicide	222.11 > 207.10	15	223.11 > 208.10	15	14.26

Table 5. *Cont.*

Pesticide	Application	Quantitation m/z	CE (eV)	Confirmation m/z	CE (eV)	RT (min)
Metalaxyl	Fungicide	249.13 > 190.10	10	249.13 > 249.13	5	12.56
Metamitron	Herbicide	202.09 > 174.07	5	202.09 > 186.08	10	10.42
Methabenzthiazuron	Herbicide	164.05 > 136.04	12	164.05 > 164.05	10	9.84
Methamidophos	Acaricide	141.00 > 95.00	10	141.00 > 126.00	5	5.77
Methidathion	Insecticide	124.98 > 98.99	22	144.98 > 84.99	10	14.65
Methiocarb	Acaricide	168.06 > 109.04	15	168.06 > 153.06	15	12.98
Metribuzin	Herbicide	198.08 > 82.03	20	198.08 > 110.05	20	12.46
Mevinphos	Acaricide	127.03 > 109.02	10	192.04 > 127.03	12	7.32
Monocrotophos	Acaricide	127.03 > 95.03	20	127.03 > 109.03	25	9.94
Omethoate	Acaricide	110.01 > 79.01	15	156.02 > 110.01	10	9.05
Penconazole	Fungicide	248.06 > 157.04	25	248.06 > 192.04	15	14.09
Pendimethalin	Herbicide	252.12 > 162.08	12	252.12 > 191.09	12	13.86
Phosalone	Acaricide	181.99 > 111.00	15	181.99 > 138.00	10	18.56
Phosphamidon	Insecticide	227.05 > 127.03	15	264.06 > 193.04	15	11.88
Pirimicarb	Insecticide	166.10 > 96.06	10	238.14 > 166.10	15	11.95
Probenfos	Insecticide	204.07 > 122.04	15	218.89 > 182.91	15	11.72
Procymidone	Fungicide	283.02 > 96.01	15	283.02 > 255.02	10	14.56
Profenofos	Insecticide	138.98 > 96.98	8	338.94 > 268.95	20	15.37
Propachlor	Herbicide	176.06 > 120.04	10	196.07 > 120.04	10	9.45
Propanil	Herbicide	217.01 > 161.00	10	219.01 > 163.00	10	12.16
Propargite	Acaricide	135.06 > 107.05	15	350.16 > 201.09	10	17.24
Propoxur	Acaricide	110.06 > 64.03	10	152.08 > 110.06	10	9.02
Pyrimethanil	Fungicide	198.11 > 158.09	30	198.11 > 183.10	15	11.28
Pyriproxyfen	Juvenile hormone mimics	226.10 > 186.10	12	136.10 > 96.00	10	10.45
Pyridaben	Acaricide	147.10 > 117.10	20	147.10 > 132.10	12	11.35
Quinalphos	Acaricide	146.03 > 118.02 1	15	157.03 > 129.02	13	14.29
Spiromesifen	Insecticide	371.24 > 273.15	15	371.24 > 255.64	25	18.42
Spiroxamine	Fungicide	100.09 > 58.05	15	100.09 > 72.06	15	12.89
Tefluthrin	Insecticide	177.02 > 127.02	20	197.03 > 141.02	15	11.27
Tetradifon	Acaricide	226.93 > 198.94	18	353.88 > 158.95	15	18.56
Tolclofos-methy	Fungicide	264.96 > 92.99	20	264.96 > 249.96	15	12.34
Triazophos	Acaricide	161.03 > 134.03	10	257.05 > 162.03	10	16.55
Trifluralin	Herbicide	264.09 > 160.05	15	306.10 > 264.09	15	9.87
Vinclozolin	Fungicide	100.09 > 58.05	15	100.09 > 72.06	15	12.35

CV = cone voltage; CE = collision energy; Rt: Retention Time.

4. Conclusions

High consumption of fruits and vegetables contaminated with pesticide residues above the MRL leads to a threat to the population's health, and this is due to the poor handling practices for pests and disease control that also do not follow the pre-harvest interval (PHI) for pesticides. Therefore, it is important to update the data on the population's real consumption value to obtain a true estimate of the risk of actual exposure to pesticides. It is impotent to continue with the pesticide residues program to reduce exposure to residues that cause long-term effects or immediate serious illness.

Author Contributions: Each author participated actively in conducting analyses, editing, and approving the final, submitted version. Conceptualization; methodology and software, M.H.E.-S.; validation, M.T.S., M.M.A. and H.I.S.; formal analysis, M.H.E.-S.; investigation, M.M.A. and H.I.S.; resources, M.H.E.-S., M.T.S. and M.M.A.; data curation, All authors writing original draft preparation, M.H.E.-S. writing—review and editing, M.T.S., M.M.A. and H.I.S.; visualization and supervision, M.M.A.; administration. All authors have read and agreed to the published version of the manuscript.

Funding: This research was supported by the Deputy-Ministry of Agriculture, Ministry of Environment, Water and Agriculture, Kingdom of Saudi Arabia.

Institutional Review Board Statement: Not applicable.

Informed Consent Statement: Not applicable.

Data Availability Statement: Not applicable.

Acknowledgments: The researchers extend their thanks to the Deputy-Ministry of Agriculture, Ministry of Environment, Water and Agriculture, Kingdom of Saudi Arabia for their support of this research, which greatly impacted its completion and publication in a peer-reviewed scientific prestigious journal.

Conflicts of Interest: The authors declare no conflict of interest.

Sample Availability: Not applicable.

References

1. Chen, C.; Qian, Y.; Chen, Q.; Tao, C.; Li, C.; Li, Y. Evaluation of pesticide residues in fruits and vegetables from Xiamen, China. *Food Control* **2011**, *22*, 1114–1120. [CrossRef]
2. Claeys, W.L.; Schmit, J.F.; Bragard, C.; Maghuin-Rogister, G.; Pussemier, L.; Schiffers, B. Exposure of several Belgian consumer groups to pesticide residues through fresh fruit and vegetable consumption. *Food Control* **2011**, *22*, 508–516. [CrossRef]
3. Na, T.W.; Rahman, M.M.; Park, J.H.; Yang, A.; Park, K.H.; Abd El-Aty, A.M.; Shim, J.-H. Residual pattern of acequinocyl and hydroxyl acequinocyl in perilla leaf grown under greenhouse conditions using ultra performance liquid chromatography–photo diode array detector with tandem mass confirmation. *J. Korean Soc. Appl. Biol. Chem.* **2012**, *55*, 657–662. [CrossRef]
4. European Commission. Directive 91/414/EEC, concerning the placing of plant protection products on the market. *Off. J. Eur. Commun.* **1991**, *230*.
5. Erney, D.R.; Gillespie, A.M.; Gilvydis, D.M.; Poole, C.F. Explanation of the matrix-induced chromatographic response enhancement of organophosphorus pesticides during open tubular column gas chromatography with splitless or hot on-column injection and flame photometric detection. *J. Chromatogr. A* **1993**, *638*, 57–63. [CrossRef]
6. Regulation (EC) No 396/2005 of the European Parliament and the Council of 23 February 2005 on Maximum Residue Levels of Pesticides in or on Food and Feed of Plant and Animal Origin and Amending Council Directive 91/414/EEC. Available online: http://ec.europa.eu/sanco_pesticides/public/?event=commodity.resultat (accessed on 14 April 2014).
7. Hernando, M.D.; Ferrer, C.; Ulaszewska, M.; García-Reyes, J.F.; Molina-Díaz, A.; Fernández-Alba, A.R. Application of high-performance liquid chromatography–tandem mass spectrometry with a quadrupole/linear ion trap instrument for the analysis of pesticide residues in olive oil. *Anal. Bioanal. Chem.* **2007**, *389*, 1815–1831. [CrossRef]
8. Díez, C.; Traag, W.; Zommer, P.; Marinero, P.; Atienza, J. Comparison of an acetonitrile extraction/partitioning and "dispersive solid-phase extraction" method with classical multi-residue methods for the extraction of herbicide residues in barley samples. *J. Chromatogr. A* **2006**, *1131*, 11–23. [CrossRef]
9. Anastassiades, M.; Lehotay, S.J.; Štajnbaher, D.; Schenck, F.J. Fast and Easy Multiresidue Method Employing Acetonitrile Extraction/Partitioning and "Dispersive Solid-Phase Extraction" for the Determination of Pesticide Residues in Produce. *J. AOAC Int.* **2003**, *86*, 412–431. [CrossRef]
10. Hajšlová, J.; Zrostlíková, J. Matrix effects in (ultra)trace analysis of pesticide residues in food and biotic matrices. *J. Chromatogr. A* **2003**, *1000*, 181–197. [CrossRef]
11. Anastassiades, M.; Tasdelen, B.; Scherbaum, E.; Stajnbaher, D. Recent developments in QuEChERS methodology for pesticide multiresidue analysis. In *Pesticide Chemistry: Crop Protection, Public Health, Environmental Safety*; Ohkawa, H., Miyagawa, H., Lee, P.W., Eds.; Wiley-VCH: Weinheim, Germany, 2007.
12. *EN 15662:2008*; Foods of Plant Origin: Determination of Pesticide Residues Using GC–MS and/or LC–MS (/MS) Following Acetonitrile Extraction/Partitioning and Cleanup by Dispersive SPE–QuEChERS Method. European Committee for Standardization: Brussels, Belgium, 2007.

13. Pizzutti, I.R.; Kok, A.; Zanella, R.; Adaime, M.B.; Hiemstra, M.; Wickert, C.; Prestes, O.D. Method validation for the analysis of 169 pesticides in soya grain, without clean up, by liquid chromatography-tandem mass spectrometry using positive and negative electrospray ionization. *Chromatogr. A* **2007**, *1142*, 123. [CrossRef] [PubMed]
14. Luke, M.; Froberg, J.E.; Masumoto, H.T. Extraction and Cleanup of Organochlorine, Organophosphate, Organonitrogen, and Hydrocarbon Pesticides in Produce for Determination by Gas-Liquid Chromatography. *J. AOAC Int.* **1975**, *58*, 1020–1026. [CrossRef]
15. Specht, W.; Pelz, S.; Gilsbach, W. Gas-chromatographic determination of pesticide residues after clean-up by gel-permeation chromatography and mini-silica gel-column chromatography. *Fresenius' J. Anal. Chem.* **1995**, *353*, 183–190. [CrossRef]
16. Fillion, J.; Sauvé, F.; Selwyn, J. Multiresidue Method for the Determination of Residues of 251 Pesticides in Fruits and Vegetables by Gas Chromatography/Mass Spectrometry and Liquid Chromatography with Fluorescence Detection. *J. AOAC Int.* **2000**, *83*, 698–713. [CrossRef] [PubMed]
17. Cook, J.; Beckett, M.P.; Reliford, B.; Hammock, W.; Engel, M. Quick, Easy, Cheap, Effective, Rugged, and Safe Approach for Determining Pesticide Residues. *J. AOAC Int.* **1999**, *82*, 1419–1435. [CrossRef] [PubMed]
18. Andersson, A.; Palsheden, H. Comparison of the efficiency of different GLC multi-residue methods on crops containing pesticide residues. *Fresenius' J. Anal. Chem.* **1991**, *339*, 365–367. [CrossRef]
19. Lehotay, S.J.; Son, K.A.; Kwon, H.; Koesukwiwat, U.; Fu, W.; Mastovska, K.; Hoh, E.; Leepipatpiboon, N. Comparison of QuEChERS sample preparation methods for the analysis of pesticide residues in fruits and vegetables. *J. Chromatogr. A* **2010**, *1217*, 2548–2560. [CrossRef] [PubMed]
20. European Commission. *Guidance Document on Analytical Quality Control and Method Validation Procedures for Pesticide Residues and Analysis in Food and Feed, Document SANTE/11813/2017, 21–22 November 2017 Rev.0*; European Commission: Brussels, Belgium, 2019.
21. FAO/WHO. Codex Alimentarius Commission. 2004. Available online: http://app.fao.org (accessed on 23 January 2023).
22. FAO (Food and Agriculture Organization). *Submission and Evaluation of Pesticide Residues Data for the Estimation of Maximum Residue Levels in Food and Feed*; FAO: Rome, Italy, 2002; pp. 1–279.
23. Farag, R.S.; Latif, A.M.S.; El-Gawad, A.E.; Dogheim, S.M. Monitoring of pesticide residues in some Egyptian herbs, fruits and vegetables. *Int. Food Res. J.* **2011**, *18*, 659–665.
24. Arain, M.; Kazi, T.; Baig, J.A.; Jamali, M.; Afridi, H.; Shah, A.; Jalbani, N.; Sarfraz, R. Determination of arsenic levels in lake water, sediment, and foodstuff from selected area of Sindh, Pakistan: Estimation of daily dietary intake. *Food Chem. Toxicol.* **2009**, *47*, 242–248. [CrossRef] [PubMed]
25. Amin, N.-U.; Hussain, A.; Alamzeb, S.; Begum, S. Accumulation of heavy metals in edible parts of vegetables irrigated with waste water and their daily intake to adults and children, District Mardan, Pakistan. *Food Chem.* **2013**, *136*, 1515–1523.

molecules

Article

Development of a Novel LC-MS/MS Multi-Method for the Determination of Regulated and Emerging Food Contaminants Including Tenuazonic Acid, a Chromatographically Challenging *Alternaria* Toxin

Ádám Tölgyesi [1,*], Attila Cseh [1], Andrea Simon [1] and Virender K. Sharma [2,*]

[1] Mertcontrol Hungary Ltd., 2144 Kerepes, Hungary
[2] Program for the Environment and Sustainability, Department of Environmental and Occupational Health, School of Public Health, Texas A&M University, 1266 TAMU, College Station, TX 77843, USA
* Correspondence: atolgyesi@mertcontrol.com (Á.T.); vsharma@tamu.edu (V.K.S.)

Abstract: The regulation of food contaminants in the European Union (EU) is comprehensive, and there are several compounds in the register or being added to the recommendation list. Recently, European standard methods for analysis have also been issued. The quick analysis of different groups of analytes in one sample requires a number of methods and the simultaneous use of various instruments. The aim of the present study was to develop a method that could analyze several groups of food contaminants: in this case, 266 pesticides, 12 mycotoxins, 14 alkaloid toxins, and 3 *Alternaria* toxins. The main advantage of the herein described approach over other methods is the simultaneous analysis of tenuazonic acid (TEA) and other relevant food contaminants. The developed method unites the newly published standard methods such as EN 15662:2018, EN 17194:2019, EN 17256:2019, EN 17425:2021, EN 17521:2021, which describes the analysis of both regulated and emerging contaminants. The developed method is based on a QuEChERS sample preparation, followed by LC-MS/MS analysis under alkaline mobile phase conditions. The pH of the aqueous eluent was set to 8.3, which resulted in baseline separation among ergot alkaloids and their corresponding epimers, a symmetric chromatographic peak shape for analyzing TEA and fit-for-purpose sensitivity for MS/MS detection in both positive and negative ionization modes. Those compounds, which possess the corresponding isotopically labeled internal standards (ISTD), allowed for direct quantification by the developed method and no further confirmation was necessary. This was proven by satisfactory analyses of a number of quality control (QC), proficiency test (PT), and validation samples.

Keywords: pesticides; toxins; cereals; LC-MS/MS; screening; validation

Citation: Tölgyesi, Á.; Cseh, A.; Simon, A.; Sharma, V.K. Development of a Novel LC-MS/MS Multi-Method for the Determination of Regulated and Emerging Food Contaminants Including Tenuazonic Acid, a Chromatographically Challenging *Alternaria* Toxin. *Molecules* **2023**, *28*, 1468. https://doi.org/10.3390/molecules28031468

Academic Editor: Miguel Ángel González-Curbelo

Received: 17 January 2023
Revised: 31 January 2023
Accepted: 31 January 2023
Published: 2 February 2023

1. Introduction

Contaminants in this study are substances which are either intentionally used in agriculture (e.g., pesticides) or which result from environmental contamination (e.g., plant toxins). Contamination may also occur during packaging, transport, or holding of foodstuffs, which causes a negative impact on the quality of food, thus risking human health. Therefore, the European Union (EU) has established maximum levels for several contaminants [1–8]. The pesticides are well-known groups of food contaminants. In the EU, the European Food Safety Authority (EFSA) assesses the safety of consumers based on the toxicity of pesticides and proposes a maximum residue limit (MRL) for their presence in food [7]. MRLs have been applied to more than 300 fresh products and to the same products after processing. Currently, the legislation covers more than 1000 pesticides recently or formerly used in agriculture worldwide. The MRL concentrations for pesticides set by the EU are summarized in regulation EC 396/2005 [6].

Mycotoxins have been regulated in the EU since 2006, beginning with well-known compounds such as aflatoxins, ochratoxin A or deoxynivalenol (DON) [2]. Subsequently, other compounds such as T-2 and HT-2 mycotoxins or citrinin have come into focus, and EU recommendations for these toxins are in force now [4]. In addition to these toxins, some other compounds known to be toxic have appeared in the EU regulations and recommendations. These so-called emerging toxins are the ergot and tropane alkaloids and the *Alternaria* toxins [1,3,5]. Our laboratory has been accredited for the standards listed above, and our aim in the present study was to combine all current methods into one novel multi-method. Hence, the simultaneous analysis of these groups of food contaminants is the focus of our current paper.

In the 1990s, the number of food contaminants analyzed simultaneously by HPLC was restricted due to the optical (HPLC-UV/FLD) or single-stage mass spectrometric (LC-MS) detection. With the widespread use of tandem mass spectrometric (MS/MS) detection in the early 2000s, a broader range of compounds could be separated in a single run, and multi-methods in food analysis have become popular [9,10]. One well-known LC-MS/MS multi-toxin method was published by Sulyok et al. in 2006 [11]. This method described the determination of 39 components employing a simple dilute-and-shoot approach. The extraction solvent for multi-mycotoxin analysis (acetonitrile-water-acetic/formic acid, 79/20/1, *v*/*v*/*v*) recommended by Sulyok has become a general extraction medium in control laboratories, and the recently published standard method (EN 17194:2019) included this solvent composition [12].

In addition to mycotoxins, the single pesticide group-based (e.g., chlorinated or phosphorated pesticide) methods have also been modified to multi-methods using both LC-MS/MS and GC-MS/MS techniques [13–15]. This required a general sample manipulation that can be used for all pesticides. The QuEChERS (Quick, Easy, Cheap, Efficient, Rugged, and Safe) sample preparation, which utilized acetonitrile-based extraction and phase partition, allowed for the extraction of a number of medium-polar or non-polar molecules. This extraction is commonly used for pesticides [13].

Thereafter, QuEChERS was tested and introduced as the sample preparation protocol for other LC-MS/MS methods as well [16–18]. Recently published papers and standards describe QuEChERS employed in the analysis of mycotoxins, ergot alkaloid or *Alternaria* toxin [19–23]. Although the extraction of *Alternaria* toxins from food is preferably done with a methanolic medium, QuEChERS can successfully be applied for their extraction in various food samples as well [22]. Thus, QuEChERS has become a general method for sample preparation before LC-MS/MS analyses [24].

Even though the extraction of compounds having different structures and hydrophobicity could be carried out with QuEChERS, the simultaneous HPLC analysis of the target compounds needs thorough optimization because some toxins (e.g., ergot alkaloids or tenuazonic acid (TEA)) require alkaline pH conditions in the mobile phase [19,22,25,26]. However, the HPLC separation of acidic (e.g., ochratoxin, *Alternaria* toxins) or basic compounds (e.g., alkaloids) generally requires an acidic pH condition to obtain appropriate peak shape and resolution [27]. Furthermore, MS detection in the positive ionization mode yields better sensitivity with an acidic mobile phase composition as the precursor ions are generally protonated molecules ($[M+H]^+$). Consequently, multi-methods published earlier focused on compounds that could be separated with acidic or neutral eluents [10–13,28–31] and excluded those compounds that required alkaline conditions.

Six ergot alkaloids and their corresponding six epimers are currently regulated in the EU [1]. The simultaneous analysis of the twelve compounds requires an alkaline mobile phase pH to obtain baseline separation between the ergot alkaloids and their corresponding epimers [19,25]. TEA belongs to the *Alternaria* toxin group [3]; it is a chelating compound and forms complexes at acidic pH with metal ions occurring in the eluent [32,33]. Therefore, its LC-MS separation needs either pre-column derivatization or alkaline (pH > 8) conditions in the eluent [32–36] to include in a multi-method. Consequently, a multi-method that includes TEA and all regulated ergot alkaloids along with other mycotoxins and pesticides has not been published yet.

The aim of the present study was to develop for the first time a multi-method that allows for the analysis of food contaminants such as pesticides and toxins as well as alkaloids and *Alternaria* toxins including TEA. Therefore, the pH of the mobile phase has been optimized so that the HPLC separation allows fit-for-purpose chromatographic resolution for analyzing ergot alkaloids together with their epimers a functional peak shape for challenging TEA. In addition, appropriate sensitivity for MS/MS detection carried out with polarity switching had to be optimized based on the mobile phase condition. A further goal of the paper was to verify the method with validation at low concentration levels and to evaluate the accuracy of the method involving a number of QC and PT sample analyses. Finally, the results of the multi-method on QC samples were compared to those obtained after analysis of the samples with official standard methods.

2. Results and Discussion

2.1. LC-MS/MS Method

In the analysis of pesticides and mycotoxins, the HPLC separation is generally done under acidic pH or sometimes at neural pH conditions [12,37,38]. In contrast, the separation of alkaloids and the *Alternaria* toxins requires alkaline pH conditions in the eluent recommended by the standards [19,25,26]. Hence, the optimal pH condition must be obtained at a weak alkaline pH to achieve fit-for-purpose separation of all compounds in the developed method. The pH conditions between 8.0 and 8.8 were therefore tested since the TEA gives a distorted peak shape below pH 8.0 using an HPLC column packed with C18 material, and the pH limit of the HPLC column utilized was at pH 9.0.

The EU standard methods recommend pH 10.0 to analyze ergot alkaloids in order to obtain an appropriate peak resolution between ergot alkaloids and their corresponding epimers, otherwise, peak interference may occur due to isobaric ion transitions [19,25]. On the other hand, an alkaline pH can decrease the sensitivity of those compounds ionized in the protonated molecule form. With these limitations in mind, the pH of the aqueous mobile phase was increased stepwise (in 0.1 unit increments) from 8.0 to 8.8. At pH 8.8, some pesticides displayed low intensity, e.g., cypermethrin, cyprodinil, pendimethalin and permethrin. However, this pH produced better resolution for ergot alkaloids. The lowest limit of pH in which the baseline separation could be achieved between ergot alkaloids and their corresponding epimers was at 8.3 (Figure 1). This alkaline pH did not considerably influence the retention and sensitivity of mycotoxins. Only ochratoxin A (OTA) and the fumonisins (FB1, FB2 and FB3) had retention time shifts between pH 8.0 and 8.8. The sensitivity of the detection of mycotoxins, carried out in a positive ionization mode, did not decrease under alkaline pH conditions (Figures 2 and 3) compared with the acidic conditions detailed in the standard method [12].

In our earlier studies, we found that the mobile phase did not require acidic conditions to obtain high sensitivity for the analysis of pesticides and mycotoxins using LC-MS/MS separation and employing positive ionization [37,39]. The response of DON, aflatoxins and some pesticides (e.g., chlorpyrifos-ethyl/methyl), detected as a protonated molecule ion, slightly increased at alkaline pH in comparison with acidic conditions (Figure 2). This may be caused by the sodium content of the utilized HPLC water. The higher sodium level in water yields sodium formate in the eluent when formic acid is used for acidification, and this salt can decrease the ionization of protonated molecules in the ion source. Again, the non-acidified eluent caused sensitivity drops for only a few compounds, but rather enhanced the sensitivity for most of the molecules, which resulted in fit-for-purpose sensitivity for all compounds pursued for analysis in the developed method.

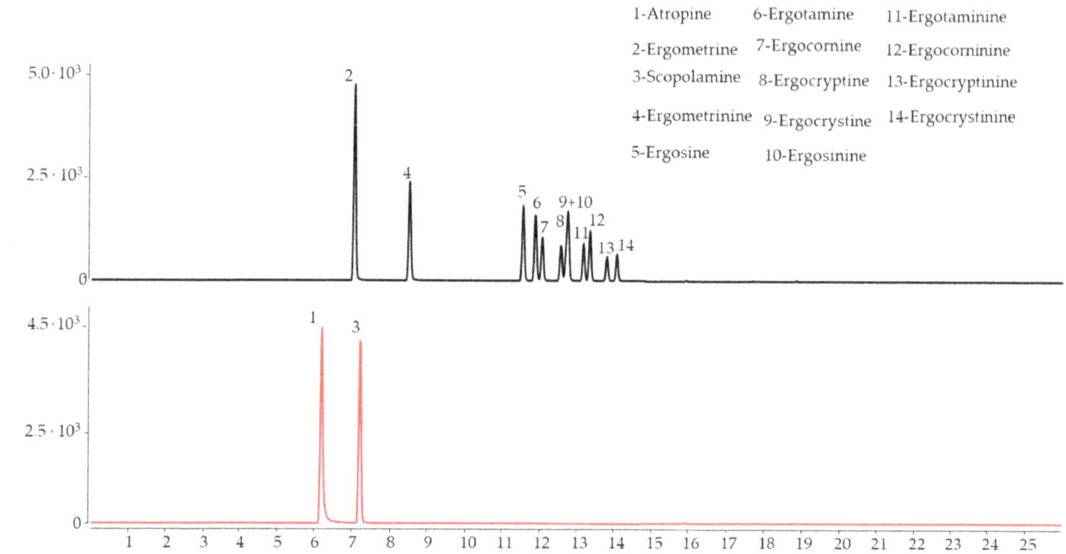

Figure 1. Separation of ergot (10 µg/kg) and tropane (5 µg/kg) alkaloids in spiked wheat samples using the optimized method.

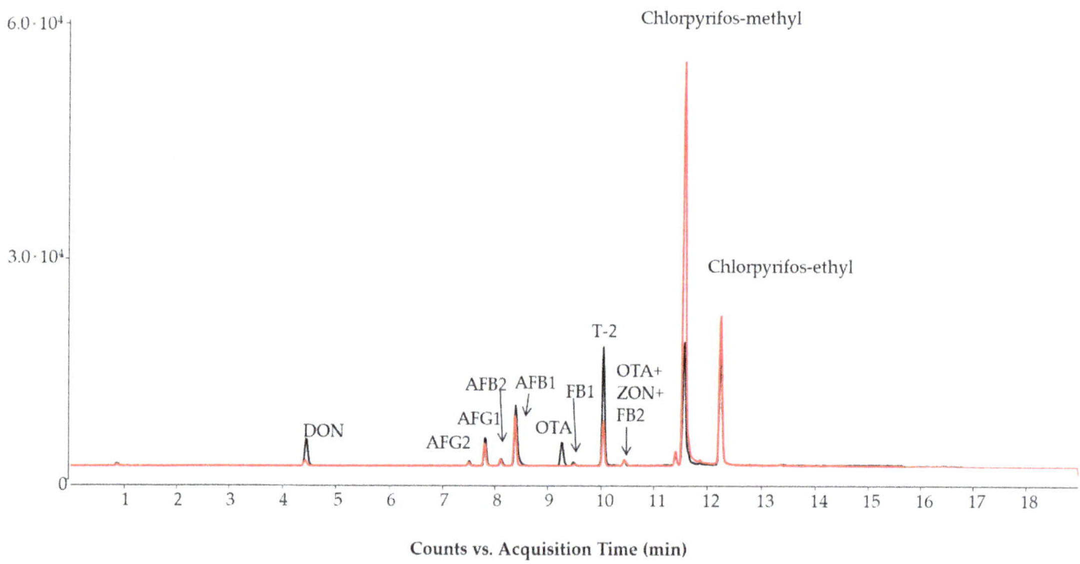

Figure 2. Total ion chromatograms of mycotoxins and chlorpyrifos-ethyl/methyl in standard solution recorded under acidic (pH 3, red line) and alkaline (pH 8.3, black line) mobile phase conditions. Concentrations: chlorpyrifos-ethyl/methyl, DON, FB1, FB2, 10 ng/mL; AFB1, AFG1, OTA, 1 ng/mL; AFB2, AFG2, 0.25 ng/mL; HT-2, T-2, ZON, 5 ng/mL. The chromatograms were recorded in an earlier stage of the method development.

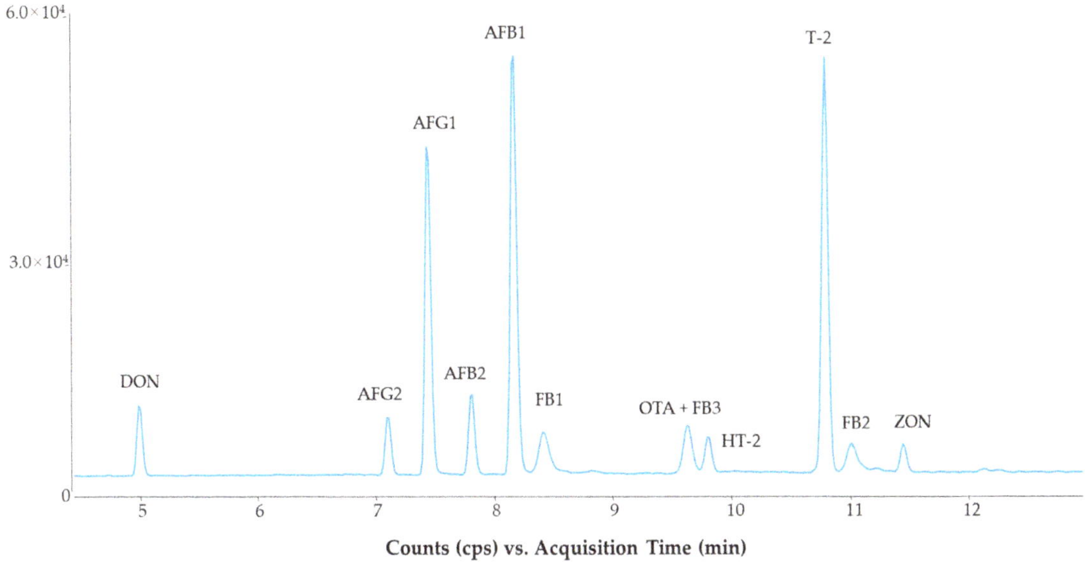

Figure 3. Separation of mycotoxins in wheat samples using the optimized method. Concentrations: DON, FB1, FB2, FB3, 50 µg/kg; AFB1, AFG1, OTA, 5 µg/kg; AFB2, AFG2, 1.25 µg/kg; HT-2, T-2, ZON, 25 µg/kg.

The other aspect of the method was the sample preparation. The QuEChERS approach described for maize and wheat samples in the standard pesticide method was tested since the QuEChERS is also used for ergot alkaloids in the EN 17425:2021 standard method [19]. Moreover, Bessaire et al. published a collaborative trial using the QuEChERS-LC-MS/MS method for analyzing mycotoxins [23] and Mujahid et al. proposed QuEChERS for Alternaria toxins [22]. The modification to the standard pesticide method involved the extraction time, which was adjusted to 30 min. Even though QuEChERS has been tested for analysis of Alternaria toxins [22], the acetonitrile-based extraction was not recommended for the extraction of Alternaria toxins earlier because methanol is preferable [40]. We also found that the absolute recovery of polar TEA, using QuEChERS extraction, was lower than 50%. The other two Alternaria toxins, AME and AOH, had higher recoveries (>80%) due to their lipophilic structures. Hence, the Alternaria ISTD solution (AME-d3, AOH-d2, TEA-^{13}C2) was added to the sample before extraction to enhance the recovery. This was also recommended in previous papers [22,40]. The isotope dilution considerably improved the recovery of TEA, but the 50% loss increased its limit of quantification (LOQ). Furthermore, the injection solution consisted of acetonitrile, which caused peak distortion of TEA when the injection volume was higher than 2.0 µL. The separation of Alternaria toxins at pH 8.3 in a cereal matrix gave fit-for-purpose LC-MS/MS analysis (Figure 4), but the high (>50%) ion suppression of AME, described also in earlier methods [34–36,40], was also seen. Therefore, isotope dilution was needed for appropriate quantification.

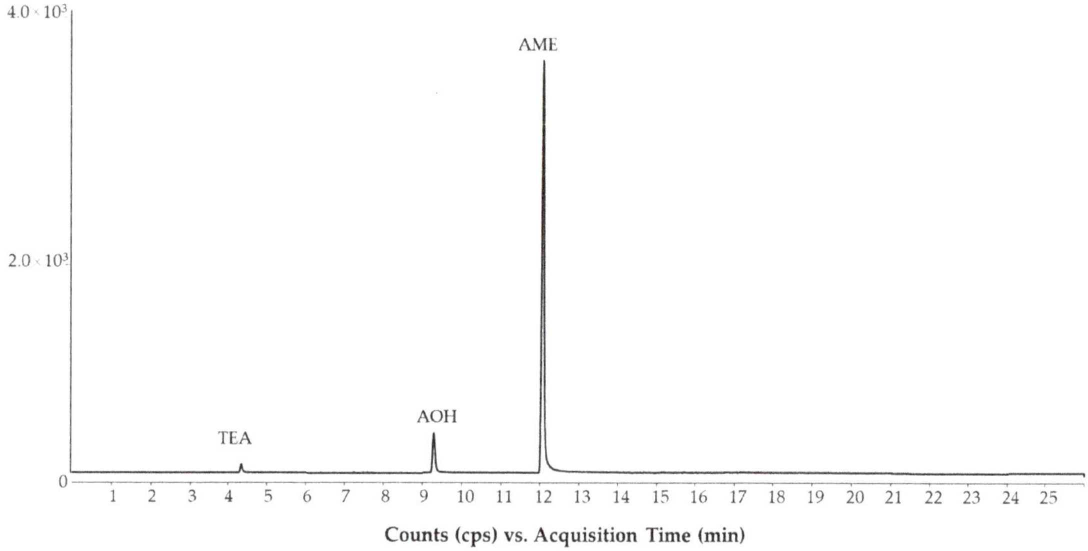

Figure 4. Separation of *Alternaria* toxins in wheat samples using the optimized method. Concentrations: AOH and AME, 2 µg/kg; TEA 100 µg/kg.

2.2. Method Evaluation

During the method evaluation, twenty-three QC and PT samples (see Supplementary Table S1) were analyzed along with validation samples (spiked blanks). The validation samples involved maize and wheat matrices (Table 1). These blank samples were spiked at two levels with 15 replicates (see Section 3.6) to evaluate the recovery and precision (Table 1). The method evaluation was done only for those compounds possessing corresponding isotopically labeled ISTD. In the case of pesticides (excluding chlorpyrifos-ethyl) and ergot alkaloids, their signals were not compensated by ISTD, so only screening and semi-quantitative analysis could be done. Even though the EN 15662:2018 and EN 17425:2021 standards allow for quantification of pesticides and ergot alkaloids using neat solvent calibration, our experience was that this leads to considerable overestimation of the pesticide concentration in spiked samples. When analyzing ergot alkaloids, low recovery was observed. The higher and lower recovery was caused by ion enhancement and ion suppression, respectively. However, the QC sample analysis gave acceptable results for both pesticides and ergot alkaloids due to their broader satisfactory range. In agreement with the SANTE 11312/2021 guideline [41], standard addition is the appropriate quantification approach for pesticides and this approach is also suggested by the EN 17256:2019 standard method for ergot and tropane alkaloids. Confirmatory analyses have been performed according to standard methods using the standard addition approach, and the results were satisfactory (Table 2). In this validation, the screening detection limit (SDL) was set for pesticides and ergot alkaloids [41]. The SDL was established as the lowest spiking level (10 µg/kg or 50 µg/kg) at which the signal-to-noise ratio (SNR > 10) and the ion ratios (within the 30% tolerance range) are acceptable. For all pesticides and ergot alkaloids, 10 µg/kg as SDL was appropriate.

The recovery calculated for those compounds listed in Table 1 was not lower than 67.1% (TEA at 500 µg/kg level) and generally ranged between 70.0% and 111%. According to the standard guideline for the determination of mycotoxins [42], recovery between 50% and 120% is acceptable with precision below 30%. These satisfactory ranges are also applicable for the analysis of alkaloid toxins and Alternaria toxins. The validation data met the standard. The validation results for analyzing chlorpyrifos-ethyl also met the SANTE requirements [41].

Table 1. Validation results for maize and wheat samples. Spiking levels are summarized in Section 3.6.

Components	Repeatability RSD% (n = 5)		Reproducibility RSD% (n = 15)		Recovery% (n = 15)		LOQ(µg/kg)	Linearity	
	Level 1	Level 2	Level 1	Level 2	Level 1	Level 2		Equation	R^2
AFB1	4.79	4.21	8.30	6.24	88.9	85.8	0.20	0.8850x − 0.0985	0.9992
AFB2	11.1	11.0	23.8	19.9	111	89.3	0.05	0.6217x − 0.0876	0.9999
AFG1	11.9	4.48	13.7	16.7	88.7	94.3	0.20	0.6388x + 0.0125	0.9998
AFG2	19.1	13.6	26.5	10.4	119	101	0.05	0.4611x − 0.00985	0.9988
AME	5.31	6.05	5.32	4.82	87.4	86.3	0.20	0.0745x − 0.0253	0.9992
AOH	21.3	14.6	23.0	14.6	108	103	0.20	0.1455x − 0.0325	0.9999
Atropine	4.02	2.18	4.11	2.18	78.6	83.9	0.20	0.6052x + 0.0897	0.9992
Chlorpyrifos-ethyl	11.3	3.86	11.3	15.6	97.1	72.7	0.20	0.0407x + 0.0014	0.9994
DON	21.1	6.71	20.5	8.86	86.0	82.9	10.0	0.04475x − 0.0014	0.9995
FB1	10.1	5.7	12.3	9.7	83.1	73.9	10.0	0.1118x − 0.00547	0.9979
FB2	9.9	5.2	17.2	9.8	91.4	81.5	10.0	0.0954x − 0.0145	0.9999
FB3	7.7	6.4	13.0	10.1	95.3	90.5	10.0	0.0954x − 0.0145	0.9999
HT-2	17.7	7.89	26.8	12.8	102.6	92	5.0	0.0051x + 0.0011	0.9998
OTA	21.7	4.25	21.7	10.5	101	70.6	1.00	0.1045x + 0.0745	0.9983
Scopolamine	3.49	2.80	5.57	2.80	74.9	76.6	0.20	0.3750x−0.0455	0.9975
T-2	8.86	3.44	11.9	13.6	96.1	89.2	1.00	0.04459x + 0.0084	0.9994
TEA	12.1	11.8	24.2	28.2	100	67.1	200	0.0153x + 0.00632	0.9988
ZON	13.8	9.87	14.6	13.3	95.4	88.2	1.00	0.0397x − 0.00754	0.9998

AFB1: aflatoxin B1; AFB2: aflatoxin B2; AFG1: aflatoxin G1; AFG2: aflatoxin G2; AME: alternariol monomethyl ether; AOH: alternariol; DON: deoxynivalenol; FB1: fumonisin B1; FB2: fumonisin B2; FB3: fumonisin B3; OTA: ochratoin A; TEA: tenuazonic acid; ZON: zearalenone.

In Table 2, we summarized the results obtained after the application of the multi-method on several naturally contaminated or spiked QC and PT samples. The concentrations of contaminants in these samples were also evaluated using the individual EU standard methods. Based on the assigned/reference values and their target standard deviations (Supplementary Table S1), the Z-score for the concentrations evaluated with the multi-method was calculated. Generally, the Z-score is satisfactory between −2 and +2.

In total, there were 11 samples analyzed for mycotoxins, of which 4 were PT samples. The PT samples had maize and wheat matrices in which aflatoxins, DON, fumonisins, HT-2, T-2, OTA and ZON could be detected, so all mycotoxins involved in the method was found at least in one PT sample. The evaluations of aflatoxins were successful at both low (below µg/kg) and medium level (sub-µg/kg) concentrations. The quantification of other mycotoxins was also satisfactory. The results obtained with the multi-method were close to those obtained by the standard methods [12]. This was also true for the seven QC samples. The quantification using the multi-method, which utilizes an alkaline mobile phase separation, QuEChERS sample preparation and isotope dilution with ^{13}C labeled analogs, resulted in satisfactory analysis for all mycotoxins in the naturally contaminated samples. In total, forty-six Z-scores were evaluated, and they were found to be between −1.67 and 1.96. Generally, the alkaline condition did not influence the analysis of mycotoxins. Retention shifts for fumonisins and OTA were observed, but the quantification was not affected by the different background. Isotope dilution with ^{13}C labeled standards further improved the quantification of mycotoxins.

Two QC (wheat and kidney bean) and two wheat PT samples were analyzed for pesticides (Figure 5). The QC samples contained multi-residues while the PT sample was contaminated with only chlorpyrifos-ethyl. The quantification of chlorpyrifos-ethyl was successfully carried out by the multi-method. The standard method used isotope dilution. However, the QC sample analysis showed underestimation of flufenoxuron and isofenphos-methyl in kidney bean QC and a questionable concentration of dimethoate and pirimiphos-methyl in wheat QC. This is caused by the pure solvent calibration, which could not compensate for the background and the recovery. The standard addition approach used for the confirmatory analysis gave satisfactory data for analyzing pesticides. In its current form, the multi-method utilizing neat solvent calibration can only be used as a screening approach.

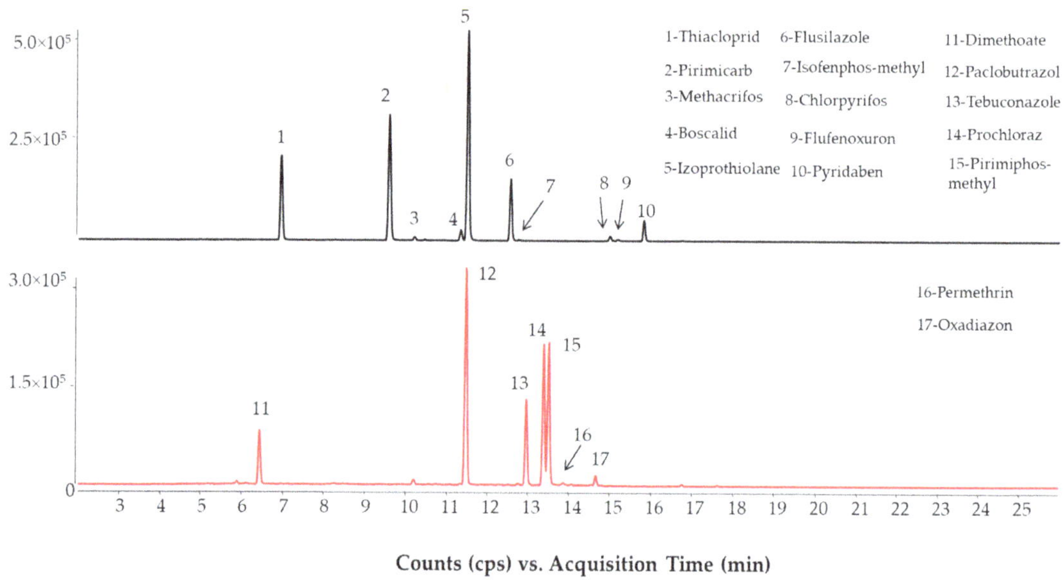

Figure 5. Separation of pesticides in kidney bean (T09133QC, above) and wheat flour (T09140QC, below) FAPAS QC samples using the multi-method. Concentrations are detailed in Table 2.

In the case of alkaloids, five QC samples were analyzed. Two rye samples were spiked with ergot alkaloids, while the three cereal samples were naturally contaminated with tropane alkaloids. The samples were evaluated with both the multi-method and the standard method (EN 17256:2019). The alkaline mobile phase condition (pH 8.3) allowed for the baseline separation of the ergot alkaloids and their corresponding epimers. Therefore, all 12 compounds could be separated in all samples. In the separation of tropane alkaloids (atropine and scopolamine), isotope dilution produced satisfactory results for all samples. However, the pure solvent calibration used in the multi-method resulted in two non-satisfactory data sets in the two QC samples for analyzing ergot alkaloids (ergosine/inine and ergotaminine). Again, the standard addition approach used in the standard method for the analysis of ergot alkaloids produced successful results (Table 2).

In the case of *Alternaria* toxins (AME, AOH and TEA), the accuracy must be improved by spiking the isotopically labeled ISTDs at the beginning of sample preparation. This is critically needed because of the use of the acetonitrile-based extraction in QuEChERS. The TEA had mostly low recovery compared to the standard method using methanolic extraction. Two naturally contaminated wheat and a sunflower seed sample were analyzed using the multi-method and the standard method (EN 17521:2021). The detected concentrations compensated by isotope dilution gave satisfactory concentrations with both methods in all samples (Table 2). However, the standard method may be preferable over the method presented here because it gives better LOQ due to the higher injection volume.

Table 2. The results of QC and PT sample analysis using the multi-method and EU standard methods.

Sample Code	Matrix	Detected Compounds	Detected Concentrations (μg/kg)	Calculated Z-Score	Evaluation	Detected Concentrations with Standard Method (Reference Value, μg/kg)	Evaluation
GAFTA PT 2022-M2	Maize	AFB1	1.77	0.49	Satisfactory	1.79 (1.60)	Satisfactory
		AFB2	0.507	0.0		0.524 (0.50)	
		AFG1	2.05	1.67		1.86 (1.50)	
		AFG2	0.69	1.90		0.552 (0.50)	
		Total Aflatoxins	5.05	1.61		4.73 (3.73)	
GAFTA PT 2022-M1	Wheat	HT-2	7.8	0.31	Satisfactory	9.2 (7.3)	Satisfactory
		T-2	30.1	−1.42		38.1 (43.8)	
Romer PT M22411 AF	Maize	AFB1	8.35	−0.22	Satisfactory	9.44 (8.79)	Satisfactory
		AFB2	0.576	−0.38		0.450 (0.63)	
		AFG1	0.706	1.96		0.554 (0.49)	
		AFG2	–	–		–	
		Total Aflatoxins	9.73	0.0		10.4 (9.72)	
		FB1	1136	−1.56		1414 (1425)	
		FB2	296	−1.21		402 (387)	
		FB3	121	−1.31		168 (168)	
		Total Fumonisins	1553	−1.55		1984 (1911)	
Romer PT M22161 DZO	Wheat	DON	2032	0.78	Satisfactory	1694 (1826)	Satisfactory
		OTA	30.3	0.79		27.5 (25.9)	
		ZON	702	1.81		519 (545)	
Romer QC M21161DZO	Wheat	DON	2597	−0.96	Satisfactory	2802 (2841)	Satisfactory
		OTA	25.0	−0.84		28.7 (30.7)	
		ZON	200	0.62		195 (177)	
EURL QC 2016 O161	Oat	HT-2	161	0.33	Satisfactory	98 (150)	Satisfactory
		T-2	63.8	−0.41		58.8 (70.3)	
EURL QC 2017 A004	Wheat	DON	434	−0.97	Satisfactory	388 (551)	Satisfactory
EURL QC 2016 C257	Maize	AFB1	10.9	0.13	Satisfactory	9.10 (10.6)	Satisfactory
		DON	553	−0.43		454 (618)	
		FB1	501	−1.58		653 (768)	
		FB2	237	0.27		246 (224)	
		ZON	210	1.33		147 (162)	
Romer QC DZO10006460	Wheat	DON	618	−1.67	Satisfactory	859 (825)	Satisfactory
		OTA	7.4	−1.24		7.2 (10)	
		ZON	34.6	−0.05		34.4 (34.9)	
Romer QC 10003613	Maize	AFB1	8.5	−0.59	Satisfactory	8.6 (9.5)	Satisfactory
		AFB2	1.93	−0.49		2.5 (2.1)	
Trilogy QC TQC MMF11-100	Maize	AFB1	20.8	0.54	Satisfactory	17.6 (18.6)	Satisfactory
		AFB2	1.32	0.07		1.20 (1.30)	
		Total Aflatoxins	22.1	0.50		18.8 (19.9)	
		DON	1932	0.16		1758 (1900)	
		FB1	1168	−1.55		1276 (1400)	
		FB2	442	1.2		366 (400)	
		FB3	95	−0.2		93 (100)	
		Total Fumonisins	1705	−0.68		1735 (1900)	
		HT-2	121.9	−0.41		149 (127)	
		T-2	104.5	1.22		92.5 (94.8)	
		OTA	17.5	−0.31		22.6 (18.5)	
		ZON	374	0.35		373 (360)	

Table 2. *Cont.*

Sample Code	Matrix	Detected Compounds	Detected Concentrations (µg/kg)	Calculated Z-Score	Evaluation	Detected Concentrations with Standard Method (Reference Value, µg/kg)	Evaluation
FAPAS QC T09133QC	Kidney Beans (Dried)	Boscalid	81	−1.12	Questionable	94 (107)	Satisfactory
		Chlorpyrifos	143	1.74		68 (103)	
		Flufenoxuron	26	−2.38		38.5 (55)	
		Flusilazole	119	−1.09		152 (155)	
		Isofenphos-methyl	25	−3.06		80 (77)	
		Isoprothiolane	154	−0.06		142 (156)	
		Methacrifos	70	−1.88		118 (119)	
		Pirimicarb	96.4	−0.21		82 (101)	
		Pyridaben	32.3	−1.79		69 (53)	
		Thiacloprid	67.2	−0.95		62 (85)	
FAPAS QC T09140QC	Wheat flour	Dimethoate	47.1	2.10	Questionable	41.1 (32.1)	Satisfactory
		Oxadiazon	101	1.35		65.4 (77.8)	
		Paclobutrazol	134	1.02		112 (98.3)	
		Permethrin	28.1	−2.57		66.8 (64.7)	
		Pirimiphos-methyl	104	0.00		99.3 (104)	
		Prochloraz	166	0.40		111 (153)	
		Tebuconazole	79	0.44		101 (87.3)	
PT, Chlorpyrifos-ethyl	Wheat	Chlorpyrifos-ethyl	28.0	−0.68	Satisfactory	27.0 (33.0)	Satisfactory
PT, Chlorpyrifos-ethyl	Wheat	Chlorpyrifos-ethyl	17.0	0.0	Satisfactory	19.6 (16.9)	Satisfactory
EURL 2017 QC EA047	Rye	Ergocornine/inine	294	0.00	Questionable	280 (295)	Satisfactory
		α-Ergocryptine/inine	304	+1.60		337 (231)	
		Ergocrystine/inine	676	−0.55		651 (752)	
		Ergometrine/inine	92.1	−1.20		114 (116)	
		Ergosine/inine	136	−2.65		222 (242)	
		Ergotamine/inine	641	−0.45		606 (695)	
FAPAS QC 22180	Rye	Ergocornine	45.2	1.79	Questionable	40.7 (32.4)	Satisfactory
		Ergocorninine	12.6	0.07		15.8 (12.4)	
		α-Ergocryptinine	13.9	−1.30		15.7 (19.5)	
		Ergocrystine	85.6	−0.91		141 (107)	
		Ergocrystinine	20.8	−1.95		23.6 (36.4)	
		Ergometrine	27.1	0.21		32 (25.9)	
		Ergometrinine	4.30	0.11		4.64 (4.2)	
		Ergosine	16.9	−1.08		19.1 (22.2)	
		Ergotamine	34.5	−1.42		41.6 (50.2)	
		Ergotaminine	6.03	−2.29		18.0 (13.6)	
		Total Ergot Alkaloides	338	−1.06		353 (419)	
EURL QC 2016 C029	Cereal	Atropine	0.81	−1.37	Questionable	1.11 (1.16)	Satisfactory
		Scopolamine	0.111	−1.85		0.169 (0.183)	
FAPAS QC 22179	Cereal	Atropine	6.5	−1.53	Satisfactory	8.82 (9.8)	Satisfactory
		Scopolamine	3.6	−0.37		4.83 (3.88)	
EURL QC 2016 E087	Cereal	Atropine	6.7	−0.45	Satisfactory	6.29 (7.44)	Satisfactory
		Scopolamine	0.63	−1.77		0.76 (1.03)	
QC 2018T15	Wheat	AME	4.0	−0.95	Satisfactory	4.08 (5.06)	Satisfactory
		AOH	3.3	−1.63		5.96 (5.11)	
		TEA	76.0	2.1		71.0 (52.0)	
QC 2018 B56	Wheat	AME	0.78	−1.53	Satisfactory	0.69 (1.17)	Satisfactory
		AOH	1.51	−1.28		2.51 (2.1)	
		TEA	180	−1.95		314 (297)	

Table 2. *Cont.*

Sample Code	Matrix	Detected Compounds	Detected Concentrations (μg/kg)	Calculated Z-Score	Evaluation	Detected Concentrations with Standard Method (Reference Value, μg/kg)	Evaluation
QC 2018 X06	Sunflower seed	AME	1.68	−0.91	Satisfactory	2.01 (2.1)	Satisfactory
		AOH	1.57	−1.06		1.32 (2.06)	
		TEA	87	−1.84		102 (146)	

The outlier results are highlighted with red.

3. Materials and Methods

3.1. Reagents and Samples

Dried-down analytical standards such as *Alternaria* toxins (100 μg), ergot alkaloids (500 μg), tropane alkaloids (100 μg), mycotoxin stock solutions and ^{13}C isotopically labeled stock solutions were obtained from Romer Labs (Tulln, Austria). Stock solutions were prepared by adding 1.0 mL methanol (*Alternaria* toxins), 1.0 mL acetonitrile (tropane alkaloids), 5.0 mL acetonitrile (ergot alkaloids) to the vial and standards were redissolved in the solvent to obtain a concentration of 100 μg/mL. Stock solutions were kept at −18 °C for a year. Deuterated isotopically labeled standards (*Alternaria* toxins: AME-d3, AOH-d2, TEA-$^{13}C2$; tropane alkaloids: atropine-d5 and scopolamine-$^{13}C1$-d3; pesticide: chlorpyrifos-d10), Pesticide Mixture 167 and Pyrethroide Pesticide Mixture 153 were acquired from LGC (Wesel, UK). Piperonyl butoxide (a synergistic component), permethrin and diphenylamine individual standards were purchased from the Merck-Sigma group (Schnelldorf, Germany). An LC-MS comprehensive pesticide mixture containing 253 compounds was purchased from Agilent Technologies (Waldbronn, Germany). Stock solutions (1 mg/mL) for individual standards were prepared and stored by following the procedure given in the pesticide database [43].

Methanol, acetonitrile, ammonium formate, formic acid, ammonia solution (25%), (either LC-MS or HPLC grade) and the Ascentis Express C18 HPLC column (100 mm × 3 mm, 2.7 μm) were purchased from the Merck-Sigma group (Schnelldorf, Germany). The EN 15662:2018 QuEChERS extraction salt (4 g MgSO4, 1 g NaCl, 1 g Na-citrate × $2H_2O$ and 0.5 g) and HPLC pre-column holders and C18 pre-column cartridges (4 mm × 3 mm; 5 μm) were obtained from Phenomenex (Torrance, CA, USA). Honeywell HPLC-gradient-grade water was acquired from Thomasker (Debrecen, Hungary). The final aqueous mobile phase (solvent A, pH 8.3) was prepared by adding 65 μL ammonia solution to 1 L HPLC water containing 5 mM ammonium formate.

PT and QC samples (23 in total) were obtained from various companies and details are provided in Supplementary Table S1.

3.2. Instrumentation

LC-MS/MS analyses were carried out using an Agilent 6470B triple quad consisting of an Agilent 1260 liquid chromatograph coupled to a 6470B MS detector equipped with an Agilent JetStream ion source (Agilent Technologies (Waldbronn, Germany)). Data acquisition and evaluation were performed with the Masshunter software version 10.1.

3.3. Sample Preparation

Samples were ground (<1 mm) before extraction and thoroughly mixed to assure adequate homogeneity. The sample preparation was based on the standard QuECHERS approach for regular pesticide residue analysis in cereals [13], with mechanical shaking modification to obtain appropriate extraction for the toxins. Samples (5.0 g) were weighed in a 50 mL plastic centrifuge tube and 100 μL Alternaria ISTD solution (25 μg/mL TEA-$^{13}C2$, 5.0 μg/mL AOH-d2 and AME-d3 in methanol) was spiked into the sample. Then, 10.0 mL distilled water was added to the samples, followed by 9.9 mL acetonitrile. The extraction was carried out with a laboratory shaker (CAT S50, CAT M. Zipperer GmbH, Ballrechten-Dottingen, Germany) at full speed (600 min^{-1}) for 30 min. After the extraction, EN 15662:2018 QuEChERS salt mixture was added and the samples were hand-shaken for 1.0 min, followed by centrifugation at 2300× *g* (Thermo Megafuge 16, Unicam Kft, Budapest, Hungary) at ambient temperature. Then, 470 µL of the upper layer and 30 µL of ISTD solution (AFB1-^{13}C17, AFB2-^{13}C17, AFG1-^{13}C17, AFG2-^{13}C17, OTA-^{13}C20 in 15 ng/mL; T-2-^{13}C24, HT-2-^{13}C22, ZON-^{13}C18 in 150 ng/mL; DON-^{13}C15, FB1-^{13}C34, FB2-^{13}C34, FB3-^{13}C34, chlorpyrifos-ethyl-d10 in 300 ng/mL; atropine-d5, scopolamine-^{13}C1-d3 in 100 ng/mL in 50% acetonitrile) were mixed in a 2.0 mL screw-cap HPLC vial and vortexed prior to injection into the LC-MS/MS instrument.

3.4. LC-MS/MS Separation

Compounds were separated on an Ascentis Express C18 HPLC column (100 × 3 mm, 2.7 µm) equipped with a C18 guard column (4 mm × 3 mm, 5 µm) Merck-Sigma group (Schnelldorf, Germany). The binary gradient elution mode was applied with solvent A containing 5 mM ammonium formate in water (pH 8.3) and solvent B containing methanol. The mobile phase gradient consisted of 5% B at 0 min; 5% B at 0.5 min; 40% B at 3.0 min; 100% B at 15 min; 100% B at 19 min; 5% B at 19.1 min; 5% B at 26.0 min; flow rate was set to 0.5 mL/min. The column thermostat and autosampler were maintained at 39 °C and at 18 °C, respectively. The injection volume was 2.0 µL. Compounds were detected using positive/negative ionization mode and dynamic multiple reaction monitoring (dMRM) scan mode. Ion transitions for 295 compounds are presented in Supplementary Table S2. The MRM time window was 60 s, and the cycle time was 1000 ms. The Agilent Jet Stream ion source parameters were as follows: drying gas temperature, 300 °C; sheath gas temperature, 350 °C; nebulizer, 35 psi; gas flow, 7 L/min; sheath gas flow, 11 L/min; capillary voltage, ± 3500 V; and nozzle voltage, +0, −1000 V. The HPLC effluent was directed into waste from 0 to 2.0 min.

3.5. Quantification

Calibrants in 50% acetonitrile were prepared from the native working standard mixture along with ISTD solutions, considering the dilution factor (2.13×) of the sample preparation. The calibration levels, expressed in µg/kg, are detailed in Table 3. For some pesticides, the lowest calibration level was 0.2 µg/kg, however, for most of them, 1 µg/kg could be used as the starting point of the calibration. Only those compounds that possessed the corresponding isotopically labeled analogs could be appropriately quantified. These were the mycotoxins, the tropane alkaloids, the Alternaria toxins, and chlorpyrifos-ethyl. Even though the pesticide (EN 15662:2018) and ergot alkaloid (EN 17425:2021) standard methods allow the quantification with neat calibrants [13,19], the presented method works only as a screening approach for them in the absence of ISTD. In the case where compounds from their group are identified, a further quantification using the standard addition approach is needed, in accordance with the SANTE 11312/2021 guidelines and EN 17256:2019 standard method [25,41].

The concentrations of analytes could be directly obtained from the equations of linear calibration weighted with the factor of 1/x. The determination coefficients obtained under the validation study were not lower than 0.9950.

Table 3. Calibration levels.

Compounds	Cal 1 (µg/kg)	Cal 2 (µg/kg)	Cal 3 (µg/kg)	Cal 4 (µg/kg)	Cal 5 (µg/kg)	Cal 6 (µg/kg)
AFB1	0.2	1	2	10	20	50
AFB2	0.05	0.25	0.5	2.5	5	12.5
AFG1	0.2	1	2	10	20	50

Table 3. *Cont.*

Compounds	Cal 1 (µg/kg)	Cal 2 (µg/kg)	Cal 3 (µg/kg)	Cal 4 (µg/kg)	Cal 5 (µg/kg)	Cal 6 (µg/kg)
AFG2	0.05	0.25	0.5	2.5	5	12.5
AME	0.2	1	2	10	20	50
AOH	0.2	1	2	10	20	50
Atropine/scopolamine	0.2	1	2	10	20	50
DON	10	50	100	500	1000	2500
Ergot alkaloids	0.2	1	2	10	20	50
Fumonisins	10	50	100	500	1000	2500
HT-2/T-2	1	5	10	50	100	250
OTA	1	5	10	50	100	250
Pesticides	0.2	1	2	10	20	50
TEA	100	500	1000	5000	10,000	25,000
ZON	1	5	10	50	100	250

3.6. *Validation*

The confirmatory validation was performed for the mycotoxins, the tropane alkaloids, the *Alternaria* toxins and chlorpyrifos-ethyl. The recovery and precision were calculated from the analysis of spiked maize and wheat samples. The fortified samples were prepared on three different days at two concentration levels (level 1 and level 2) by the operators (Table 1). In total, fifteen samples were analyzed. The levels were: AFB2, AFG2—0.25 µg/kg and 1.25 µg/kg; AFB1, AFG1, atropine and scopolamine—1.0 µg/kg and 5.0 µg/kg; AME and AOH—2 µg/kg and 10 µg/kg; HT-2, T-2 and ZON—5 µg/kg and 25 µg/kg; chlorpyrifos-ethyl, DON, FB1, FB2 and FB3—10 µg/kg and 50 µg/kg; TEA—200 µg/kg and 1000 µg/kg.

Ergot alkaloids and pesticides were also spiked into the samples along with the other compounds mentioned above. Their levels were 10 µg/kg and 50 µg/kg; however, appropriate quantification could not be performed due to the absence of background compensation with ISTD. Hence, the validation of these samples was performed as a screening validation, and the SDL (either 10 µg/kg or 50 µg/kg) was evaluated. The LOQ was set as the lowest calibration point.

4. Conclusions

A novel LC-MS/MS multi-method has been developed for analyzing toxins and pesticides together. The sample preparation is the modification of the QuEChERS-based approaches described in the pesticide and ergot alkaloid standard method or developed by Mujahid et al. (2020) or Bessaire et al. (2019). The chromatographically challenging TEA could be included in the method along with all regulated ergot alkaloids by using alkaline mobile phase conditions. The pH of the eluent did not influence the analysis of pesticides and mycotoxins. The method was evaluated by analyzing several QC and PT samples. Moreover, results obtained with the multi-method was compared with those data obtained by the individual EU standard methods. Even though the results with the standard methods are better, similarly good data can be obtained with the multi-method, which covers 295 compounds and unites five standard methods.

Supplementary Materials: The following supporting information can be downloaded at: https://www.mdpi.com/article/10.3390/molecules28031468/s1, Table S1: Quality control and proficiency test materials; Table S2: The scheduled MRM ion transitions of the tested compounds.

Author Contributions: Conceptualization, Á.T. and A.C.; methodology, Á.T. and A.C.; validation, A.C., Á.T. and A.S.; formal analysis, A.C. and Á.T.; investigation, Á.T. and A.C.; data curation, A.C.

and Á.T.; writing—original draft preparation, Á.T. and A.C. and V.K.S.; writing—review and editing, Á.T. and A.C. and V.K.S. All authors have read and agreed to the published version of the manuscript.

Funding: This research received no external funding.

Institutional Review Board Statement: Not applicable.

Informed Consent Statement: Not applicable.

Data Availability Statement: Not applicable.

Acknowledgments: Authors wish to thank Mary Sohn, Edgár Tóth and the anonymous reviewers for their comments that improved the paper greatly.

Conflicts of Interest: The authors declare no conflict of interest.

Sample Availability: Samples of the compounds are not available from the authors.

References

1. Commission Regulation (EU) 2021/1399 of 24 August 2021 Amending Regulation (EC) No 1881/2006 as Regards Maximum Levels of Ergot Sclerotia and Ergot Alkaloids in Certain Foodstuffs. Available online: https://eur-lex.europa.eu/legal-content/EN/TXT/PDF/?uri=CELEX:32021R1399&rid=2 (accessed on 16 October 2022).
2. Commission Regulation (EC) No 1881/2006 of 19 December 2006 Setting Maximum Levels for Certain Contaminants in Foodstuffs. Available online: https://eur-lex.europa.eu/legal-content/EN/TXT/PDF/?uri=CELEX:32006R1881&from=HU (accessed on 16 October 2022).
3. Commission Regulation (EU) 2022/553 of 5 April 2022 on Monitoring the Presence of Alternaria Toxins in Food. Available online: https://eur-lex.europa.eu/legal-content/EN/TXT/PDF/?uri=CELEX:32022H0553&from=EN (accessed on 16 October 2022).
4. Commission Regulation of 27 March 2013 on the Presence of T-2 and HT-2 Toxin in Cereals and Cereal Products. Available online: https://eur-lex.europa.eu/LexUriServ/LexUriServ.do?uri=OJ:L:2013:091:0012:0015:EN:PDF (accessed on 16 October 2022).
5. Commission Regulation (EU) 2021/1408 of 27 August 2021 Amending Regulation (EC) No 1881/2006 as Regards Maximum Levels of Tropane Alkaloids in Certain Foodstuffs. Available online: https://eur-lex.europa.eu/legal-content/EN/TXT/PDF/?uri=CELEX:32021R1408&from=EN (accessed on 16 October 2022).
6. Regulation (EC) No 396/2005 of the European Parliament and of the Council of 23 February 2005 on Maximum Residue Levels of Pesticides in or on Food and Feed of Plant and Animal Origin and Amending Council Directive 91/414/EEC. Available online: http://data.europa.eu/eli/reg/2005/396/2016-05-13 (accessed on 18 September 2022).
7. Pesticide Evaluations: Overview and Procedure. Available online: https://www.efsa.europa.eu/en/applications/pesticides (accessed on 18 September 2022).
8. Pesticide Residue Monitoring Program, Fiscal Year 2020 Pesticide Report, U.S. Food and Drug Administration. Available online: https://www.fda.gov/media/160464/download (accessed on 18 September 2022).
9. De Girolamo, A.; Lippolis, V.; Pascale, M. Overview of Recent Liquid Chromatography Mass Spectrometry-Based Methods for Natural Toxins Detection in Food Products. *Toxins* **2022**, *14*, 328. [CrossRef] [PubMed]
10. Sulyok, M.; Stadler, D.; Steiner, D.; Krska, R. Validation of an LC-MS/MS-based dilute-and-shoot approach for the quantification of > 500 mycotoxins and other secondary metabolites in food crops: Challenges and solutions. *Anal. Bioanal. Chem.* **2020**, *412*, 2607–2620. [CrossRef] [PubMed]
11. Sulyok, M.; Berthiller, F.; Krska, R.; Schuhmacher, R. Development and validation of a liquid chromatography/ tandem mass spectrometric method for the determination of 39 mycotoxins in wheat and maize. *Rapid Commun. Mass Spectrom.* **2006**, *20*, 2649–2659. [CrossRef] [PubMed]
12. EN 17194:2019. Animal Feeding Stuffs: Methods of Sampling and Analysis. Determination of Deoxynivalenol, Aflatoxin B1, Fumonisin B1 & B2, T-2 & HT-2 Toxins, Zearalenone and Ochratoxin A in Feed Materials and Compound Feed by LC-MS/MS. Available online: https://www.en-standard.eu/search/?q=EN+17194%3A2019 (accessed on 16 October 2022).
13. EN 15662:2018. Foods of Plant Origin—Multimethod for the Determination of Pesticide Residues Using GC- and LC-Based Analysis Following Acetonitrile Extraction/Partitioning and Clean-Up by Dispersive SPE—Modular QuEChERS-Method. Available online: https://www.en-standard.eu/bs-en-15662-2018-foods-of-plant-origin-multimethod-for-the-determination-of-pesticide-residues-using-gc-and-lc-based-analysis-following-acetonitrile-extraction-partitioning-and-clean-up-by-dispersive-spe-modular-quechers-method/ (accessed on 16 October 2022).
14. Lee, S.-H.; Kwak, S.-Y.; Sarker, A.; Moon, J.-K.; Kim, J.-E. Optimization of a Multi-Residue Analytical Method during Determination of Pesticides in Meat Products by GC-MS/MS. *Foods* **2022**, *11*, 2930. [CrossRef]
15. Maragou, N.C.; Balayiannis, G.; Karanasios, E.; Markellou, E.; Liapis, K. Targeted Multiresidue Method for the Analysis of Different Classes of Pesticides in Agro-Food Industrial Sludge by Liquid Chromatography Tandem Mass Spectrometry. *Molecules* **2021**, *26*, 6888. [CrossRef]

16. González-Gómez, L.; Morante-Zarcero, S.; Pérez-Quintanilla, D.; Sierra, I. Occurrence and Chemistry of Tropane Alkaloids in Foods, with a Focus on Sample Analysis Methods: A Review on Recent Trends and Technological Advances. *Foods* **2022**, *11*, 407. [CrossRef]
17. Martinello, M.; Borin, A.; Stella, R.; Bovo, D.; Biancotto, G.; Gallina, A.; Mutinelli, F. Development and validation of a QuEChERS method coupled to liquid chromatography and high resolution mass spectrometry to determine pyrrolizidine and tropane alkaloids in honey. *Food Chem.* **2017**, *234*, 295–302. [CrossRef] [PubMed]
18. Rausch, A.-K.; Brockmeyer, R.; Schwerdtle, T. Development and Validation of a QuEChERS-Based Liquid Chromatography Tandem Mass Spectrometry Multi-Method for the Determination of 38 Native and Modified Mycotoxins in Cereals. *J. Agric. Food Chem.* **2020**, *68*, 4657–4669. [CrossRef]
19. EN 17425:2021. Foodstuffs. Determination of Ergot Alkaloids in Cereals and Cereal Products by dSPE Clean-Up and HPLC-MS/MS. Available online: https://www.en-standard.eu/search/?q=EN+17425%3A2021 (accessed on 16 October 2022).
20. Carbonell-Rozas, L.; Gámiz-Gracia, L.; Lara, F.J.; García-Campaña, A.M. Determination of the Main Ergot Alkaloids and Their Epimers in Oat-Based Functional Foods by Ultra-High Performance Liquid Chromatography Tandem Mass Spectrometry. *Molecules* **2021**, *26*, 3717. [CrossRef]
21. González-Gómez, L.; Morante-Zarcero, S.; Pereira, J.A.M.; Câmara, J.S.; Sierra, I. Improved Analytical Approach for Determination of Tropane Alkaloids in Leafy Vegetables Based on μ-QuEChERS Combined with HPLC-MS/MS. *Toxins* **2022**, *14*, 650. [CrossRef]
22. Mujahid, C.; Savoy, M.-C.; Baslé, Q.; Woo, P.M.; Ee, E.C.Y.; Mottier, P.; Bessaire, T. Levels of *Alternaria* Toxins in Selected Food Commodities Including Green Coffee. *Toxins* **2020**, *12*, 595. [CrossRef] [PubMed]
23. Bessaire, T.; Mujahid, C.; Mottier, P.; Desmarchelier, A. Multiple Mycotoxins Determination in Food by LC-MS/MS: An International Collaborative Study. *Toxins* **2019**, *11*, 658. [CrossRef] [PubMed]
24. Kim, L.; Lee, D.; Cho, H.-K.; Choi, S.-D. Review of the QuEChERS method for the analysis of organic pollutants: Persistent organic pollutants, polycyclic aromatic hydrocarbons, and pharmaceuticals. *Trends Environ. Anal. Chem.* **2019**, *22*, e00063. [CrossRef]
25. EN 17256:2019. Animal Feeding Stuffs: Methods of Sampling and Analysis. Determination of Ergot Alkaloids and Tropane Alkaloids in Feed Materials and Compound Feeds by LC-MS/MS. Available online: https://www.en-standard.eu/search/?q=EN+17256%3A2019 (accessed on 16 October 2022).
26. EN 17521:2021. Foodstuffs. Determination of *Alternaria* Toxins in Tomato, Wheat and Sunflower Seeds by SPE Clean-Up and HPLC-MS/MS. Available online: https://www.en-standard.eu/search/?q=EN+17521%3A2021 (accessed on 16 October 2022).
27. Tölgyesi, Á.; Berky, R.; Békési, K.; Fekete, S.; Fekete, J.; Sharma, V.K. Analysis of sulfonamide residue in real honey samples using liquid chromatography with fluorescence and tandem mass spectrometry detection. *J. Liq. Chromatogr Relat. Technol.* **2013**, *36*, 1105–1125. [CrossRef]
28. Kresse, M.; Drinda, H.; Romanotto, A.; Speer, K. Simultaneous determination of pesticides, mycotoxins, and metabolites as well as other contaminants in cereals by LC-LC-MS/MS. *J. Chromatogr. B* **2019**, *1117*, 86–102. [CrossRef]
29. Reichert, B.; de Kok, A.; Pizzutti, I.R.; Scholten, J.; Cardoso, C.D.; Spanjer, M. Simultaneous determination of 117 pesticides and 30 mycotoxins in raw coffee, without clean-up, by LC-ESI-MS/MS analysis. *Anal. Chim. Acta* **2018**, *1004*, 40–50. [CrossRef]
30. Na, T.W.; Seo, H.-J.; Jang, S.-N.; Kim, H.; Yun, H.; Kim, H.; Ahn, J.; Cho, H.; Hong, S.-H.; Kim, H.J.; et al. Multi-residue analytical method for detecting pesticides, veterinary drugs, and mycotoxins in feed using liquid- and gas chromatography coupled with mass spectrometry. *J. Chromatogr. A* **2022**, *1676*, 463257. [CrossRef]
31. Eyring, P.; Tienstra, M.; Mol, H.; Herrmann, S.S.; Rasmussen, P.H.; Frandsen, H.L.; Poulsen, M.E. Development of a new generic extraction method for the analysis of pesticides, mycotoxins, and polycyclic aromatic hydrocarbons in representative animal feed and food samples. *Food Chem.* **2021**, *356*, 129653. [CrossRef]
32. Asam, S.; Liu, Y.; Konitzer, K.; Rychlik, M. Development of a stable isotope dilution assay for tenuazonic acid. *J. Agric. Food Chem.* **2011**, *59*, 2980–2987. [CrossRef]
33. Siegel, D.; Rasenko, T.; Koch, M.; Nehls, I. Determination of the Alternaria mycotoxin tenuazonic acid. in cereals by high-performance liquid chromatography–electrospray ionization iontrap multistage mass spectrometry after derivatization with 2,4-dinitrophenylhydrazine. *J. Chromatogr. A* **2009**, *1216*, 4582–4588. [CrossRef]
34. Tölgyesi, Á.; Stroka, J.; Tamosiunas, V.; Zwickel, T. Simultaneous analysis of Alternaria toxins and citrinin in tomato: An optimised method using liquid chromatography-tandem mass spectrometry. *Food Addit. Contam.* **2015**, *32*, 1512–1522. [CrossRef]
35. Tölgyesi, Á.; Kozma, L.K.; Sharma, V. Determination of *Alternaria* Toxins in Sunflower Oil by Liquid Chromatography Isotope Dilution Tandem Mass Spectrometry. *Molecules* **2020**, *25*, 1685. [CrossRef] [PubMed]
36. Tölgyesi, Á.; Farkas, T.; Bálint, M.; McDonald, T.J.; Sharma, V.K. A Dilute and Shoot Strategy for Determining *Alternaria* Toxins in Tomato-Based Samples and in Different Flours Using LC-IDMS Separation. *Molecules* **2021**, *26*, 1017. [CrossRef] [PubMed]
37. Tóth, E.; Tölgyesi, Á.; Simon, A.; Bálint, M.; Ma, X.; Sharma, V.K. An Alternative Strategy for Screening and Confirmation of 330 Pesticides in Ground- and Surface Water Using Liquid Chromatography Tandem Mass Spectrometry. *Molecules* **2022**, *27*, 1872. [CrossRef] [PubMed]
38. Lauwers, M.; De Baere, S.; Letor, B.; Rychlik, M.; Croubels, S.; Devreese, M. Multi LC-MS/MS and LC-HRMS Methods for Determination of 24 Mycotoxins including Major Phase I and II Biomarker Metabolites in Biological Matrices from Pigs and Broiler Chickens. *Toxins* **2019**, *11*, 171. [CrossRef] [PubMed]
39. Tölgyesi, Á.; Tóth Kovács, B.; Tóth, E.; Simon, A.; Bálint, M.; Sharma, V.K. Unexpected sensitivity enhancement in analysing alfatoxin M1 using LC-IDMS. *Microchem. J.* **2022**, *179*, 107469. [CrossRef]

40. Gonçalves, C.; Tölgyesi, Á.; Bouten, K.; Cordeiro, F.; Stroka, J. Determination of *Alternaria* Toxins in Food by SPE and LC-IDMS: Development and In-House Validation of a Candidate Method for Standardisation. *Separations* **2022**, *9*, 70. [CrossRef]
41. Main Changes Introduced in Document N° SANTE/11312/2021 with Respect to the Previous Version (Document N° SANTE 12682/2019). Available online: https://food.ec.europa.eu/system/files/2022-02/pesticides_mrl_guidelines_wrkdoc_2021-11312.pdf (accessed on 16 October 2022).
42. CEN/TR 16059:2010. Food Analysis—Performance Criteria for Single Laboratory Validated Methods of Analysis for the Determination of Mycotoxins. Available online: https://www.nbn.be/shop/en/standard/preview/6283/en/ (accessed on 16 October 2022).
43. EU Pesticides Database. Available online: https://ec.europa.eu/food/plants/pesticides/eu-pesticides-database_hu (accessed on 16 October 2022).

molecules

Article

Chromatographic Method for Monitoring of Pesticide Residues and Risk Assessment for Herbal Decoctions Used in Traditional Korean Medicine Clinics

Se-Mi Kang [1], Jae-Hee Won [1], Ji-Eun Han [2], Jong-Hyun Kim [3], Kyeong-Han Kim [4], Hye-In Jeong [5,*] and Soo-Hyun Sung [2,*]

1 Department of Public Infrastructure Operation, National Institute of Korean Medicine Development, Seoul 04554, Republic of Korea
2 Department of Policy Development, National Institute of Korean Medicine Development, Seoul 04554, Republic of Korea
3 Department of Medical Classics and History, College of Korean Medicine, Gachon University, Seongnam-si 13120, Republic of Korea
4 Department of Preventive Medicine, College of Korean Medicine, Woosuk University, Jeonju-si 54986, Republic of Korea
5 Department of Preventive Medicine, College of Korean Medicine, Kyung Hee University, Seoul 02447, Republic of Korea
* Correspondence: frogcream@gmail.com (H.-I.J.); koyote10010@nikom.or.kr (S.-H.S.)

Abstract: The presence of pesticide residues in herbs and the herbal products derived from them raises serious health concerns. This study was conducted to investigate the residual pesticide concentrations and assess potential human health risks from herbal medicines used in traditional Korean medicine clinics. A total of 40 samples of herbal decoctions were collected from 10 external herbal dispensaries. The pesticide residues were analyzed by the multiresidue method for 320 different pesticides using liquid chromatography tandem mass spectrometry (LC-MS/MS) and gas chromatography tandem mass spectrometry (GC-MS/MS). As a result of the monitoring, carbendazim was detected at 0.01 and 0.03 μg/g in eight samples and no pesticide was detected in the other herbal decoctions. Carbendazim was set for each individual item as less than 0.05 μg/g in Paeoniae radix, less than 0.05 μg/g in Cassiae semen, less than 2.0 μg/g in Lycii fructus, and less than 10 μg/g in Schisandrae fructus (dried). Therefore, the results of this study suggested that the detected pesticide residues in herbal decoctions could not be considered as posing a serious health risk.

Keywords: herbal decoction; traditional Korean medicine; pesticide residues; risk assessment

Citation: Kang, S.-M.; Won, J.-H.; Han, J.-E.; Kim, J.-H.; Kim, K.-H.; Jeong, H.-I.; Sung, S.-H. Chromatographic Method for Monitoring of Pesticide Residues and Risk Assessment for Herbal Decoctions Used in Traditional Korean Medicine Clinics. *Molecules* **2023**, *28*, 3343. https://doi.org/10.3390/molecules28083343

Academic Editors: Xingang Liu and Miguel Ángel González-Curbelo

Received: 31 January 2023
Revised: 29 March 2023
Accepted: 3 April 2023
Published: 10 April 2023

1. Introduction

In a rapidly growing population, various pesticides are used to increase overall agricultural production [1]. Because pesticides tend to remain in harvested crops and these residues can harm humans, many countries use a variety of methods to detect pesticide residues [2]. Lists of pesticides that require testing in accordance with the pharmacopoeia of each country and the national food safety standards have been established [3,4]. Moreover, there has been a growing interest in analyzing more pesticides and developing new pretreatment methods [5,6].

Herbal medicines are grown as agricultural products before being washed, cut, dried, and packaged in facilities with good manufacturing practices; therefore, they are exposed to various pesticides [7]. In Korea, Monograph Part 2 of the Korean Pharmacopoeia presents individual standard specifications for herbal medicines and preparations [8]. Standards for pesticide residues, including α-BHC, β-BHC, δ-BHC, γ-BHC, aldrin, dieldrin, endrin, P.P′-DDD, P.P′-DDE, O.P′-DDT, and P.P′-DDT, ranging from 11 to 31, are set and presented depending on the product. The test method involves using gas chromatography equipped with an electron capture detector, nitrogen-phosphorus detector, etc., or high-performance liquid chromatography (HPLC) after pretreatment.

However, the number of pesticides registered every year for the cultivation of agricultural products is increasing [9] because herbal medicines are imported from various countries, and it is becoming difficult to secure the safety of herbal medicines using only the herbal medicine test method No. 30 and the standards in the Korean Pharmacopoeia. In particular, in the case of herbal decoctions, since patients take them within approximately 24 h after dispensing, test results should be obtained as quickly as possible [10]. However, various types of herbal medicines are used according to the prescriptions, and the procedure to analyze pesticide residues in the Korean Pharmacopoeia is so complicated that it takes approximately 2–7 days to determine the results [11]. Therefore, developing an effective analysis method that can shorten the test time and increase the accuracy is necessary.

In Korea, the QuEChERS sampling method has been applied to the preprocessing stage in the food sector for a long time, and many pesticide multicomponent analysis methods have been developed using HPLC, gas chromatography (GC), gas chromatography tandem mass spectrometry (GC-MS/MS), and liquid chromatography tandem mass spectrometry (LC-MS/MS) [12]. Representatively, there are multiclass pesticide multiresidue methods [13] in the Korea Food Code and methods of analyzing harmful substances, such as the agricultural products of the Agricultural Products Quality Management Service [14]. In this study, the pretreatment process was supplemented with a single test method to overcome the limitations of human resources and time, in accordance with the limitations of the analysis of pesticide components in the existing analysis methods (multiclass pesticide multiresidue methods), the limitations of the diversity of pretreatment methods, and instrumental analysis. The specificity, linearity, accuracy, detection limit, and quantification limit of the pesticide analysis methods LC-MS/MS and GC-MS/MS were verified, and the pesticide detection results of 40 prepared decoctions using the test methods are presented.

2. Results

2.1. Test Analysis Method Verification Result

2.1.1. Specificity

GC-MS/MS Target Pesticide

The specificity was confirmed under multiple reaction monitoring (MRM) conditions for 113 standard products for GC-MS/MS analysis to confirm the selectivity of the measured analytes without interference from other components (Figure 1).

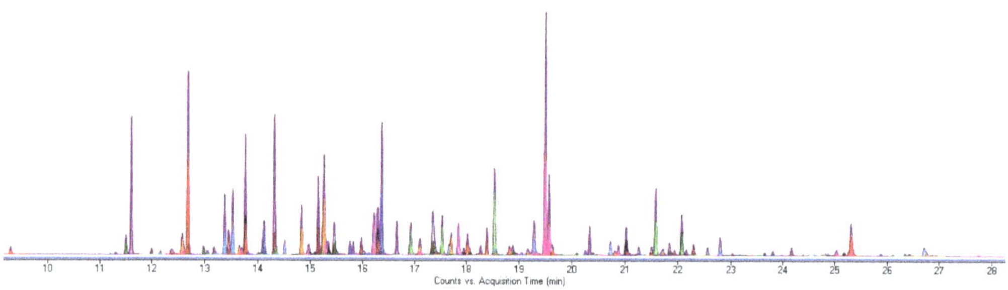

Figure 1. GC-MS/MS standard solution chromatogram. GC: gas chromatography; MS: mass spectrometry.

LC-MS/MS Target Pesticide

To confirm the selectivity to measure analytes without interference from other components, specificity was confirmed under MRM conditions for 207 standard products for LC-MS/MS analysis (Figure 2).

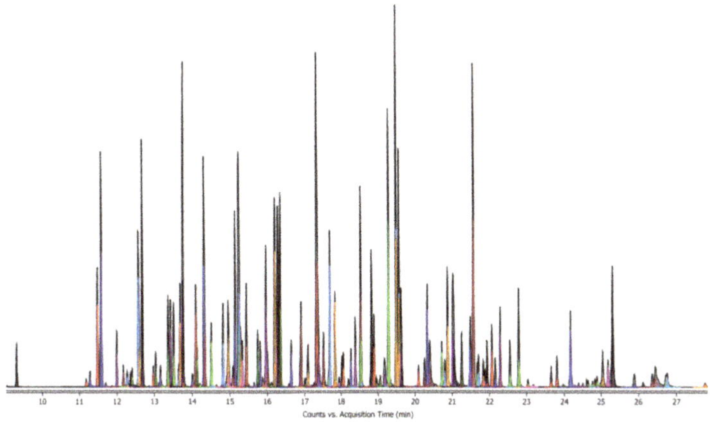

Figure 2. LC-MS/MS standard solution chromatogram. LC: liquid chromatography; MS: mass spectrometry.

2.1.2. Linearity

The linearity between instrument signals according to pesticide concentration was evaluated in the dilution range of 1−200 μg/kg of the calibration curve standard solution. As a result of examining the linearity of 113 diluted pesticide mixture standard solutions for GC-MS/MS analysis in 5 stages at concentrations of 5, 10, 20, 100, and 200 μg/kg, most showed good linearity with $R^2 \geq 0.99$. As a result of examining the linearity of the 207 diluted standard pesticide mixtures for LC-MS/MS analysis at concentrations of 1, 5, 10, 100, and 200 μg/kg in 5 stages, all showed good linearity with $R^2 \geq 0.99$ (Supplementary File S1, Figures S1 and S2).

2.1.3. Accuracy

Accuracy, the degree of agreement between the measurement result and the standard value, was measured by recovery. The accuracy of the simultaneous analysis of pesticide residues in the decoction was confirmed by adding a standard solution to Galgeun-tang. For the recovery rate test, standard solutions were added to Galgeun-tang at concentrations of 10, 50, and 100 μg/kg, extracted using the above pretreatment method, and the test solution was then analyzed using LC-MS/MS and GC-MS/MS. The recovery rate results are shown in Table S3 and S4 (See the Supplementary File S2). Galgeun-tang is the herbal decoction most commonly prescribed to patients in traditional Korean medicine (TKM) and was selected after expert consultation to measure accuracy [15]. Herbal medicines (*Pueraria lobata* Ohwi, *Cinnamomum cassia* Presl, *Ephedra sinica* Stapf, *Paeonia lactiflora* Pallas, *Glycyrrhiza uralensis* Fischer, *Zingiber officinale* Roscoe, and *Zizyphus jujuba* Miller) that make up Galgeun-tang are all manufactured in an hGMP manufacturing facility licensed by the Korea Ministry of Food and Drug Safety [16,17].

The recovery rates of the 207 pesticide components subjected to LC-MS/MS analysis were 65−161% in the case of low concentrations; 201 pesticides with standard deviations within 15% consisted of 201 species; and 201 components were qualitative. Six types of pesticides, cyazofamid, cyflufenamid, fenoxaprop-ethyl, gibberellic acid, propaquizafop, and pyroquilon, were outside the recovery rate of 70−125%. The recovery rates at high concentrations were 79−119%, and the standard deviation was within 15% for the 207 pesticides, all of which could be analyzed at high concentrations.

The recovery rates of the 113 pesticide components analyzed by GC-MS/MS were 90−1349% at low concentrations. There were 111 pesticides whose recovery rates were 70−125%, with standard deviations within 15%, and there were 111 types of ingredients that could be quantified. The recovery rate was higher than 125% for two types of pesticides (prochloraz and indanofan).

The recovery rates at high concentrations were 97–189%. The recovery rates of 112 pesticides were 70–125%; the standard deviation was within 15%, and 112 components were available for qualitative treatment. Indanofan had a recovery rate of 189%, and the standard deviation was 20%, making it qualitatively difficult.

2.1.4. Limit of Detection and Quantification

The detection and quantification limits were calculated using the Mass Hunter program (version 11.2; Agilent Technologies, Santa Clara, CA, USA) after repeated measurements of the lowest concentration seven times based on the calibration curve prepared to confirm linearity. It was calculated using the slope of the calibration curve and the deviation of repeated measurements. The limit of detection (LOD) of each pesticide was 0.05 to 5.0 μg/kg in the LC-MS/MS analysis and 0.2 to 7.0 μg/kg in the GC-MS/MS analysis. The limit of quantification was in the range of 0.15 to 15 μg/kg in the LC-MS/MS analysis and 0.6 to 20 μg/kg in the GC-MS/MS, and trace amounts of pesticide components could be detected at the level of 10 μg/kg contained in the sample.

2.2. Pesticide Residues in Analyzed Samples

As a result of the analysis of 40 herbal decoctions prepared in the outpatient bathroom, no pesticides other than carbendazim were detected. Carbendazim was detected in the range of 0.01 to 0.03 μg/g in 8 samples of 40 herbal decoctions.

3. Discussion

The existing method (QuEChERS sample pretreatment method) is a test method for solids such as food and agricultural products, and there is a difference in the detection concentration depending on the volume of solids and the type of sample. Therefore, most tests using the existing method are used only as monitoring test methods (detection checks), and the quantitative results at the time of detection are obtained by conducting individual experiments for each pesticide component. However, it was confirmed that this test method is a stable method with accuracy and precision that can be widely applied without large deviations for liquid samples with certain properties, such as the decoction obtained by first hot water extraction of the sample. This study presented an improved test method that can be applied to liquid types with similar properties to decoction samples and that is valuable in that accuracy and precision were verified through validation.

The possibility of simultaneous multicomponent analysis of 320 pesticides using LC-MS/MS and GC-MS/MS was investigated to analyze the pesticide residues in herbal decoctions. The validity of the test method was verified by applying a preprocessing method modified from the existing QuEChERS method.

In AOAC and Codex, the suitability of the analysis method used in a study is judged by a 70–125% recovery rate and a 15% relative standard deviation. The recovery rate of the analytical method using LC-MS/MS and GC-MS/MS was at least 70%, and the relative standard deviation was less than 15%, meeting international standards, while eight pesticides at 10 μg/kg concentration, four pesticides at 50 μg/kg concentration, and one pesticide at 100 μg/kg concentration were excluded.

Therefore, the analysis method applied in this study is considered applicable to the analysis of multicomponent pesticides remaining in herbal decoctions.

Among the 320 species to be analyzed, it is considered necessary to apply and develop additional test methods to improve the preprocessing methods and increase the efficiency of the analysis of indanofan components that cannot be quantified beyond the range of recovery rates [18].

The multicomponent analysis method for various pesticides using LC-MS/MS and GC-MS/MS reviewed in this study is expected to be applicable for monitoring herbal decoctions.

Currently, no distinct regulations exist for managing harmful substances, including residual pesticides, in the herbal decoctions utilized by TKM clinics. This is due to the fact that such decoctions are not classified as products approved by the MFDS but rather as traditional medicines prepared through herb decoction at TKM clinics. Nonetheless, the Herbal Medicine Test Method of the Korean Pharmacopoeia applies specific residue limits for various pesticides in herbal medicines and extracts, including total DDT (sum of p,p′-DDD, p,p′-DDE, o,p′-DDT, and p,p′-DDT) at 0.1 ppm or less, dieldrin at 0.01 ppm or less, total BHC (the sum of α, β, γ, and δ-BHC) at 0.2 ppm or less, aldrin at 0.01 ppm or less, and endrin at 0.01 ppm or less [8]. Furthermore, the Korean Pharmacopoeia applies an interim standard of 0.01 ppm or less for pesticides that lack established standards in the herbal medicine test method in processed foods [8]. Although this study's validation of the test method demonstrated that all LODs were below 0.01 ppm, the LOQs for five pesticides, including bifenox, exceeded 0.01 ppm, which is higher than the residual pesticide limits set by the Korea MFDS for herbal extracts and processed foods.

Carbendazim is commonly used to control fungal diseases in vegetables and fruits [19,20]. It was used worldwide before toxicological evidence was detected, and the U.S. Environmental Protection Agency canceled the registration. Although it has carcinogenicity and reproductive toxicity and can damage organs, such as the liver, it is still used in agriculture in some countries. China is a representative country that uses carbendazim in agriculture. Since most of the herbal medicines imported to Korea are from China, decoctions made from herbal medicines may be contaminated with carbendazim [21,22].

According to the herbal medicine standard of Korea, carbendazim should be no more than 0.05 mg/kg in peony, 2.0 mg/kg in goji berry, 4.0 mg/kg in jujube, 10 mg/kg in Schisandra chinensis, 0.05 mg/kg in hemp, 2.0 mg/kg in raspberry, 0.5 mg/kg in ginseng, and 0.05 mg/kg in ginger. For agricultural products, the standard specifications are set for 123 items, e.g., eggplant, tangerine, potato, and mustard, and the permissible standard is set differently, from low to high concentration, depending on the item: less than 0.03 mg/kg of potato and less than 50 mg/kg of kale.

In this study, we detected 0.01 and 0.03 mg/kg in eight samples. According to a previous study, in the case of mushrooms, washing, drying, and heating can reduce carbendazim residues. In particular, there were no carbendazim residues after the boiling process [23]. Herbal medicines are also submitted to washing and drying processes, and herbal decoctions are the result of boiling herbal medicines; therefore, the carbendazim concentration should be detected at very low levels. We compared previous studies conducted in other countries because there are no pesticide residue standards for herbal decoctions in Korea.

Fan et al., analyzed 77 Fragaria and 74 Myrica rubra sold in Hangzhou, China. They detected prochloraz and carbendazim mostly with detection rates of 71.6% and 68.9% in Myrica rubra, and the mean concentration of carbendazim was 0.149 mg/kg (0.0110−1.02 mg/kg) [24]. In Zhang's et al. [25] study, 99 Chuanxiong Rhizoma samples were analyzed. As a result, carbendazim and prometryn were the pesticides the most frequently detected, with a 100% detection rate. Carbendazim was found at 38.92 ± 83.68 μg/kg (0.38−343.55 μg/kg), which exceeded the Chinese Pharmacopoeia standard by 20 times [25]. In the study by Xiao, carbendazim was the most widely used pesticide (>85%) [26]. Since our study focused on herbal decoction and comparative studies targeted herbal medicine, which is the raw material for herbal decoction, it was not possible to individually compare the amount and the ratio of detection. However, it cannot be said that the level of pesticide residue contamination in herbal decoction is high.

This study has several limitations. The collection of 4 frequently used prescriptions in 10 external herbal dispensaries (EHD) was a valuable component of this study. However, the prescription names and composition of 40 samples could not be included in the thesis as the EHDs refused to disclose this information for business reasons. Consequently, it was not possible to provide them as a supplementary file. It is important to note that in Korea, national surveys are regularly conducted on the use of herbal medicines in TKM clinics, and these surveys suggest the frequently used prescriptions. Disclosure of the name and composition of the prescription may be possible in future studies if a standard prescription is selected and a dispensing request is made.

4. Materials and Methods

4.1. Sample Collection

From March to April 2020, the research team collected 40 herbal decoctions in a portable refrigerator from the 10 EHDs and used them for the experiment. Herbal decoction pouches were collected for 4 prescriptions used frequently in 10 EHDs. An EHD is a type of pharmacy that provides various types of herbal medicines to other TKM institutions in Korea [27].

4.2. Standards and Reagents

The pesticide standards used in the analysis were purchased from AccuStandard® (New Haven, CT, USA) at a concentration of 1000 mg/L. Each pesticide standard was mixed and used, and acetonitrile (Merk, Rahway, NJ, USA) was used as a dilution solvent for each concentration. QuEChERS kits (mixed with anhydrous magnesium sulfate 4 g, sodium chloride 1 g, sodium citrate 1 g, disodium hydrogen citrate sesquihydrate 0.5 g, anhydrous magnesium sulfate 150 mg, and primary secondary amine 25 mg) were used with BEKOlut® (Bruchmühlbach-Miesau, Germany) for pretreatment. The solvents used in the analysis were acetonitrile (hyper grade for LC-MS, Merk, Rahway, NJ, USA), formic acid (for LC-MS 98–100%, Merk, USA), and ammonium acetate (for mass spectrometry, Sigma-Aldrich, St. Louis, MI, USA).

A total of 113 pesticides subjected to GC-MS/MS analysis were prepared by mixing and diluting a standard product prepared at a concentration of 1000 mg/kg to a concentration of 5 mg/kg and then diluted to 5, 10, 20, 100, and 200 µg/kg using acetonitrile for analysis. A total of 207 pesticides subjected to LC-MS/MS analysis were prepared by mixing and diluting a standard product prepared at a concentration of 1000 mg/kg to a concentration of 2 mg/kg and then diluted to 1, 5, 10, 100, and 200 µg/kg using acetonitrile for analysis. Because the standard solution for analysis shows a matrix-induced chromate graphic response enhancement effect, in which the response value of the standard solution of pesticides increases as a result of the matrix, the extract of the nonpesticide decoction was mixed 1:1 with the standard solution for analysis.

4.3. Pretreatment of Samples

The pretreatment method for analyzing residual pesticides in decoction was as follows:

(1) Precisely add 10 mL of decoction to a 50 mL centrifuge tube along with 10 mL of acetonitrile containing the internal standard (0.1 mg/kg triphenylphosphate), shake it, and extract it for 1 to 2 min.

(2) Add 4 g of MgSO4, 1 g of NaCl, 1 g of trisodium citrate dihydrate, and 0.5 g of disodium hydrogen citrate sesquihydrate to the centrifuge tube of (1) and shake it for 1 min. After centrifugation (3000 rpm/min, 5 min) to separate the acetonitrile layer and the water layer, mix the acetonitrile layer with the buffer solution, filter it through a PTFE filter (0.2 µm), and analyze it by LC-MS/MS.

(3) In the case of GC-MS/MS, add 1 mL of acetonitrile extract from (2) to a powdered solid phase extraction tube containing 150 mg of $MgSO_4$ and 25 mg of PSA, shake it for 1 min, and centrifuge it (10,000 rpm/min, 2 min). Then, analyze it by GC-MS/MS.

In this study, the possibility of a multicomponent simultaneous analysis method using LC-MS/MS and GC-MS/MS for pesticide residue analysis in a decoction was confirmed by applying a modified pretreatment method of the QuEChERS sample pretreatment method (Figure 3).

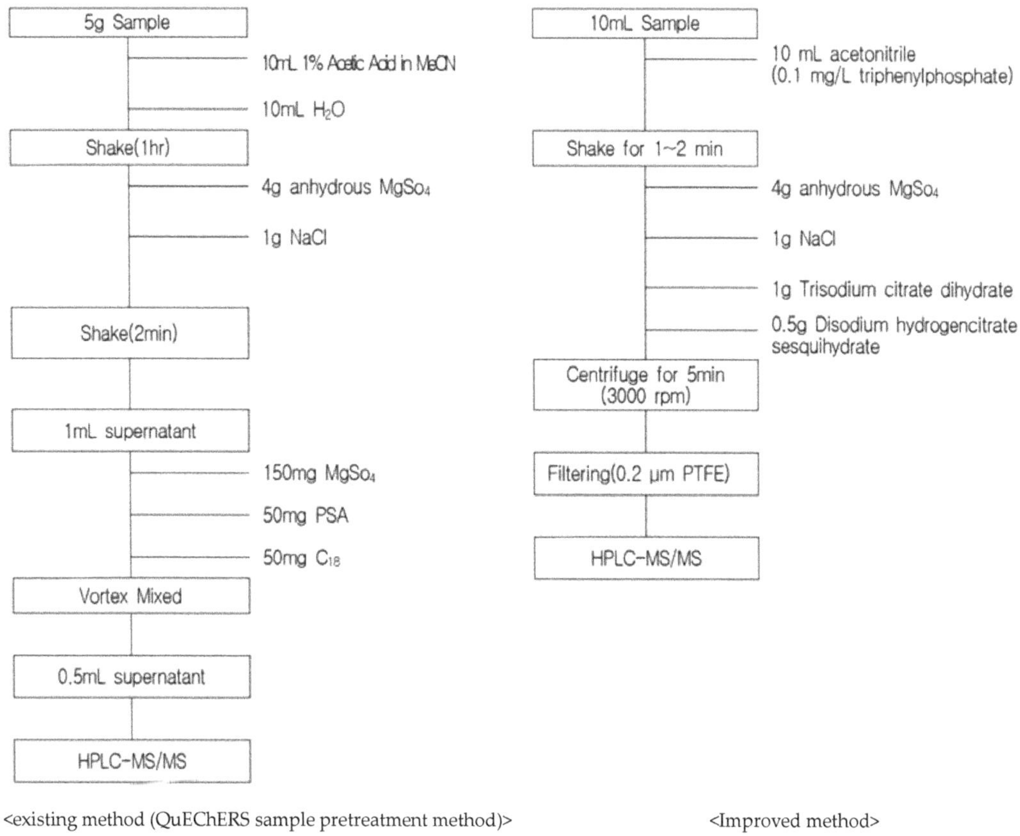

Figure 3. Pretreatment method for analyzing residual pesticides of herbal decoctions. PTFE: polytetrafluoroethylene; GC: gas chromatography; LC: liquid chromatography; MS: mass spectrometry.

Analysis of Instrument and Instrument Conditions

A total of 320 pesticide residue analysis methods of decoction were extracted and purified using salt-containing acetonitrile/powder phase solid phase extraction (QuEChERS); 207 species were analyzed by LC-MS/MS, and 113 species were analyzed by GC-MS/MS.

The MRM conditions for the 207 pesticides subjected to LC-MS/MS analysis and the 113 pesticides subjected to GC-MS/MS analysis are provided in Supplementary File S2 (Tables S1 and S2). The slope and correlation (R^2) of the analytical pesticide standard solution calibration curve of LC-MS/MS and GC-MS/MS are presented in Supplementary File S2 (Tables S1 and S2).

4.4. Validation of the Test Method

The validity of the analysis method was verified using specificity, linearity, accuracy, precision, detection limit, and quantification limit.

Accuracy was confirmed by a recovery experiment for the standard solution, and the standard treatment concentrations were 10, 50, and 100 μg/kg, including the quantification limit for each pesticide component. the results were confirmed after three repetitions.

The detection and quantification limits were calculated using an analytical instrument program after 7 repeated measurements of 10 μg/kg concentration based on a calibration curve prepared at concentration levels of 1 to 200 μg/kg.

5. Conclusions

A total of 320 pesticide residues in 40 decoctions were analyzed using an improved LC-MS/MS and GC-MS/MS analysis. As a result of the monitoring, carbendazim was detected at 0.01 and 0.03 μg/g in eight samples, and no pesticide was detected in the other herbal decoctions. In addition, this study verified the specificity, linearity, accuracy, detection limit, and quantification limit of pesticides using improved LC-MS/MS and GC-MS/MS analysis methods compared with existing methods (multiclass pesticide multiresidue methods). Therefore, our results provide a framework for pesticide residue management in countries with traditional medicines.

Supplementary Materials: The following supporting information can be downloaded at https://www.mdpi.com/article/10.3390/molecules28083343/s1. Supplementary File S1: GC-MS/MS and LC-MS/MS standard solution calibration curve, Supplementary File S2: GC-MS/MS and LC-MS/MS analysis pesticide standard solution slope and correlation (R^2).

Author Contributions: Conceptualization, S.-M.K., H.-I.J. and S.-H.S.; methodology, H.-I.J., J.-H.K., K.-H.K. and S.-H.S.; software, S.-M.K. and J.-H.W.; validation, S.-M.K. and J.-H.W.; formal analysis, S.-M.K. and J.-H.W.; investigation, S.-M.K. and J.-H.W.; resources, S.-M.K. and J.-H.W.; data curation, S.-M.K. and J.-H.W.; writing—original draft preparation, S.-M.K., H.-I.J., J.-E.H. and S.-H.S.; writing—review and editing, J.-H.K. and K.-H.K.; visualization, H.-I.J.; supervision, J.-H.W.; project administration, J.-H.W. and S.-H.S.; funding acquisition, S.-H.S. All authors have read and agreed to the published version of the manuscript.

Funding: This research was supported by a grant from the Project of National Development Institute of Korean Medicine, Accreditation of External Herbal Dispensaries of Traditional Korean Medicine Clinics, funded by the Korean Ministry of Health and Welfare.

Institutional Review Board Statement: Not applicable.

Informed Consent Statement: Not applicable.

Data Availability Statement: The data will be made available upon reasonable request.

Acknowledgments: The authors thank traditional Korean medicine clinics for providing the herbal decoction samples.

Conflicts of Interest: The authors declare no conflict of interest.

Abbreviations

AOAC: Association of Official Agricultural Chemists, EHD: external herbal dispensary, EPRW: European Pesticide Residue Workshop, GC: gas chromatography, HPLC: high-performance liquid chromatography, LC: liquid chromatography, LOD: limit of detection, LOQ: limit of quantification, MRM: multiple reaction monitoring, MS: mass spectrometry, QuEChERS: quick, easy, cheap, effective, rugged, and safe, TKM: traditional Korean medicine, U.S.: United States.

References

1. Nicolopoulou-Stamati, P.; Maipas, S.; Kotampasi, C.; Stamatis, P.; Hens, L. Chemical Pesticides and Human Health: The Urgent Need for a New Concept in Agriculture. *Front. Public Health* **2016**, *4*, 148. [CrossRef] [PubMed]
2. Ntzani, E.E.; Ntritsos, G.; Chondrogiorgi, M.; Evangelou, E.; Tzoulaki, I. Literature Review on Epidemiological Studies Linking Exposure to Pesticides and Health Effects. *EFSA Support. Publ.* **2013**, *10*, 497E. [CrossRef]
3. Wang, Y.; Han, J.; Zhang, J.; Li, X.; Bai, R.; Hu, F. A Monitoring Survey and Health Risk Assessment for Pesticide Residues on Codonopsis Radix in China. *Sci. Rep.* **2022**, *12*, 8133. [CrossRef] [PubMed]
4. Xu, X.; Xu, H.; Shang, Y.; Zhu, R.; Hong, X.; Song, Z.; Yang, Z. Development of the General Chapters of the Chinese Pharmacopoeia 2020 Edition: A Review. *J. Pharm. Anal.* **2021**, *11*, 398–404. [CrossRef]
5. Li, M.; Sun, W.; Zhou, R.; Huang, Y.; Meng, F.; Luo, L.; Yan, Z.; Shi, X.; Huang, F.; Fan, C.; et al. Non-Pharmaceutical Treatments to Relieve Pain or Reduce Opioid Analgesic Intake and Improve Quality of Life after Total Hip Replacement: A Meta Analysis. *Am. J. Transl. Res.* **2022**, *14*, 6828–6845.
6. Li, S.; Yu, P.; Zhou, C.; Tong, L.; Li, D.; Yu, Z.; Zhao, Y. Analysis of Pesticide Residues in Commercially Available Chenpi Using a Modified QuEChERS Method and GC-MS/MS Determination. *J. Pharm. Anal.* **2020**, *10*, 60–69. [CrossRef]

7. Tudi, M.; Daniel Ruan, H.; Wang, L.; Lyu, J.; Sadler, R.; Connell, D.; Chu, C.; Phung, D.T. Agriculture Development, Pesticide Application and Its Impact on the Environment. *Int. J. Environ. Res. Public Health* **2021**, *18*, 1112. [CrossRef] [PubMed]
8. Ministry of Food and Drug Safety. *The Korean Pharmacopoeia*; Ministry of Food and Drug Safety: Seoul, Republic of Korea, 2022.
9. Carvalho, F.P. Pesticides, Environment, and Food Safety. *Food Energy Secur.* **2017**, *6*, 48–60. [CrossRef]
10. Luo, H.; Li, Q.; Flower, A.; Lewith, G.; Liu, J. Comparison of Effectiveness and Safety between Granules and Decoction of Chinese Herbal Medicine: A Systematic Review of Randomized Clinical Trials. *J. Ethnopharmacol.* **2012**, *140*, 555–567. [CrossRef]
11. Lu, X.-Y.; Ouyang, Y.-Q.; Zeng, W.-Y.; Lin, C.-Q.; Xiao, L.-H.; Luo, G.-H.; Zhan, R.-T.; Yan, P. Effect of Pretreatment on Detection of 37 Pesticide Residues in Chrysanthemum Indicum. *J. Anal. Methods Chem.* **2021**, *2021*, 8854025. [CrossRef]
12. Zahler, M.; Filippova, N.A.; Morel, P.C.; Gothe, R.; Rinder, H. Relationships between Species of the Rhipicephalus Sanguineus Group: A Molecular Approach. *J. Parasitol.* **1997**, *83*, 302–306. [CrossRef]
13. Ministry of Food and Drug Safety. Food and Food Additives Codex. Available online: https://www.foodsafetykorea.go.kr/foodcode/01_03.jsp?idx=404 (accessed on 29 January 2023).
14. Hazardous Substance Analysis Method for Agricultural Products, Etc. Available online: https://www.law.go.kr/%ED%96%89%EC%A0%95%EA%B7%9C%EC%B9%99/%EB%86%8D%EC%82%B0%EB%AC%BC%EB%93%B1%EC%9D%98%EC%9C%A0%ED%95%B4%EB%AC%BC%EC%A7%88%EB%B6%84%EC%84%9D%EB%B2%95/(2021-27,20210325) (accessed on 29 January 2023).
15. Lee, J.; Kim, J.; Shin, H. Systematic Studies on Galgeun-Tang(Gegen-Tang, Kakkon-to) for Establishment of Evidence Based Medicine. *Herb. Formula Sci.* **2011**, *19*, 103–111.
16. Ministry of Health and Welfare. The 2017 Korean Medicine Utilization and Herbal Medicine Consumption Survey. Available online: https://nikom.or.kr/koms/index.do (accessed on 16 March 2023).
17. Ministry of Food and Drug Safety. *Ministry of Food and Drug Safety White Paper 2015*; Ministry of Food and Drug Safety: Seoul, Republic of Korea, 2016.
18. Indanofan | C20H17ClO3—PubChem. Available online: https://pubchem.ncbi.nlm.nih.gov/compound/Indanofan (accessed on 29 January 2023).
19. Mazellier, P.; Leroy, É.; Legube, B. Photochemical Behavior of the Fungicide Carbendazim in Dilute Aqueous Solution. *J. Photochem. Photobiol. A Chem.* **2002**, *153*, 221–227. [CrossRef]
20. McCarroll, N. A Survey of EPA/OPP and Open Literature on Selected Pesticide Chemicals III. Mutagenicity and Carcinogenicity of Benomyl and Carbendazim. *Mutat. Res./Rev. Mutat. Res.* **2002**, *512*, 1–35. [CrossRef] [PubMed]
21. Sharma, M.; Maheshwari, N.; Khan, F.H.; Mahmood, R. Carbendazim Toxicity in Different Cell Lines and Mammalian Tissues. *J. Biochem. Mol. Toxicol.* **2022**, *36*, e23194. [CrossRef]
22. Zhang, X.; Huang, Y.; Harvey, P.R.; Li, H.; Ren, Y.; Li, J.; Wang, J.; Yang, H. Isolation and Characterization of Carbendazim-Degrading Rhodococcus Erythropolis Djl-11. *PLoS ONE* **2013**, *8*, e74810. [CrossRef]
23. Xia, F.; Tao, W.; Yao, X.; Wang, J.; Tang, F. Effects of Processing on Carbendazim Residue in *Pleurotus ostreatus*. *Food Sci. Nutr.* **2016**, *4*, 645–650. [CrossRef]
24. Ji-Cai, F.; Hua-Li, H.; Shao-Ying, L.; Ren, R.; Shu-Ting, W. Investigation of Pesticide Residues in Fragaria and Myrica Rubra Sold in Hangzhou. *J. Food Prot.* **2022**, *85*, 534–538. [CrossRef]
25. Zhang, D.; Gou, Y.; Yu, X.; Wang, M.; Yu, W.; Zhou, J.; Liu, W.; Li, M. Detection and Risk Assessments of Multi-Pesticides in Traditional Chinese Medicine Chuanxiong Rhizoma by LC/MS-MS and GC/MS-MS. *Molecules* **2022**, *27*, 622. [CrossRef]
26. Xiao, J.; Xu, X.; Wang, F.; Ma, J.; Liao, M.; Shi, Y.; Fang, Q.; Cao, H. Analysis of Exposure to Pesticide Residues from Traditional Chinese Medicine. *J. Hazard. Mater.* **2019**, *365*, 857–867. [CrossRef]
27. Sung, S.H.; Han, J.E.; Ryu, J.Y.; Sung, A.D.M.; Park, J.Y.; Ha, I.H.; Kim, K.H.; Park, J.K.; Shin, B.C. Current Status and Future Perspective of External Herbal Dispensaries Preparing Traditional Herbal Medicine in South Korea: The First National-Wide Survey Results. *BMC Complement. Med. Ther.* **2020**, *20*, 354. [CrossRef] [PubMed]

molecules

MDPI

Article

Pesticide Residues in Mandarins: Three-Year Monitoring Results

Emrah Gormez [1], Ozgur Golge [2], Miguel Ángel González-Curbelo [3] and Bulent Kabak [4,5,*]

1 Pia Frucht Food Control Laboratory, Alaşehir 45600, Turkey
2 Department of Gastronomy and Culinary Arts, Faculty of Tourism, Alanya Alaaddin Keykubat University, Alanya 07425, Turkey
3 Departamento de Ciencias Básicas, Facultad de Ingeniería, Universidad EAN, Calle 79 n° 11-45, Bogotá 110221, Colombia
4 Department of Food Engineering, Faculty of Engineering, Hitit University, Corum 19030, Turkey
5 Biotechnology Laboratory, Machinery and Manufacturing Technology Application and Research Center, Hitit University, Corum 19030, Turkey
* Correspondence: bulentkabak@hitit.edu.tr

Abstract: The demand of plant production product use has increased because of the current system of citrus production, which prioritizes high agricultural yields. Therefore, the monitoring of pesticide residues in citrus fruits and other agricultural products and their impacts on human health and food security are of great concern. This study aims to determine multi-class pesticides including highly polar residues in satsuma mandarins. A total of 226 mandarin samples were collected over three consecutive harvesting years from 2019 to 2021 in the Izmir region of Turkey. Targeted compounds included pesticides and metabolites with European Union (EU) regulatory levels, plus other non-approved residues and highly polar compounds. The residues excluding highly polar substances were analyzed by applying the quick, easy, cheap, effective, rugged, and safe (QuEChERS) extraction and liquid chromatography–triple quadrupole mass spectrometry (LC-MS/MS) determination for 434 analytes and gas chromatography–triple quadrupole mass spectrometry (GC-MS/MS) determination for 71 analytes. For six highly polar pesticides, sample preparation was based on Quick Polar Pesticides (QuPPe) extraction. The polar residues were determined by LC-MS/MS using internal standards. Forty different residues, including two highly polar substances, were recorded in mandarin samples through three harvesting years. In 8.4% of the samples, no quantifiable residues were detected, whereas 207 samples contained at least one residue. The maximum residue level (MRL) exceedances were recorded for 22.1% of the samples. The two most frequently found pesticides were phosphonic acid and spirotetramat, with an incidence rate of 48.7% and 46.5%, respectively. The concentration of phosphonic acid and spirotetramat in mandarin samples varied from 0.026 to 39.386 mg kg^{-1} and from 0.010 to 1.485 mg kg^{-1}, respectively. The results will enable researchers and regulatory authorities to assess the extent of pesticide presence, identify potential risks, and take necessary measures to ensure the safety of satsuma mandarins for consumers.

Keywords: chromatography; food safety; mass spectrometry; pesticides; polar pesticides; QuEChERS; QuPPe

Citation: Gormez, E.; Golge, O.; González-Curbelo, M.Á.; Kabak, B. Pesticide Residues in Mandarins: Three-Year Monitoring Results. *Molecules* **2023**, *28*, 5611. https://doi.org/10.3390/molecules28145611

Academic Editor: Jesus Simal-Gandara

Received: 13 July 2023
Revised: 20 July 2023
Accepted: 21 July 2023
Published: 24 July 2023

1. Introduction

Mandarins, also known as tangerines in some parts of the world, are the second most commonly cultivated citrus type, with 38 million tons (22.4% of global citrus production), after oranges but ahead of lemons and grapefruit. In the 2020 season, the global mandarin crop totaled over 38 million tons. China is the global supply leader with over 23 million metric tons, accounting for more than 60% of the world's mandarin crops in 2020/2021. Spain and Turkey came in second and third, with 5.6% and 4.1% of the global market share, respectively. In 2020, Spain was the top exporter of mandarins with over 1.3 million tons, accounting for 23.4% of the global exports. Turkey and China were the second- and third-largest exporters of mandarins, accounting for 15% and 12.8% of the global market, respectively. In 2020, the Russian Federation was the leading importer of mandarins in the world, with a 16.9% share of global imports, followed by the United States (7.4%), Germany (7.3%), France (6.7%), and the United Kingdom (5.9%) [1].

Carbohydrates, mainly sucrose, glucose, and fructose, and dietary fiber are the principal macronutrients in mandarins. They are also a well-known source of many valuable substances such as organic acids (mainly citric and malic acids), carotenoids (β-cryptoxanthin), polyphenols (flavonoids and phenolic acids), vitamin C, and minerals (mainly potassium) [2]. When compared to other citrus fruits such as lemons, oranges, or grapefruits, mandarins are generally not suited to long-term storage.

Many different pathogens, including insect pests such as the Mediterranean fruit fly (*Ceratitis capitata*), affect mandarins and other citrus fruits, causing diseases with adverse effects in orchards worldwide. Mandarins are susceptible to various citrus diseases, including citrus blast caused by the bacterial pathogen *Pseudomonas syringae*, citrus canker caused by *Xanthomonas* spp., anthracnose (*Colletotrichum* spp.), green mold caused by *Penicillium digitatum*, blue mold caused by *Penicillium italicum*, collar rot caused by *Phytophthora citrophthora*, sour rot caused by *Geotrichum citri-aurantii*, *Alternaria* brown spot caused by *Alternaria* spp., gray mold caused by *Botrytis cinerea*, and *Mucor* rot caused by *Mucor piriformis* [3,4].

In spite of the rising consumer resistance to the presence of chemical residues on products, the utilization of pesticides remains the prevailing practice for the prevention of pre-harvest and post-harvest infestations. Very small amounts of pesticides called residues may remain in or on fruits and vegetables and might pose a potential risk to human health due to their sub-acute and long-term toxicity. For this reason, it is very important to control and regulate pesticide use in agricultural production and to monitor their levels in fruits and vegetables [5]. National and international organizations establish a maximum residue level (MRL) for each agricultural and other product, aiming to establish benchmarks for food safety and promote global trade. In Turkey, the Ministry of Agriculture and Forestry bears the responsibility of assessing the permissible levels of pesticide residues in agricultural and other products [6], adhering to the regulations set forth by European Union (EU) legislation [7], to ascertain the levels of residues.

Pesticides are widely used in fruit growing and in the treatment of citrus fruits for pre-harvest and post-harvest protection by the citrus farmers in Turkey and around the world. The Commission's Rapid Alert System for Food and Feed (RASFF) shows that in 2022, a total of 293 notifications on fruits and vegetables from Turkey were transmitted through the system, 23 of which (7.85%) concerning mandarins [8]. Following an increase in the number of interceptions of Turkish citrus fruits that do not meet requirements on pesticide residues, the European Commission has decided to temporarily increase by 20% the frequency of physical checks on citrus fruits, including mandarin and clementine imports from Turkey [9].

Methods used for the analysis of pesticides vary widely. However, liquid chromatography and gas chromatography coupled with triple quadrupole mass spectrometry (LC-MS/MS and GC-MS/MS) methods are the most powerful techniques in the determination of pesticides [10–13]. While there are many sample-extraction methods including solid-phase extraction [14,15], solid-phase microextraction [16], liquid–liquid extraction [17], liquid-phase microextraction [18], pressurized liquid extraction [12], accelerated solvent extraction [19], ultrasonic solvent extraction [12], supercritical fluid extraction [20,21], ultrasonic solvent extraction [12], matrix solid-phase dispersion [22,23], and microwave-assisted extraction [24], the quick, easy, cheap, effective, rugged, and safe (QuEChERS) extraction method developed by Anastassiades et al. [25] has increasingly being used in combination with LC-MS/MS and/or GC-MS/MS for the detection of multi-class residues in agricultural and other products. QuEChERS has gained significant popularity in pesticide residue analysis due to its simplicity, time saving, cost-effectiveness, high throughput, and minimal solvent requirement. In the QuEChERS extraction method, the process comprises two steps: extraction and clean-up. In the first step, the residues are extracted from the matrix with acidified acetonitrile and salts/buffers. To reduce interferences, sugars, fatty acids, organic acids, lipids, and polar pigments are removed in the clean-up step by the use of primary–secondary amine (PSA). However, highly polar pesticides have been excluded for a long time from the routine scope of laboratory investigations because they are not amenable to extraction via QuEChERS. Recently, the Quick Polar Pesticides (QuPPe) extraction method for the simultaneous analysis of highly polar substances has been developed by the EU Reference Laboratory for Pesticides Single Residue Methods (EURL-SRM). With this technique, many polar substances are extracted with acidified methanol from the various matrices without a sample clean-up process [26].

The main purpose of this study was to monitor the residual concentration of pesticides in Turkish satsuma mandarins (*Citrus unshiu* Marcovitch) intended for export during three harvesting years. The methodologies involved the QuEChERS and QuPPe sample preparation approaches for the determination of 505 non-polar/medium-polar and six highly polar residues, respectively. The QuEChERS extracts were analyzed by LC-MS/MS and GC-MS/MS, whereas only LC-MS/MS in electrospray negative ionization (ESI-) mode was used in the determination stage for QuPPe extracts.

2. Results and Discussion

2.1. Method Validation Data

In-house method validation involving LC-MS/MS and GC-MS/MS was conducted to establish method performance characteristics for the detection and quantification of target compounds in the matrix of high acid content and high water content. The method validation data for the detected residues are shown in Table S1. The LC-MS/MS and GC-MS/MS methods demonstrated satisfactory selectivity. No visible interfering peaks were evident at or close to the expected retention times of the target analytes. Linearity of response was acceptable (coefficient of determination (R^2) > 0.99) for the majority of the compounds (except for five residues: dinobuton, hexachlorobenzene, tefluthrin, tralkoxydim, and vinclozolin) including highly polar substances. Method limits of quantification (LOQ) for all target compounds were lower than 0.01 mg kg^{-1}. All recovery values are compliant with provisions set in SANTE 11312/2021 Guideline [27], which recommends a recovery rate of 70–120%. A recovery range of 73.9 to 113.5% was observed after a spiking matrix with detected analytes at 0.01 mg kg^{-1}. After extraction of the higher level compound-spiked matrix, recoveries of detected residues fell within the range of 86.5 to 109.9%. For the blank matrix spiked with residues at 0.01 and 0.05 mg kg^{-1}, the repeatability (RSD_r, %) was found to be from 0.20 to 16.31% and from 0.27 to 8.41%, respectively, for detected residues. The within-laboratory method reproducibility (RSD_R) data were found to be in the range of 0.62 to 13.56% and 1.73 to 6.65% at 0.01 and 0.05 mg kg^{-1} spiking levels, respectively. The measurement uncertainties for detected residues were between 8.5 and 42.8%.

2.2. Pesticide Residues in Mandarins

Between 2019 and 2021, a total of 226 mandarin samples cultivated in the Izmir region, Turkey, were monitored for the presence of 511 pesticide residues. Monitoring pesticide residues over multiple years provides a more robust and representative dataset, as it helps account for potential variations in pesticide use and fruit quality across different harvest seasons. Agricultural practices, including pesticide application, can vary from year to year based on factors such as weather conditions, pest pressure, and farmer practices. By sampling satsuma mandarins over three years, the study can better capture the overall trend and consistency of pesticide contamination in the region.

At least one pesticide residue was detected in 91.6% of the analyzed samples, while no pesticide residue was found in 19 samples. Forty different pesticides were detected in the mandarin samples, including 23 insecticides, 14 fungicides, two acaricides, and one insect growth regulator. Among the 40 active substances recorded in mandarin samples, 12 of them were non-approved in the EU. Only fosetyl and phosphonic acid were detected in mandarins among the six highly polar residues (chlorate, ethephon, fosetyl, glyphosate, perchlorate, and phosphonic acid).

In 2019, 29 mandarin samples were analyzed. Table 1 shows the quantified pesticide residues and their concentrations. Only three samples (10.3%) were free of quantifiable residues. In total, 89.7% of mandarin samples contained at least one pesticide residue, but only two of them (6.9%) exceeded the MRL. These two exceedances were related to residues of buprofezin and propiconazole.

Table 1. The presence and quantification of pesticide residues in mandarins in 2019.

Pesticide	Type of Residue	EU MRL (mg kg^{-1})	% of Samples <LOQ	% of Samples between LOQ-MRL	% of Samples >MRL	Range (mg kg^{-1})	
						Min.–Max.	Mean
Acetamiprid	IN	0.9	89.7	10.3	-	0.010–0.129	0.056
Buprofezin	IN	0.01	96.6	-	3.4	0.013	0.013
Carbendazim *	FU	0.7	96.6	3.4	-	0.038	0.038
Chlorpyrifos-methyl *	IN/AC	0.01	96.6	3.4	-	0.010	0.010
Deltamethrin	IN	0.04	96.6	3.4	-	0.015	0.015
Fludioxonil	FU	10	96.6	3.4	-	0.627	0.627
Fluopyram	FU	0.9	96.6	3.4	-	0.013	0.013
Fosetyl **	FU	150	93.1	6.9	-	0.006–0.351	0.179
Imazalil	FU	5	96.6	3.4	-	0.801	0.801
Imidacloprid *	IN	0.9	96.6	3.4	-	0.012	0.012
Lambda-cyhalothrin	IN	0.2	93.1	6.9	-	0.029–0.190	0.110
Malathion	IN	2	65.5	34.5	-	0.014–0.716	0.141
Phosmet *	IN/AC	0.5	96.6	3.4	-	0.014	0.014
Phosphonic acid **	FU	150	27.6	72.4	-	0.028–3.835	0.826
Pirimicarb	IN	3	96.6	3.4	-	0.037	0.037
Propiconazole *	FU	0.01	96.6	-	3.4	1.008	1.008
Pyrimethanil	FU	8	96.6	3.4	-	0.701	0.701
Pyriproxyfen	IN	0.6	96.6	3.4	-	0.028	0.028
Spirotetramat	IN	0.5	82.8	17.2	-	0.017–0.112	0.052
Sulfoxaflor	IN	0.8	82.8	17.2	-	0.010–0.015	0.012
Tau-fluvalinate	IN	0.4	89.7	10.3	-	0.063–0.228	0.135
Thiophanate-methyl *	FU	6	96.6	3.4	-	0.022	0.022

IN: insecticide; FU: fungicide; AC: acaricide; * not approved in the EU; ** sum of fosetyl, phosphonic acid, and their salts expresses as fosetyl.

In mandarin samples from 2019, 22 different residues were detected in quantifiable concentrations. While the majority of recorded residues (16 pesticides) relate to approved pesticides, six non-approved pesticides (carbendazim, chlorpyrifos-methyl, imidacloprid, phosmet, propiconazole, and thiophanate-methyl) were found in different mandarin samples. In total, 31% of mandarin samples contained only one residue, while multiple residues were quantified in 17 samples (58.7%); mandarin samples were recorded with up to seven different residues (Figure 1).

The most frequent residue detected in mandarins from 2019 was phosphonic acid, with a detection rate of 72.4%. The presence of this compound can be attributed to the utilization of fungicides such as fosetyl and phosphonic acid salts, as well as the prior application of growth enhancers. Notably, phosphonic acid is encompassed within the permissible MRL for fosetyl-aluminium (fosetyl-Al), considering the cumulative amount of fosetyl, phosphonic acids, and their respective salts, calculated as fosetyl [28]. The concentration of phosphonic acid varied from 0.028 to 3.835 mg kg^{-1} (0.038–5.49 mg kg^{-1} for fosetyl, sum), with a mean level of 0.826 mg kg^{-1} (1.12 mg kg^{-1} for fosetyl, sum). Only two mandarin samples from 2019 contained fosetyl per se in concentrations of 0.006 and 0.351 mg kg^{-1}.

Residues of malathion (34.5%; range = 0.014–0.716 mg kg^{-1}), spirotetramat (17.2%; 0.017–0.112 mg kg^{-1}), sulfoxaflor (17.2%; 0.010–0.015 mg kg^{-1}), acetamiprid (10.3%; 0.010–0.129 mg kg^{-1}), and tau-fluvalinate (10.3%; 0.063–0.228 mg kg^{-1}) were found in more than 10% of the mandarin samples. While lambda-cyhalothrin was detected in two mandarin samples (0.029–0.190 mg kg^{-1}) from 2019 in measurable concentrations, fourteen pesticides, namely buprofezin (0.013 mg kg^{-1}), carbendazim (0.038 mg kg^{-1}), chlorpyrifos-methyl (0.010 mg kg^{-1}), deltamethrin (0.015 mg kg^{-1}), fludioxonil (0.627 mg kg^{-1}), fluopyram (0.013 mg kg^{-1}), imazalil (0.801 mg kg^{-1}), imidacloprid (0.012 mg kg^{-1}), phosmet (0.014 mg kg^{-1}), pirimicarb (0.037 mg kg^{-1}), propiconazole (1.008 mg kg^{-1}), pyrimethanil (0.701 mg kg^{-1}), pyriproxyfen (0.028 mg kg^{-1}), and thiophanate-methyl (0.022 mg kg^{-1}), were quantified only in one sample.

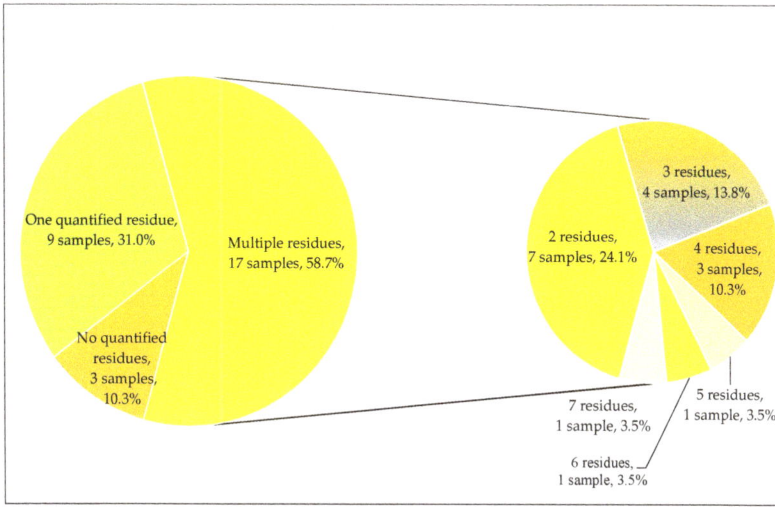

Figure 1. Number of quantified residues in satsuma mandarins from 2019.

In 2020, 93 mandarin samples were analyzed. Table 2 shows the quantified pesticide residues and their concentrations in mandarin samples in 2020. The quantification rate of pesticides in mandarin samples from 2020 (95.7%) was slightly increased compared to the 2019 results (89.7%). In 4.3% of mandarin samples, no measurable residues were recorded. A total of 67 mandarin samples (72%) contained at least one detectable pesticide within the legally permitted concentrations, whereas the MRL exceedances were recorded in 21 samples (22.6% of the analyzed samples in 2020). Compared with 2019, the MRL exceedance rate for mandarin went up. Residues exceeding the MRL were related to five residues (buprofezin, chlorpyrifos, fenbutatin-oxide, malathion, and propiconazole).

In the 2020 monitoring year, 24 different residues were found at measurable concentrations in mandarin samples. While 20 detected residues in mandarins were approved pesticides, the remaining four residues (propiconazole, fenbutatin-oxide, spirodiclofen, and chlorpyrifos) were non-approved. While 12 mandarin samples contained only one residue, multiple residues were detected in 82.8% (77 samples) of the samples; up to eight pesticides were found in individual mandarin samples (Figure 2). It should be noted that the multiple-residue rate significantly increased from 55.2% in 2019 to 82.8% in 2020.

Among the residues, phosphonic acid (58.1%, range = 0.039–39.386 mg kg^{-1}; 0.052–52.777 mg kg^{-1} for fosetyl, sum), spirotetramat (55.9%, 0.011–0.324 mg kg^{-1}), fludioxonil (46.2%, 0.011–0.648 mg kg^{-1}), imazalil (46.2%, 0.408–1.006 mg kg^{-1}), pyrimethanil (46.2%, 0.329–1.200 mg kg^{-1}), and 2-phenylphenol (44.1%, 0.584–2.667 mg kg^{-1}) were the most frequently detected pesticides present in more than 40% of the mandarin samples. Compared to the 2019 results, the quantification rate was more than tenfold higher for the pesticides fludioxonil, imazalil, and pyrimethanil. The fungicide 2-phenylphenol, which was frequently detected in mandarins in 2020, was not recorded in 2019.

Table 2. The presence and quantification of pesticide residues in mandarins in 2020.

Pesticide	Type of Residue	EU MRL (mg kg^{-1})	% of Samples <LOQ	% of Samples between LOQ-MRL	% of Samples >MRL	Range (mg kg^{-1})	
						Min.–Max.	Mean
2-Phenylphenol	FU	10	55.9	44.1	-	0.584–2.667	0.993
Acetamiprid	IN	0.9	89.2	10.8	-	0.019–0.318	0.067
Azoxystrobin	FU	15	98.9	1.1		0.010	0.010
Buprofezin	IN	0.01	90.3	2.2	7.5	0.010–0.109	0.042
Chlorpyrifos *	IN/AC	0.01	94.6	-	5.4	0.038–0.418	0.149
Cypermethrin	IN	2	98.9	1.1	-	0.512	0.512
Deltamethrin	IN	0.04	97.8	2.2	-	0.012–0.037	0.025
Difenoconazole	FU	0.6	98.9	1.1	-	0.378	0.378
Esfenvalerate	IN	0.02	98.9	1.1	-	0.017	0.017
Fenbutatin-oxide *	AC	0.01	92.5	1.1	6.5	0.010–0.047	0.028
Fludioxonil	FU	10	53.8	46.2	-	0.011–0.648	0.131
Imazalil	FU	5	53.8	46.2	-	0.408–1.006	0.675
Lambda-cyhalothrin	IN	0.2	98.9	1.1	-	0.111	0.111
Malathion	IN	2	82.8	16.1	1.1	0.011–2.855	0.493
Phosphonic acid **	FU	150	41.9	58.1	-	0.039–39.386	2.917
Pirimicarb	IN	3	83.9	16.1	-	0.040–0.165	0.084
Propiconazole *	FU	0.01	91.4	-	8.6	0.020–0.171	0.044
Pyrimethanil	FU	8	53.8	46.2	-	0.329–1.200	0.588
Pyriproxyfen	IN	0.6	89.2	10.8	-	0.021–0.140	0.053
Spirodiclofen *	AC	0.4	93.5	6.5	-	0.010–0.166	0.049
Spirotetramat	IN	0.5	44.1	55.9	-	0.011–0.324	0.061
Sulfoxaflor	IN	0.8	89.2	10.8	-	0.011–0.131	0.040
Tau-fluvalinate	IN	0.4	91.4	8.6	-	0.029–0.385	0.156
Tetraconazole	FU	0.02	97.8	2.2		0.016–0.019	0.018

FU: fungicide; IN: insecticide; AC: acaricide; * not approved in the EU; ** sum of fosetyl, phosphonic acid, and their salts expresses as fosetyl.

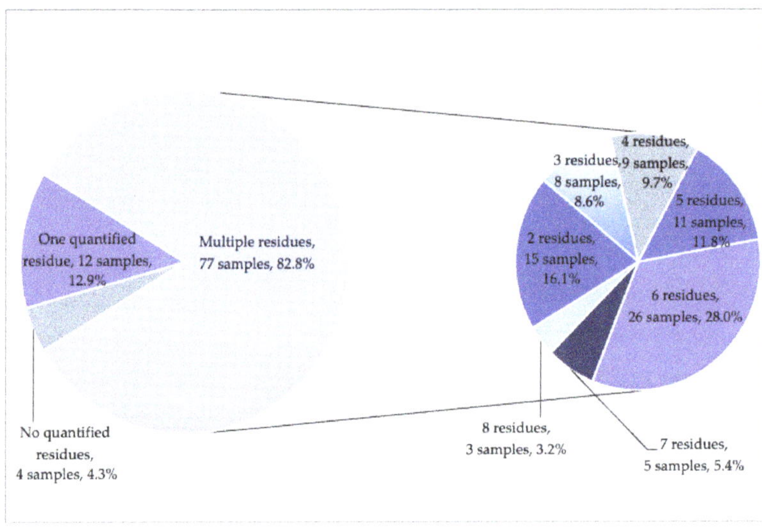

Figure 2. Number of quantified residues in satsuma mandarins from 2020.

In 2021, 104 mandarin samples were analyzed. Table 3 shows the distribution of pesticide residue contents in mandarin samples from 2021. In 12 samples (11.54%) no pesticide residues were quantified, whereas 92 samples contained one or several pesticides in measurable concentrations. For 27 samples (25.96% of the analyzed mandarin samples), the residue concentrations exceeded the MRL. These exceedances were mainly related to buprofezin residue (16 samples), followed by propiconazole (7 samples), fenbutatin-oxide (5 samples), and spirotetramat (4 samples). Among the 39 individual determinations that exceeded the MRL, 18 determinations were observed for residues that are currently non-approved.

Table 3. The presence and quantification of pesticide residues in mandarins from 2021.

Pesticide	Type of Residue	EU MRL (mg kg^{-1})	% of Samples <LOQ	% of Samples between LOQ-MRL	% of Samples >MRL	Range (mg kg^{-1})	
						Min.–Max.	Mean
2-Phenylphenol	FU	10	88.5	11.5	-	0.809–1.258	0.979
Acetamiprid	IN	0.9	75.0	25.0	-	0.010–0.121	0.032
Bifenthrin *	IN	0.05	99.0	1.0	-	0.017	0.017
Boscalid	FU	2	99.0	1.0	-	0.012	0.012
Buprofezin	IN	0.01	84.6	-	15.4	0.011–0.164	0.063
Chlorpyrifos *	IN/AC	0.01	98.1	1.0	1.0	0.010–0.013	0.012
Chlorpyrifos-methyl *	IN/AC	0.01	99.0	-	1.0	0.012	0.012
Cyantraniliprole	IN	0.9	97.1	2.9	-	0.010–0.106	0.044
Cypermethrin	IN	2	96.2	3.8	-	0.011–0.024	0.019
Difenoconazole	FU	0.6	92.3	7.7	-	0.010–0.119	0.062
Etoxazole	IN	0.1	97.1	2.9	-	0.013–0.032	0.021
Fenbutatin-oxide *	AC	0.01	95.2	-	4.8	0.013–0.359	0.102
Flonicamid	IN	0.15	99.0	1.0	-	0.018	0.018
Fludioxonil	FU	10	87.5	12.5	-	0.201–0.413	0.313
Fosetyl **	FU	150	94.2	5.8	-	0.110–0.164	0.135
Imazalil	FU	5	89.4	10.6	-	0.436–0.702	0.605
Imidacloprid *	IN	0.9	98.1	1.9	-	0.015–0.029	0.022
Malathion	IN	2	80.8	19.2	-	0.010–1.596	0.256
Novaluron *	IGR	0.01	98.1	-	1.9	0.019–0.111	0.065
Phosphonic acid **	FU	150	66.3	33.7	-	0.026–5.342	1.844
Pirimicarb	IN	3	95.2	4.8	-	0.015–0.182	0.073
Propiconazole *	FU	0.01	93.3	-	6.7	0.031–0.086	0.054
Pyridaben	IN/AC	0.3	86.5	12.5	1.0	0.011–0.318	0.113
Pyrimethanil	FU	8	88.5	11.5	-	0.271–0.601	0.473
Pyriproxyfen	IN	0.6	78.8	21.2	-	0.010–0.166	0.072
Spinosad	IN	0.3	99.0	1.0	-	0.012	0.012
Spirodiclofen *	AC	0.4	85.6	14.4	-	0.018–0.385	0.064
Spirotetramat	IN	0.5	53.8	42.3	3.8	0.010–1.485	0.155
Sulfoxaflor	IN	0.8	93.3	6.7	-	0.012–0.231	0.066
Tau-fluvalinate	IN	0.4	79.8	20.2	-	0.010–0.358	0.090
Thiacloprid *	IN	0.01	98.1	-	1.9	0.013–0.033	0.023
Thiophanate-methyl *	FU	6.0	99.0	1.0	-	0.017	0.017

FU: fungicide; IN: insecticide; AC: acaricide; IGR: insect growth regulator; * not approved in the EU; ** sum of fosetyl, phosphonic acid, and their salts expresses as fosetyl.

Compared with 2019 and 2020, a higher number of residues were found in mandarins in 2021. In total, 32 different pesticides were recorded in concentrations equal to or above the LOQ in mandarin samples from 2021. In 14.4% of the samples, only one residue was found in quantifiable concentrations. Multiple residues were recorded in 74.1% of the samples; up to nine residues were detected in individual mandarin samples from 2021 (Figure 3). Among the 77 samples that contained more than one residue, 22.1% of which (17 samples) had two residues, 18.2% (14 samples) three residues, 19.5% (15 samples) four residues, 22.1% (17 samples) five residues, 13% (10 samples) six residues, 1.3% (one sample) seven residues, 1.3% (one sample) eight residues, and 2.6% (two samples) nine residues.

The most frequently detected pesticide was spirotetramat in mandarin samples from 2021, with an incidence rate of 46.2% (48 samples). The samples contained spirotetramat concentrations ranging from 0.010 up to 1.485 mg kg^{-1}, with a mean concentration of 0.155 mg kg^{-1}. The MRL of 0.5 mg kg^{-1} for spirotetramat was exceeded for only four mandarin samples. The insecticide spirotetramat, derived from tetramic acid, has been widely used in citrus orchards in Turkey for the control of sucking insects, including *Planococcus citri*, *Aonidiella citrina*, *Aonidiella aurantia*, *Aphis gossypi*, and *Aphis citricola* [29]. It acts as an acetyl-coA carboxylase inhibitor and interrupts the biosynthesis of lipids in insects. After the foliar application of spirotetramat, it enters the plant and transforms into its metabolite enol, along with the metabolites -enol-glucoside and -ketohydroxy, which are the three main products of degradation [30,31]. Its derivatives are included in the current MRL for spirotetramat (sum of spirotetramat and their derivatives, spirotetramat-enol, spirotetramat-enol-glucoside, spitotetramet-monohydroxy, and spirotetramat-ketohydroxy, calculated as spirotetramat, sum) [32]. For spirotetramat, an acceptable daily intake (ADI)

processor (Retsch GmBh, GM 300, Haan, Germany) to achieve consistent and small particle sizes. Each analytical result was derived from a single laboratory sample taken from each lot.

3.3. Sample Preparation

For extraction of multi-class pesticide residues except for polar pesticides, the Association of Official Analytical Chemists (AOAC) version of the QuEChERS method [39] was used with slight modifications. The QuEChERS sample preparation methodology was summarized in Figure 4. Briefly, 15 g of homogenized mandarin sample was placed into a 50 mL polypropylene extraction tube, and 100 µL of TPP solution (internal standard, 10 µg mL^{-1}) and 15 mL of acetonitrile containing 1% acetic acid were added. After shaking the tube vigorously for 1 min, the QuEChERS salt extraction packet (containing 6 g of $MgSO_4$ and 1.5 g of sodium acetate) for AOAC 2007 method was added. The tube was shaken on a platform shaker (Collomix GmbH, VIBA 330, Gaimersheim, Germany) for 2 min and centrifuged (Hettich, Rotofix 32A, Tuttlingen, Germany) for 1 min at 5000 rpm. After extraction, 8 mL of acetonitrile layer (supernatant) was transferred to a clean-up dispersive tube containing 900 mg $MgSO_4$ and 150 mg PSA to remove residual water and further remove matrix interferences (sugars, organic acids, and polar pigments) from the sample. The tube was shaken on a platform shaker for 2 min and centrifuged (4000 rpm, 3 min). The supernatant was then filtered using a 0.20 µm cellulose syringe filter and analyzed by LC-MS/MS or GC-MS/MS.

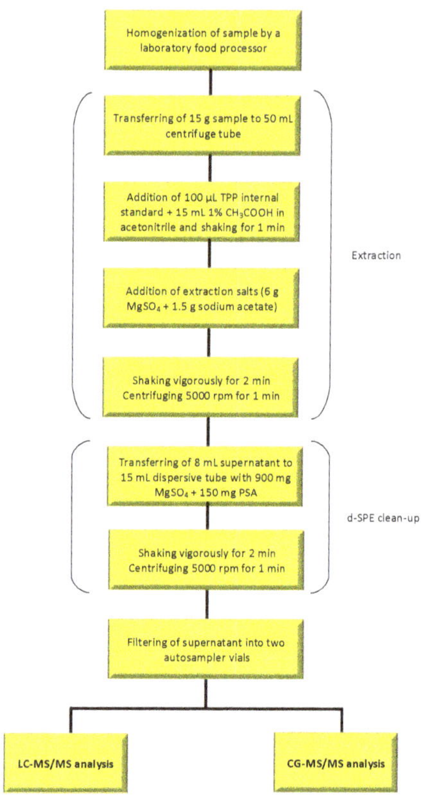

Figure 4. The schematic diagram of the QuEChERS sample preparation.

For extraction of six highly polar pesticides (chlorate, ethephon, fosetyl, glyphosate, perchlorate, and phosphonic acid) from mandarin samples, the QuPPe method developed by the EURL-SRM [26] was employed as shown in Figure 5. For mandarins, 10 g of homogenized sample was taken in a 50 mL centrifuge tube, and 1.5 mL of ultrapure water was added to adjust the total extract volume. Before the extraction with 10 mL of acidified methanol (containing 1% formic acid), 50 μL of ILISs solution (40 μg mL^{-1}) was added to the tube. After shaking and centrifugation (4000 rpm for 3 min) steps, the methanol layer (supernatant) from the QuPPe extract was filtered through a 0.20 μm cellulose syringe filter and analyzed by LC-MS/MS.

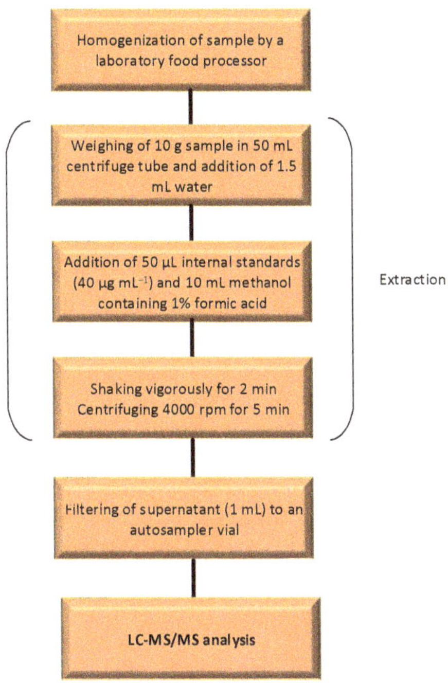

Figure 5. Workflow diagram for QuPPe sample preparation.

3.4. LC-MS/MS Analysis

LC-amenable of 440 pesticides separation was conducted using an Agilent 1290 HPLC system (Agilent, Santa Clara, CA, USA). This was equipped with an autosampler, a degasser module, a binary pump, and temperature-controlled column oven. Via a jet stream electrospray ionization (ESI) source, the LC was coupled to an Agilent 6470 QQQ triple quadrupole mass spectrometer (MS/MS, Agilent, Santa Clara, CA, USA).

Chromatographic separation of LC-amenable pesticides (434 substances) except for polar substances was achieved using an InfinityLab Poroshell 120 SB-C18 column (3 × 100 mm, 2.7 μm particle size) (Agilent Technologies, Santa Clara, CA, USA). The column temperature was set at 45 °C, and the flow rate was set at 0.66 mL min^{-1}. Eluent A was water, containing 5 mM ammonium formate, and eluent B was 100% methanol. Gradient elution was applied as follows: 0–0.5 min 40% B, 0.5–3.5 min: 40–60% B, 3.5–7 min: 60–98% B, 7–8.7 min 98% B, 8.7–8.8 min: 98–40% B, and 8.8–11 min: 40% B.

A porous graphitic carbon-based Hypercarb 2.1 × 100 mm column with 5 μm particle size (Thermo Scientific™, Waltham, MA, USA) was used for the separation of six polar compounds at 40 °C. The mobile phase for QuPPe extracts was composed of 94:5:1 water-methanol-acetic acid ($v/v/v$) as eluent A and methanol-acetic acid at the ratio of 99:1 (v/v) as eluent B. The gradient started at 0 min, 0% B, and increased linearly to 30% B in 10 min at a flow rate of 0.2 mL min^{-1}. The 30% B was kept for 8 min and then increased linearly to 90% B in 1 min at a flow rate of 0.4 mL min^{-1}. The 90% B was kept for 3 min and returned to 0% B within 0.1 min at the initial flow rate of 0.2 mL min^{-1} and held for 10 min.

Electrospray negative ionization (ESI-) was used for the analysis of QuPPe extract. The ionization conditions of the ESI source were as flows: gas temperature of 230 °C, gas flow of 10 L min^{-1}, nebulizing gas pressure of 45 psi, sheath gas temperature of 300 °C, sheath gas flow of 11 L min^{-1}, capillary of 3500 V, and nozzle voltage of 500 V. Nitrogen was used as the collision gas. Data acquisition was performed using Agilent MassHunter software (Version B.07.01).

3.5. GC-MS/MS Analysis

A total of 71 GC-amenable pesticides were analyzed using an Agilent (Santa Clara, CA, USA) 7890A GC system equipped with an Agilent 7693 autosampler, interfaced to an Agilent 7000B triple quadrupole mass spectrometer. An Agilent HP-5MS Ultra Inert analytical column (30 × 0.25 mm, 0.25 µm) was used in the residue separation, with helium as a carrier gas at a constant flow rate of 1.25 mL min^{-1}. The GC oven was operated under the following conditions: initial temperature of 75 °C held for 2.5 min, 50 °C min^{-1} rate to 150 °C, then 20 °C min^{-1} rate to 200 °C, and finally 16 °C min^{-1} rate to 310 °C and held for 15 min. The injection port temperature was 280 °C and 5 µL volume was injected with a multimode inlet in programmable temperature vaporizer (PTV) mode.

The triple quadrupole mass spectrometer was operated in electron ionization (EI) mode with an ionization voltage of 35 eV, ion source temperature of 230 °C, quadrupole temperature of 150 °C, and transfer line temperature of 300 °C, scanning from m/z 50 to 500 at 2.5 s per scan, solvent delay 3.75 min. Default instrument settings of collision gas flow of N_2 at 1.5 mL min^{-1} and quench gas of He at 2.35 mL min^{-1} were used. Agilent MassHunter software was used for acquisition, data handling, and reporting.

3.6. Validation Studies

The validation of the analytical methods was implemented according to SANTE 11312/2021 guidelines [27]. Method performance for LC-amenable and GC-amenable residues was verified, including parameters such as linearity, LOQs, recovery, precision, and measurement uncertainties. The validation procedures were extensively described in our previous papers [5,11,13,40,41]. Five levels (0.002, 0.005, 0.01, 0.05, and 0.1 mg kg^{-1}) of matrix-matched calibrations were prepared for each target analyte.

For recovery, replicate homogenates (n = 5) were spiked at two levels of concentrations: an upper level of fortification of 0.05 mg kg^{-1} and a lower level of sample spiking with residue concentration of 0.01 mg kg^{-1}. The repeatability of the method was assessed through the relative standard deviations (RSD_r, %) associated with measurements of target compounds performed during recovery analyses on the same day. Over a one-week period, the within-laboratory method reproducibility (RSD_R, %) was assessed. This involved two laboratory analysts performing matrix homogenate spiking, extraction, and analysis on different days. Each operator extracted and analyzed a batch of fortified homogenates (n = 10). To determine the expanded measurement uncertainty for each analyte, trueness (bias) and within-laboratory reproducibility uncertainties were taken into account.

4. Conclusions

This study has focused on the determination of 511 pesticide residues including widely used pesticides in the citrus industry, non-approved residues, and six highly polar substances in satsuma mandarin samples. Two sample extraction methods, QuEChERS and QuPPe, have been successfully applied for the analysis of non-polar/medium-polar and highly polar substances, respectively. This three-year monitoring study showed that 91.6% of 226 satsuma mandarin samples collected from the Izmir region, Turkey, contained one or multiple residues, up to nine residues. Forty different residues comprising 23 insecticides, 14 fungicides, two acaricides, and one insect growth regulator were detected in mandarin samples during the three harvesting years. While one residue was found in 15.9% of mandarin samples, two or more residues were recorded in 75.7% of samples. In 22.1% of the mandarin samples, the residue concentrations exceeded the MRLs. Among the residues, phosphonic acid (48.7%), spirotetramat (46.5%), fludioxonil (25.2%), pyrimethanil (24.8%), imazalil (24.3%), and 2-phenylphenol (23.5%) were the most frequently found pesticides in satsuma mandarins. The increase in the use of active ingredients in mandarin farming can be attributed to a combination of factors, including the need to manage pests and diseases effectively, meet market demands, improve crop quality, address environmental conditions, adopt sustainable practices, and adhere to regulatory requirements.

These results showed that official citrus monitoring programs should be conducted routinely by governments. Moreover, more strictly controlled measures for hormone-disrupting pesticides such as imazalil should be enacted to protect consumers. The influence of various processing techniques including washing, peeling, and juicing on pesticide residues in mandarins and other citrus fruits should also be investigated. Conducting toxicity studies on the potential synergistic or additive effects of multiple pesticide residues found in citrus fruits will provide valuable information for risk assessment. Furthermore, the cumulative dietary exposure of consumers to detected residues should be analyzed to formulate appropriate risk management measures and establish revised MRLs.

Supplementary Materials: The following supporting information can be downloaded at: https://www.mdpi.com/article/10.3390/molecules28145611/s1, Table S1: MRM transitions and in-house validation data for 40 residues.

Author Contributions: Conceptualization, E.G. and B.K.; methodology, E.G., O.G. and B.K.; validation, E.G. and O.G.; formal analysis, E.G.; investigation, E.G. and B.K.; writing—review and editing, M.Á.G.-C. and B.K.; supervision, B.K.; funding acquisition, E.G. and M.Á.G.-C. All authors have read and agreed to the published version of the manuscript.

Funding: The APC was funded by Universidad EAN.

Institutional Review Board Statement: Not applicable.

Informed Consent Statement: Not applicable.

Data Availability Statement: Not applicable.

Conflicts of Interest: The authors declare no conflict of interest.

References

1. FAO (Food and Agricultural Organization). *FAO Statistical Databases and Data Sets*; FAO (Food and Agricultural Organization): Rome, Italy, 2022. Available online: https://www.fao.org/faostat/en/#data/QC/visualize (accessed on 25 March 2023).
2. Lado, J.; Cuellar, F.; Rodrigo, M.J.; Zacarías, L. Nutritional composition of mandarins. In *Nutritional Composition of Fruit Cultivars*; Simmonds, M.S.J., Preedy, V.R., Eds.; Academic Press: Cambridge, MA, USA, 2016; pp. 419–443.
3. Saito, S.; Xiao, C.L. Prevalence of postharvest diseases of mandarin fruit in California. *Plant Health Prog.* **2017**, *18*, 204–210. [CrossRef]
4. Turgutoğlu, E. Mandarin yetiştiriciliği T.C. Tarım ve Orman Bakanlığı Tarımsal Araştırmalar ve Politikalar Genel Müdürlüğü Batı Akdeniz Tarımsal Araştırma Enstitüsü, Antalya. 2020. Available online: https://arastirma.tarimorman.gov.tr/batem/Belgeler/Kutuphane/Teknik%20Bilgiler/Mandarin%20Yetiştiriciliği.pdf (accessed on 26 October 2022).
5. Aslantas, S.; Golge, O.; González-Curbelo, M.Á.; Kabak, B. Determination of 355 pesticides in lemon and lemon juice by LC-MS/MS and GC-MS/MS. *Foods* **2023**, *12*, 1812. [CrossRef] [PubMed]
6. Resmi Gazete. *Türk Gıda Kodeksi Pestisitlerin Maksimum Kalıntı Limitleri Yönetmeliği*; sayı: 31611; Resmi Gazete: Ankara, Turkey, 2021.
7. European Commission. Regulation (EC) no 396/2005 of the European Parliament and of the council of 23 February 2005 on maximum residue levels of pesticides in or on food and feed of plant and animal origin and amending Council Directive 91/414/EEC. *Off. J. Eur. Union* **2005**, *L70*, 1–16.
8. RASFF (Rapid Alert System for Food and Feed). RASFF Portal. Available online: https://webgate.ec.europa.eu/rasff-winndow/screen/search (accessed on 20 February 2023).

9. European Commission. Commission implementing regulation (EU) 2019/1793 of 22 October 2019 on the temporary increase of official controls and emergency measures governing the entry into the Union of certain goods from certain third countries implementing Regulations (EU) 2017/625 and (EC) No 178/2002 of the European Parliament and of the Council and repealing Commission Regulations (EC) No 669/2009, (EU) No 884/2014, (EU) 2015/175, (EU) 2017/186 and (EU) 2018/1660. *Off. J. Eur. Union* **2019**, *L277*, 89–129.
10. Vu Duc, N.; Nguyen-Quang, T.; Le-Minh, T.; Nguyen-Thi, X.; Tran, T.M.; Vu, H.A.; Nguyen, L.-A.; Doan-Duy, T.; Hoi, B.V.; Vu, C.-T.; et al. Multiresidue pesticides analysis of vegetables in Vietnam by ultrahigh-performance liquid chromatography in combination with high-resolution mass spectrometry (UPLC-orbitrap MS). *J. Anal. Methods Chem.* **2019**, *2019*, 3489634. [CrossRef]
11. Golge, O.; Cinpolat, S.; Kabak, B. Quantification of pesticide residues in gherkins by liquid and gas chromatography coupled to tandem mass spectrometry. *J. Food Compos. Anal.* **2021**, *96*, 103755. [CrossRef]
12. González-Curbelo, M.Á.; Varela-Martínez, D.A.; Riaño-Herrera, D.A. Pesticide-residue analysis in soils by the QuEChERS method: A review. *Molecules* **2022**, *27*, 4323. [CrossRef]
13. Deveci, B.; Golge, O.; Kabak, B. Quantification of 363 pesticides in leafy vegetables (dill, rocket and parsley) in the Turkey market by using QuEChERS with LC-MS/MS and GC-MS/MS. *Foods* **2023**, *12*, 1034. [CrossRef]
14. Yang, X.; Zhang, H.; Liu, Y.; Wang, J.; Zhang, Y.; Dong, A.; Zhao, H.; Sun, C.; Cui, J. Multiresidue method for determination of 88 pesticides in berry fruits using solid phase extraction and gas-chromatography-mass spectrometry. *Food Chem.* **2011**, *127*, 855–865. [CrossRef]
15. Iwafune, T.; Ogino, T.; Watanabe, E. Water-based extraction and liquid chromatography–tandem mass spectrometry analysis of neonicotinoid insecticides and their metabolites in green pepper/tomato samples. *J. Agric. Food Chem.* **2014**, *62*, 2790–2796. [CrossRef]
16. Correia, M.; Delerue, C.; Alves, A. Development of a SPME-GC-ECD methodology for selected pesticides in must and wine samples. *Fresenius' J. Anal. Chem.* **2001**, *369*, 647–651. [CrossRef] [PubMed]
17. Bedessa, T.; Megersa, N.; Gure, A. Salting out assisted liquid-liquid extraction for the determination of multiresidue pesticides in alcoholic beverages by high performance liquid chromatography. *Sci. J. Anal. Chem.* **2017**, *5*, 38–45. [CrossRef]
18. González-Curbelo, M.Á.; Hernández-Borges, J.; Borges-Miguel, T.M.; Rodríguez-Delgado, M.Á. Determination of organophosphorus pesticides and metabolites in cereal-based baby foods and wheat flour by means of ultrasound-assisted extraction and hollow-fiber liquid-phase microextraction prior to gas chromatography with nitrogen phosphorus detection. *J. Chromatogr. A* **2013**, *1313*, 166–174. [CrossRef] [PubMed]
19. Adou, K.; Bontoyan, W.R.; Sweeney, P.J. Multiresidue method for the analysis of pesticide residues in fruits and vegetables by accelerated solvent extraction and capillary gas chromatography. *J. Agric. Food Chem.* **2001**, *49*, 4153–4160. [CrossRef] [PubMed]
20. Rissato, S.R.; Galhaine, M.S.; Knoll, F.R.; Apon, P.M. Supercritical fluid extraction for pesticide multiresidue analysis in honey: Determination by gas chromatography with electron-capture and mass spectrometry detection. *J. Chromatogr. A* **2004**, *1048*, 153–159. [CrossRef] [PubMed]
21. Yousefi, M.; Rahimi-Nasrabadi, M.; Mirsadeghi, S.; Pourmortazavi, S.M. Supercritical fluid extraction of pesticides and insecticides from food samples and plant materials. *Crit. Rev. Anal. Chem.* **2021**, *51*, 482–501. [CrossRef]
22. Valsamaki, V.I.; Boti, V.I.; Sakkas, V.A.; Albanis, T.A. Determination of organochlorine pesticides and polychlorinated biphenyls in chicken eggs by matrix solid phase dispersion. *Anal. Chim. Acta* **2006**, *573–574*, 195–201. [CrossRef]
23. Lozowicka, B.; Ilyasova, G.; Kaczynski, P.; Jankowska, M.; Rutkowska, E.; Hrynko, I.; Mojsak, P.; Szabunko, J. Multi-residue methods for the determination of over four hundred pesticides in solid and liquid high sucrose content matrices by tandem mass spectrometry coupled with gas and liquid chromatography. *Talanta* **2016**, *151*, 51–61. [CrossRef]
24. Singh, S.B.; Foster, G.D.; Khan, S.U. Microwave-assisted extraction for the simultaneous determination of thiamethoxam, imidacloprid, and carbendazim residues in fresh and cooked vegetable samples. *J. Agric. Food Chem.* **2004**, *52*, 105–109. [CrossRef]
25. Anastassiades, M.; Lehotay, S.J.; Štajnbaher, D.; Schenck, F.J. Fast and easy multiresidue method employing acetonitrile extraction/partitioning and "dispersive solid-phase extraction" for the determination of pesticide residues in produce. *J. AOAC Int.* **2003**, *86*, 412–431. [CrossRef]
26. Anastassiades, M.; Kolberg, D.I.; Benkenstein, A.; Eichorn, E.; Zechmann, S.; Mack, D. *Quick Method for the Analysis of Numerous Highly Polar Pesticides in Foods of Plant Origin via LC-MS/MS Involving Simultaneous Extraction with Methanol (QuPPe Method)—Version 9.3*; EU Reference Laboratory for Pesticides Requiring Single Residue Methods (EURL-SRM): Stuttgart, Germany; CVUA: Stuttgart, Germany, 2017.
27. European Commission. *Guidance Document on Analytical Quality Control and Method Validation Procedures for Pesticide Residues Analysis in Food and Feed*; Document No. SANTE 11312/2021; EC Directorate-General for Health and Food Safety: Brussels, Belgium, 2021.
28. CVUA Stuttgart (Chemisches und Veterinäruntersuchungsamt). Residues and Contaminants in Fresh Fruit from Conventional Cultivation. 2018. Available online: https://www.ua-bw.de/pesticides/beitrag_en.asp?subid=1&Thema_ID=5&ID=2931&Pdf=No&lang=EN (accessed on 26 November 2022).
29. Tarım ve Köy İşleri Bakanlığı. Turunçgil Hastalık ve Zararlıları ile Mücadele. Ankara. 2011. Available online: https://cadcom.com.tr/downloads/CsAgriMedya/documents/turuncgil_hastaliklari.pdf (accessed on 21 June 2023).
30. Salazar-López, N.J.; Aldana, L.; Silveira-Gramont, M.-I. Sprirotetramat-An alternative for the control of parasitic sucking insects and its fate in the environment. In *Insecticides Resistance*; Trdan, S., Ed.; IntechOpen: London, UK, 2016; 450p.

31. EFSA (European Food Safety Authority). Modification of the existing maximum residue levels for spirotetramat in herbs and edible flowers. *EFSA J.* **2022**, *20*, 7668.
32. EFSA (European Food Safety Authority). Reasoned opinion on the review of the existing maximum residue levels for spirotetramat according to Article 12 of Regulation (EC) No 396/2005. *EFSA J.* **2020**, *18*, 5960.
33. Tağa, Ö. Determination of Pesticide Residue Levels on Citrus Fruits Grown on Aegean and Mediterranean Region in Turkey. Master's Thesis, Namık Kemal University Graduate School of Natural and Applied Science, Tekirdağ, Turkey, 2007.
34. Bakırcı, G.T.; Acay, D.B.Y.; Bakırcı, F.; Ötleş, S. Pesticide residues in fruits and vegetables from the Aegean region, Turkey. *Food Chem.* **2014**, *160*, 379–392. [CrossRef] [PubMed]
35. Al-Nasir, F.M.; Jiries, A.G.; Al-Rabadi, G.J.; Alu'datt, M.H.; Tranchant, C.C.; Al-Dalain, S.A.; Alrabadi, N.; Madanat, O.Y.; Al-Dmour, R.S. Determination of pesticide residues in selected citrus fruits and vegetables cultivated in the Jordan Valley. *LWT Food Sci. Technol.* **2020**, *123*, 109005. [CrossRef]
36. Li, Z.; Zhang, Y.; Zhao, Q.; Wang, C.; Cui, Y.; Li, J.; Chen, A.; Liang, G.; Jiao, B. Occurrence, temporal variation, quality and safety assessment of pesticide residues on citrus fruits in China. *Chemosphere* **2020**, *258*, 127381. [CrossRef]
37. EFSA (European Food Safety Authority). The 2015 European Union report on pesticide residues in food. *EFSA J.* **2017**, *15*, 4791.
38. European Commission. Commission Directive 2002/63/EC of 11 July 2002 establishing Community methods of sampling for the official control of pesticide residues in and on products of plant and animal origin and repealing Directive 79/700/EEC. *Off. J. Eur. Communities* **2002**, *2*, 30–43.
39. AOAC International. *Pesticide Residues in Foods by Acetonitrile Extraction and Partitioning with Magnesium Sulphate: AOAC Official Method 2007.01*; AOAC International: Rockville, MD, USA, 2007.
40. Gormez, E.; Golge, O.; Kabak, B. Quantification of fosetyl-aluminium /phosphonic acid and other highly polar residues in pomegranates using Quick Polar Pesticides method involving liquid chromatography-tandem mass spectrometry measurement. *J. Chromatogr. A* **2021**, *1642*, 4622038. [CrossRef] [PubMed]
41. Gormez, E.; Golge, O.; González-Curbelo, M.Á.; Kabak, B. Monitoring and exposure assessment of fosetyl aluminium and other highly polar pesticide residues in sweet cherry. *Molecules* **2023**, *28*, 252. [CrossRef] [PubMed]

molecules

Article

Simultaneous Determination of 54 Pesticides in Proso Millet Using QuEChERS with Liquid Chromatography-Tandem Mass Spectrometry (LC–MS/MS)

Chao Ding, Pengcheng Ren, Yanli Qi, Yanmei Yang and Shu Qin *

Shanxi Center for Testing of Functional Agro-Products, Shanxi Agricultural University, Taiyuan 030031, China; ymyangsuda@163.com (Y.Y.)
* Correspondence: qinshu@sxau.edu.cn

Abstract: To assess the potential risks posed to the environment and human health, analyzing pesticide residues in proso millet is important. This paper aimed to develop a modified QuEChERS method with liquid chromatography-tandem mass spectrometry (LC–MS/MS) for the analysis of 54 pesticide residues in proso millet. Parameters including the mobile phase of the instrument, the acidity of the extraction solvent, and the type of absorbents were optimized to provide satisfactory performance. The method was validated concerning linearity, limit of quantification (LOQ), matrix effect, accuracy, and precision. In detail, the linearity of the matrix-matched calibration curve was acceptable with correlation coefficients (R^2) higher than 0.99. The mean recovery was in the range of 86% to 114% with relative standard deviations (RSDs) $\leq$ 20% (n = 5). The LOQ was determined to be 0.25–10 μg/kg. The developed method was feasible for the determination of multiple pesticide residues in proso millet.

Keywords: proso millet; pesticides; determination; QuEChERS; LC–MS/MS

Citation: Ding, C.; Ren, P.; Qi, Y.; Yang, Y.; Qin, S. Simultaneous Determination of 54 Pesticides in Proso Millet Using QuEChERS with Liquid Chromatography-Tandem Mass Spectrometry (LC–MS/MS). *Molecules* **2023**, *28*, 6575. https://doi.org/10.3390/molecules28186575

Academic Editor: Miguel Ángel González-Curbelo

Received: 10 August 2023
Revised: 8 September 2023
Accepted: 9 September 2023
Published: 12 September 2023

1. Introduction

Proso millet (*Panicum miliaceum* L.) is one of the featured coarse cereals in Shanxi, China. Due to its tolerance to drought and heat conditions as well as short growing period, proso millet is planted widely. In past years, proso millet has also received extensive attention because it is rich in protein, starch, fat, vitamins, and minerals [1,2]. Besides, natural active substances in proso millet, including polyphenols, phytic acid, and alkaloids, are beneficial for regulating blood sugar and lipids, as well as resisting oxidation [3]. Previous studies indicated that intake of proso millet and its processed product could help reduce the risk of chronic diseases, such as type 2 diabetes and liver damage [4–7].

In the cultivation of proso millet, pesticides are widely utilized to control disease, insect pests, and weeds, which is beneficial for improving its production and quality [8]. However, due to its low economic benefit, registered pesticide for proso millet is limited [9]. Widely used pesticides concerning organophosphorus, carbamate, pyrethroid, and nicotinoids may be applied during the cultivation of proso millet. Moreover, pesticides remaining in the soil can be uptaken and transported by crops, such as thiamethoxam, imidacloprid, and chlorpyrifos, azoxystrobin, acetamiprid [10,11]. The residue in crops will pose a toxicological impact on the environment and human health, especially for those with high toxicity, such as phorate [12]. Thus, developing a reliable method for determining multiple pesticide residues to monitor the dietary intake risk from proso millet is crucial. However, relevant studies are still limited.

As a cheap, easy, quick, safe method, QuEChERS has been the most commonly used sample preparation method since first developed in 2003, enabling the determination of multiple pesticides simultaneously [13]. It has been used for the analysis of multi-residues in various cereals, such as rice [14], corn [15], and wheat grains [16]. Ruan et al. also developed a method for the analysis of 34 pesticides in proso millet by QuEChERS [17]. However, in this study, QuEChERS was coupled with online gel permeation chromatography–gas chromatography, which was time-consuming and required labor. In comparison, liquid chromatography-tandem mass spectrometry (LC–MS/MS) is predominant due to its good sensitivity, selectivity, and short analysis time [18,19]. Based on the above discussion, the combination of QuEChERS with LC–MS/MS is an ideal strategy for detecting multiple pesticide residues in proso millet.

The purpose of this study was to develop and validate an efficient analytic method based on QuEChERS coupled with LC–MS/MS for the determination of multi-residue pesticides. In our study, 54 widely used pesticides, some of which were high-toxic or with a long residual period, were chosen as target compounds. The type of target pesticides includes acaricide, fungicide, herbicide, insecticide, and plant growth regulator. The developed method was proven to meet the requirements for the detection of multi-residues in proso millet.

2. Results and Discussion

2.1. Optimization of Instrumental Conditions

In this study, mass spectrum acquisition parameters of 54 pesticides were firstly optimized by direct injection of 0.1 mg/L individual standard solution in ACN to a mass spectrometer. The precursor and product ion were selected using the scan type of Q1 MS and product scan, respectively. Then collision energy (CE) and declustering potential (DP) were optimized based on the intensity of the ion. Most of the pesticides were detected using the positive mode while the other four pesticides including fipronil, fipronil desulfinyl, fipronil sulfone, and fipronil sulfide gave a better response using the negative mode. Detailed parameters were exhibited in Table 1. The product ion with a higher response was selected as a quantitative ion and the other one as a confirmative ion.

Table 1. Category and mass spectrum acquisition parameters of 54 pesticides.

No.	Pesticide	Category [1]	RT	Parent Ion	Product Ion [2]	DP	CE
				Positive Mode			
1	Methamidophos	I	2.34	142.0	94.0 *; 125.0	57	19; 18
2	Phorate	I	5.54	261.0	75.0 *; 47.0	51	21; 53
3	Omethoate	I	2.76	214.0	183.0 *; 109.0	60	16; 36
4	Dichlorvos	I	4.62	221.0	109.0 *; 127.0	70	23; 27
5	Triazophos	I	5.07	314.1	162.1 *; 119.1	80	24; 50
6	Dimethoate	I	3.71	230.0	199.0 *; 125.0	56	13; 29
7	Chlorpyrifos	I	6.11	349.9	197.9 *; 97.0	75	28; 45
8	Acephate	I	2.68	184.0	143.0 *; 125.0	50	12; 25
9	Malathion	I	5.03	331.0	127.0 *; 99.0	70	16; 32
10	Phosalone	I	5.48	368.0	182.0 *; 322.0	76	20; 13
11	Phosmet	I	4.90	318.0	160.0 *; 133.0	61	17; 49
12	Isocarbophos	I	5.20	231.0	121.0 *; 109.0	100	26; 38
13	Diazinon	I	5.42	305.1	169.0 *; 153.1	90	28; 28
14	Profenofos	I	5.81	372.9	302.9 *; 344.9	80	25; 18
15	Phorate Sulfone	I	4.78	293.0	97.0 *; 115.0	65	50; 35
16	Phorate Sulfoxide	I	4.74	277.0	199.0 *; 153.0	25	13; 19
17	Tau-Fluvalinate	I	6.53	503.1	208.1 *; 181.1	61	16; 38
18	Iprodione	F	5.26	330.0	245.0 *; 288.0	30	20; 18
19	Deltamethrin	I	6.37	523.0	280.9 *; 506.0	55	23; 16
20	Fenpropathrin	I	6.11	350.2	125.1 *; 97.1	85	23; 46
21	Triadimefon	F	5.07	294.1	197.1 *; 225.1	70	21; 17
22	Aldicarb	I	4.30	116.1	89.0 *; 70.0	25	14; 14
23	Aldicarb Sulfone	I	2.90	240.1	148.0 *; 166.1	30	17; 16
24	Aldicarb Sulfoxide	I	2.82	207.1	132.0 *; 89.0	55	9; 20

Table 1. *Cont.*

No.	Pesticide	Category [1]	RT	Parent Ion	Product Ion [2]	DP	CE
25	Carbofuran	I	4.63	222.1	165.1 *; 123.0	70	16; 29
26	3-Hydroxy Carbofuran	I	3.63	238.1	181.1 *; 163.1	70	16; 18
27	Methomyl	I	3.05	163.1	88.0 *; 106.0	38	12; 14
28	Carbaryl	I	4.69	202.1	145.1 *; 127.1	56	15; 40
29	Carbendazim	F	3.06	192.1	160.1 *; 132.1	80	25; 41
30	Phoxim	I	5.41	299.1	129.0 *; 153.0	55	18; 10
31	Pyridaben	A	6.47	365.1	309.1 *; 147.1	77	17; 34
32	Pyrimethanil	F	4.97	200.1	183.1 *; 168.1	30	33; 40
33	Difenoconazole	F	5.55	406.1	251.0 *; 337.0	105	35; 24
34	Acetamiprid	I	3.64	223.1	126.0 *; 99.0	65	28; 60
35	Imidacloprid	I	3.38	256.1	175.1 *; 209.1	45	27; 22
36	Dimethomorph	F	5.01	388.1	301.1 *; 165.1	105	29; 43
37	Pendimethalin	H	6.16	282.1	212.1 *; 194.1	40	15; 28
38	Azoxystrobin	F	4.86	404.1	372.1 *; 344.1	80	20; 34
39	Thiamethoxam	I	3.10	292.0	211.1 *; 181.1	30	16; 30
40	Chlorfluazuron	I	6.32	540.0	382.9 *; 384.9	70	30; 30
41	Prochloraz	F	5.45	376.2	308.0 *; 266.0	20	15; 22
42	Chlorbenzuron	I	5.33	309.0	156.0 *; 139.0	50	18; 40
43	Diflubenzuron	I	5.24	311.0	158.0 *; 141.0	45	20; 49
44	Propamocarb	F	2.77	189.2	102.1 *; 74.0	70	24; 34
45	Forchlorfenuron	R	4.83	248.1	129.0 *; 93.0	50	23; 47
46	Etofenprox	I	6.90	394.2	177.1 *; 107.0	30	19; 59
47	Chlorantraniliprole	I	4.84	484.0	285.9 *; 452.9	45	19; 25
48	Pyraclostrobin	F	5.38	388.1	194.1 *; 163.1	50	18; 36
49	Metalaxyl	F	4.82	280.2	220.1 *; 192.1	75	18; 24
50	Paclobutrazol	R	5.02	294.1	70.0 *; 125.0	90	40; 45
				Negative Mode			
51	Fipronil	I	5.17	434.9	330.0 *; 250.0	−25	−24; −38
52	Fipronil Desulfinyl	I	5.12	387.0	351.0 *; 282.0	−30	−19; −47
53	Fipronil Sulfone	I	5.26	450.9	414.9 *; 282.0	−28	−26; −38
54	Fipronil Sulfide	I	5.20	418.9	262.0 *; 383.0	−20	−35; −22

[1] A acaricide, F fungicide, H herbicide, I insecticide, R plant growth regulator. [2], * refers to Quantitative ion.

The mobile phase of LC Chromatographic played an important role in the ionization efficiency of pesticides [20]. The formic acid (FA) and ammonium acetate studied were recommended to be commonly used as MS-compatible additives [21]. In the following study, the effect of three different mobile combinations was investigated in terms of peak area: 0.1% formic acid (FA) aqueous solution containing 4 mM ammonium acetate + methanol (mobile phase 1); 0.1% formic acid (FA) aqueous solution containing 4 mM ammonium acetate + acetonitrile (mobile phase 2); 0.1% formic acid (FA) aqueous solution + methanol (mobile phase 3); According to results exhibited in Figure 1, compared with mobile phase 2 and 3, higher peak response of most pesticides was obtained when using mobile phase 1. Thus, mobile phase 1 was utilized for the separation of 54 target compounds in 10 min. It is noted that the peak shape or retention time of basic compounds is easily affected by the reversed-phase (RP) column due to overload behavior. However, McCalley indicated that reduced column efficiency occurs with sample amounts introduced onto RP columns of greater than 50 ng [22]. And the injection amount in our experiment was far less than the above value. The effect of different mobile on peak shape or retention time was not studied herein. The corresponding MRM chromatograms of 54 pesticides (0.1 mg/L) using the positive mode and negative mode were exhibited in Figure 2a,b.

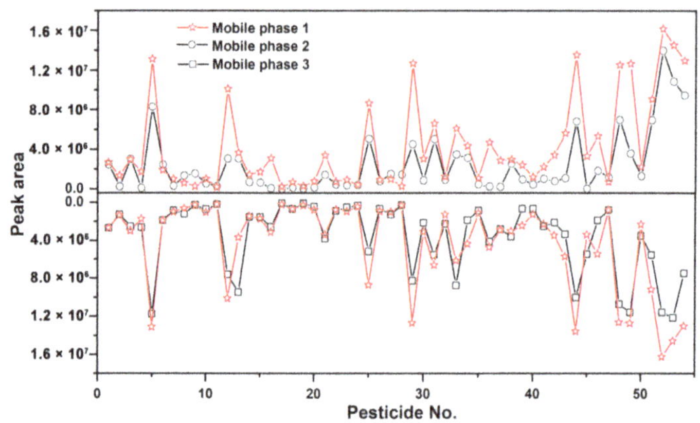

Figure 1. The peak response of 54 pesticides (0.1 mg/L) was obtained using different combinations of mobile phase.

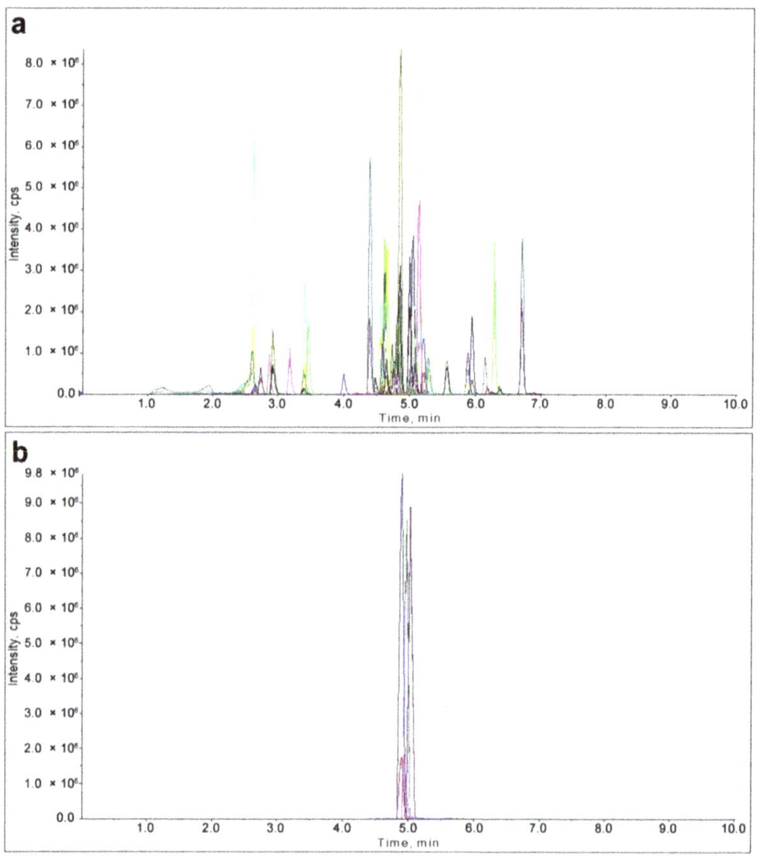

Figure 2. MRM chromatograms of 54 pesticides (0.1 mg/L) were obtained using the positive mode (**a**) and negative mode (**b**). The different colored lines correspond to different pesticides.

2.2. *Optimization of QuEChERS Method*

2.2.1. Extraction

Based on EN15662-2018, soaking the sample with water can improve the extraction efficiency of pesticides if the water content of the sample is <10% [23]. Therefore, 5 mL of water was added to homogenized proso millet samples and allowed to stand for 20 min before extraction.

The solvent is one of the main factors to affect the extraction efficiency. Due to its medium polarity, acetonitrile is the most common solvent used to determine various pesticides with different physicochemical properties [24,25]. The addition of acid could also influence the extraction ability of acetonitrile. In this section, the recovery of pure acetonitrile (ACN), acetonitrile with 1% acetic acid (1% HAc-ACN), and acetonitrile with 2% acetic acid (2% HAc-ACN) were compared. In detail, 50 μL of standard solution (10 mg/L) containing 54 pesticides was added into a blank proso millet sample. After standing for 30 min, samples were extracted with three different extract solvents. Recovery indicated that no obvious difference was shown for 48 pesticides. While for methamidophos, the recovery increased from 67% to 89% with the increase of HAc content in acetonitrile from 0% to 2%. Previous studies revealed that the increase in acid could improve the stability, which led to a significant increase in recovery [26]. An increase of acid in acetonitrile could also improve the extract efficiency of dichlorvos, phosmet, carbendazim, propamocarb, forchlorfenuron, and etofenprox. As shown in Figure 3a, the recovery of these five pesticides extracted with 2% HAc-ACN was all in the range of 80–110%, which was more satisfactory than pure ACN and 1% HAc-ACN. Thus, the concentration of acid in ACN was determined to be 2% and a further increase of acid will result in more interference, which can contaminate the instrument. In the following experiment, the type of acid was identified as the next variable to be optimized. The recovery of pesticides extracted with 2% HAc-ACN and acetonitrile with 2% formic acid (2% FA-ACN) was compared. Figure 3b shows that recoveries of 54 pesticides all ranged from 70–120% when extracted with 2% HAc-ACN and 2% FA-ACN. However, when using 2% FA-ACN as the extraction reagent, 6 of 54 pesticides had recoveries between 70–80%, whereas recoveries of 54 pesticides were all above 80% in the case of 2% HAc-ACN under the same conditions. Based on the above results, 2% HAc-ACN was selected as the best extract solvent.

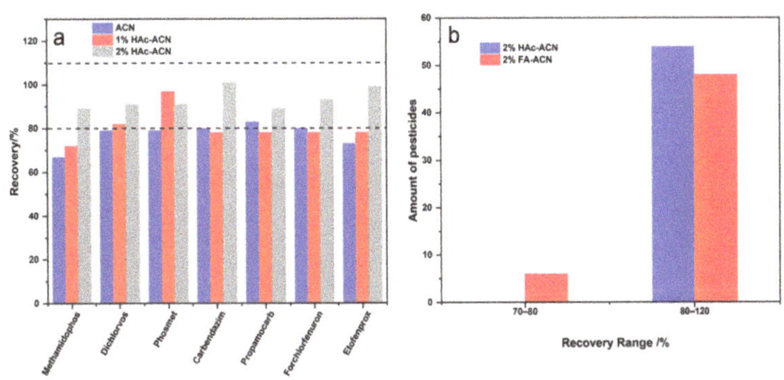

Figure 3. (**a**) Recoveries of methamidophos, dichlorvos, phosmet, carbendazim, propamocarb, forchlofenuron, and etofenprox extracted with pure ACN, 1% HAc-ACN and 2% HAc-ACN; The upper and lower dotted line indicated recoveries of 110% and 80%, respectively. (**b**) Amount of pesticides in different recovery range using 2% HAc-ACN and 2% FA-ACN as extract solvents.

2.2.2. Clean-Up

The clean-up procedure is a critical step for the determination of pesticides. The optimum absorbent should enable satisfactory recovery as well as minimum interferences. Proso millet comprises complex components, including carbohydrates, proteins, fats, and dietary fibers. Thus, the optimization of absorbents to remove interferences is essential.

Commonly used absorbents for purification include C_{18}, PSA, and GCB. C_{18} is used to remove non-polar substances such as lipids and fats due to its large surface [27]. PSA, as a weak anion exchange filler, is used for the removal of fatty acids and sugars [28]. GCB with a large surface area exhibited a good clean-up effect on pigments [29]. In the following study, the recovery of 54 pesticides purified with different combinations of absorbents (1: 37.5 mg PSA + 225 mg $MgSO_4$; 2: 37.5 mg PSA + 225 mg $MgSO_4$ + 7.5 mg GCB; 3: 37.5 mg PSA + 225 mg $MgSO_4$ + 50 mg C_{18}; 4: 37.5 mg PSA + 50 mg C_{18} +7.5 mg GCB + 225 mg $MgSO_4$) was investigated. The recovery of 54 pesticides was all within an acceptable range between 70–110% and no obvious difference was exhibited when using four kinds of absorbents [30,31]. Taking minimum interferences into consideration, a combination of 37.5 mg PSA + 50 mg C_{18} + 7.5 mg GCB + 225 mg $MgSO_4$ was chosen as the best absorbent for the final method.

2.3. Method Validation

The optimized method was validated in terms of linear range, LOQ, matrix effect, accuracy, and precision. Standard solutions were prepared using matrix extract with a concentration from 0.004–0.2 mg/L. 54 pesticides exhibited a satisfactory linear relationship between peak area and concentration with a correlation coefficient (r^2) > 0.99. LOQ was determined as the concentration with the signal-to-noise ratio (S/N) to be 10. As shown in Table 2, the calculated LOQ was in the range of 0.25 to 10 μg/kg.

Table 2. Method limit of quantification, precision of 54 pesticides in proso millet.

Pesticide	R^2	LOQ μg/kg	ME (%)	Recovery, % (RSD, %)		
				0.01 mg/kg	0.1 mg/kg	0.2 mg/kg
Methamidophos	0.9996	3.3	−5.7	86 (2)	89 (2)	89 (2)
Phorate	0.9900	2.0	6.8	95 (17)	85 (19)	105 (5)
Omethoate	0.9987	0.50	−18.0	98 (3)	100 (2)	94 (6)
Dichlorvos	0.9932	3.3	−18.5	93 (3)	91 (6)	95 (4)
Triazophos	0.9900	2.0	−36.4	103 (3)	102 (6)	107 (3)
Dimethoate	0.9975	1.7	−9.8	101 (3)	101 (2)	98 (2)
Chlorpyrifos	0.9970	4.0	−29.0	94 (5)	98 (5)	100 (5)
Acephate	0.9986	6.7	−8.0	99 (5)	92 (7)	98 (3)
Malathion	0.9905	1.0	−45.7	99 (7)	98 (17)	100 (5)
Phosalone	0.9977	2.5	−38.4	99 (11)	94 (10)	104 (6)
Phosmet	0.9939	10.0	−43.4	92 (5)	91 (11)	96 (6)
Isocarbophos	0.9943	5.0	−48.0	103 (4)	101 (3)	101 (1)
Diazinon	0.9976	3.3	−17.0	101 (4)	106 (6)	100 (2)
Profenofos	0.9994	3.3	−32.0	103 (8)	100 (7)	101 (7)
Phorate Sulfone	0.9904	0.67	−32.3	105 (3)	100 (6)	101 (7)
Phorate Sulfoxide	0.9929	0.50	−21.7	106 (8)	102 (5)	108 (1)
Tau-Fluvalinate	0.9909	0.33	−32.1	100 (11)	93 (9)	99 (11)
Iprodione	0.9989	2.0	−41.5	102 (8)	106 (2)	97 (2)
Deltamethrin	0.9960	8.3	−28.6	104 (8)	104 (5)	92 (14)
Fenpropathrin	0.9956	2.0	−29.7	97 (12)	104 (6)	104 (4)
Triadimefon	0.9994	5.0	−10.7	107 (4)	96 (4)	103 (2)
Aldicarb	0.9981	0.50	1.7	103 (11)	100 (4)	105 (2)
Aldicarb Sulfone	0.9994	3.0	−10.2	107 (7)	102 (4)	100 (6)
Aldicarb Sulfoxide	0.9972	0.033	−11.9	104 (10)	100 (9)	97 (8)
Carbofuran	0.9968	1.0	−32.4	105 (8)	109 (2)	104 (3)
3-HydroxyCarbofuran	0.9991	2.0	−26.1	103 (3)	104 (3)	100 (5)
Methomyl	0.9999	0.67	3.3	98 (5)	105 (3)	105 (1)
Carbaryl	0.9946	5.0	−30.2	95 (11)	96 (13)	105 (4)
Carbendazim	0.9995	0.67	−26.5	99 (2)	101 (2)	96 (2)
Phoxim	0.9982	10.0	−24.3	99 (3)	93 (1)	102 (3)
Pyridaben	0.9981	1.0	−43.3	102 (8)	114 (4)	107 (1)
Pyrimethanil	0.9986	2.5	−34.9	104 (8)	94 (5)	100 (4)

Table 2. *Cont.*

Pesticide	R^2	LOQ μg/kg	ME (%)	Recovery, % (RSD, %)		
				0.01 mg/kg	0.1 mg/kg	0.2 mg/kg
Difenoconazole	0.9970	2.0	−31.7	106 (5)	109 (4)	107 (3)
Acetamiprid	0.9987	0.50	−23.4	100 (2)	101 (2)	107 (3)
Imidacloprid	0.9978	2.0	−44.3	104 (3)	101 (3)	96 (3)
Dimethomorph	0.9970	3.3	−4.4	100 (6)	96 (5)	96 (4)
Pendimethalin	0.9955	10.0	30.6	95 (6)	100 (6)	94 (11)
Azoxystrobin	0.9983	2.5	−0.8	93 (10)	99 (7)	104 (4)
Thiamethoxam	0.9998	1.0	−14.1	100 (5)	105 (5)	100 (5)
Chlorfluazuron	0.9998	0.25	−39.4	106 (5)	99 (5)	95 (7)
Prochloraz	0.9976	0.50	21.0	90 (20)	99 (4)	96 (9)
Chlorbenzuron	0.9972	1.0	−40.9	102 (5)	103 (3)	103 (5)
Diflubenzuron	0.9993	2.0	−39.6	102 (7)	101 (3)	106 (2)
Propamocarb	0.9974	0.40	10.1	83 (2)	89 (3)	90 (5)
Forchlorfenuron	0.9982	2.0	−67.3	98 (3)	93 (4)	96 (2)
Etofenprox	0.9947	1.0	6.9	102 (7)	99 (5)	100 (5)
Chlorantraniliprole	0.9980	2.0	−57.2	98 (8)	104 (5)	85 (4)
Pyraclostrobin	0.9924	1.0	−29.0	103 (8)	98 (5)	101 (3)
Metalaxyl	0.9964	1.0	−40.0	105 (4)	102 (4)	100 (4)
Paclobutrazol	0.9963	3.3	−54.0	104 (7)	101 (4)	102 (3)
Fipronil	0.9946	1.0	58.6	99 (1)	103 (4)	103 (1)
Fipronil Desulfinyl	0.9935	2.0	−2.7	105 (1)	102 (2)	106 (1)
Fipronil Sulfone	0.9958	2.0	−4.9	102 (1)	105 (2)	105 (1)
Fipronil Sulfide	0.9936	1.0	23.0	106 (1)	103 (1)	106 (1)

Matrix effects (ME) were evaluated by comparing the slope of the matrix-matched and solvent-based calibration curves in triplicate. According to ME (%) value, there are two cases: (1) the matrix effect is ignored if the value is within the range from −20% to 20%; (2) the matrix effect is significant if the value is lower than -20% or higher than 20% [32]. ME (%) value of 54 pesticides are exhibited in Table 2. The results show that 19 pesticides exhibited ignored matrix effects, while 4 and 31 pesticides exhibited significant enhancement and suppression effects, respectively. The matrix effect of polar compounds (log Kow < 1) was reported to be more significant [33]. However, methamidophos, acephate, thiamethoxam, and methomyl exhibited an ignored matrix effect in our manuscript. It was suggested that the matrix effect was closely related to the sample type and extraction method [34]. To compensate for matrix effects, a matrix matrix-matched calibration curve was applied to obviate possible interferences for quantification in samples.

The accuracy and precision of the method were evaluated by spiking blank proso millet samples at 3 concentration levels with five replications: 0.01, 0.1, and 0.2 mg/kg. As shown in Table 2, recoveries of 54 pesticides were within the acceptable range from 86% to 114%. Precision was evaluated by relative standard deviation (RSDs, %), which was in the range of 1–20%. According to the Guideline for the Testing of Pesticide Residues in Crops (NY/T 788-2018) [30], the recovery criteria at spiking levels of 0.01, 0.1 and 0.2 mg/kg is 60%–120% (RSD $\leq$ 30%), 70%–120% (RSD $\leq$ 20%) and 70%–110% (RSD $\leq$ 15%), respectively. The recovery based on the EU's criteria of method validation procedures (SANTE/11312/2021) is 70%–110% with RSD $\leq$ 20% [31]. The results of recovery and RSDs could meet the requirements of NY/T 788-2018 and SANTE/11312/2021.

All of the above data demonstrated that the optimized QuEChERS was reliable for the determination of 54 pesticides in the proso millet sample.

2.4. Real Sample Analysis

The optimized were utilized to screen and quantify pesticides in 50 samples, which were collected from local farmer's fields. The results in Table 3 show that aldicarb sulfone and imidacloprid were detected in one batch of samples respectively, with concentrations of 0.022 and 0.011 mg/kg, while the other 52 pesticides were not detected in all samples. The residue of aldicarb sulfone might have resulted from the application of aldicarb during planting. Maximum residue levels (MRLs) of pesticides in proso millet are not prescribed in China [35]. However, MRLs of aldicarb and imidacloprid in other similar grains, such as millet, are 0.02–0.05 mg/kg therefore, it might pose a residue risk on the consumers of proso millet. Furthermore, no pesticide product for proso millet is registered according to China Pesticide Information Network. Thus, relevant work is suggested to monitor the dietary intake risk of pesticides on proso millet.

Table 3. Detection of pesticide residue in proso millet samples (n = 50) collected from a local field.

Sample No.	Pesticide	Concentration (mg/kg)	MRL (mg/kg)
7	Aldicarb sulfone	0.022	NA [1]
25	Imidacloprid	0.011	NA [1]

[1] Not Available.

3. Materials and Methods

3.1. Chemicals and Reagents

The certified reference materials of 54 pesticides (Table 1) in this work (≥98.0%) were purchased from the Agro-environmental Quality Supervision, Inspection & Testing Center (Tianjin, China). Methanol (HPLC grade) was obtained from Merck (Darmstadt, Germany). Acetonitrile (HPLC grade) was purchased from Tedia Company (Fairfield, OH, USA). Acetic acid (HPLC grade) and formic acid (HPLC grade) were from Fisher Regent Company (Beijing, China). Ammonium acetate (HPLC grade) was from Tianjin Guangfu Fine Chemical Research Institute (Tianjin, China), QuEChERS extraction salts pachets (4 g magnesium sulfate, 1 g sodium chloride, 1 g sodium citrate and 0.5 g disodium citrate sesquihydrate) were provided by Shimadzu Corporation. Anhydrous magnesium sulfate ($MgSO_4$) (>99.0%) was from Sinopharm Chemical Reagent Co., Ltd. (Beijing, China). Primary secondary amine (PSA, 40–60 μm), graphitized carbon black (GCB, 120–400 Mesh), and octadecylsilane (C_{18}, 50 μm, 60 A) were from Agela Technologies Inc. (Tianjin, China).

3.2. Sample Pretreatment

A 5.0 g homogenized sample was weighed into a 50 mL centrifuge tube. Then 5 mL of water was added to the sample and allowed to stand for 20 min. Afterwards, 10 mL of 2% of HAc in ACN was added and the mixture was then vortexed at 2500 rpm for 10 min. Subsequently, QuEChERS extraction salts were added, and the tubes were vortexed for another 5 min. The sample was then centrifuged at 8000 rpm for 3 min and the upper layer was collected.

1.5 mL of the upper organic layer was transferred into a 2 mL centrifuge tube containing 225 mg $MgSO_4$, 50 mg C_{18}, 7.5 mg GCB, and 37.5 mg PSA. Then the mixture was vortexed at 2500 rpm for 5 min followed by centrifugation at 5000 rpm for 2 min. Finally, 1.5 mL of the upper layer was filtered through a 0.22 μm nylon syringe filter and transferred into autosampler vials.

3.3. LC–MS/MS Analysis

Analysis of 54 pesticides was performed on Triple Quad 4500 (AB SCIEX, Framingham, MA, USA). Separation of target compounds was achieved on Waters ACQUITY UPLC®BEH C18 column (100 mm × 2.1 mm, 1.7 μm) at a temperature of 40 °C. The mobile phase consisted of 0.1% formic acid (FA) aqueous solution containing 4 mM ammonium acetate (A) and methanol (B). Gradient elution procedure was: 5–20% B at 0–1 min, 20–40% B for 0.10 min, 40–60% B for 1.9 min, 60–80% B for 0.1 min, 80–95% B for 1.9 min and holding for 2 min, 95–5% B for 2 min then holding for 1 min. The flow rate of the mobile phase was 0.30 mL/min and the injection volume was 2 μL.

Mass spectrometry analyses of pesticides were conducted in both positive mode (ESI^+) and negative mode (ESI^-). Mass spectrometry parameters were set as follows: curtain gas 35 psi, collision gas 9, ion source gas 1 (GS1) 55 psi, ion source gas 2 (GS2) 55 psi, ionspray voltage 5500 V (ESI^+) and 4500 V (ESI^-), temperature 550 °C. MRM parameters and retention time of 54 pesticides are shown in Table 1.

3.4. Method Validation

Method validation concerning linearity, matrix effect (ME), limit of quantification (LOQ), accuracy, and precision, was conducted according to SANTE/11312/2021 guidelines [31]. In detail, linearity was performed by preparing matrix-matched standards at concentrations of 0.004, 0.01, 0.02, 0.04, 0.1, and 0.2 mg/L. The ME was evaluated by comparing the slope of matrix-matched calibration curve and solvent-based calibration curve according to the following equation:

$$\text{ME}\% = \left(\frac{\text{Slope of matrix matched calibration curve}}{\text{Slope of solvent based calibration curve}} - 1 \right) \times \%$$

The accuracy and precision were evaluated by spiking the proso millet sample at three levels of 0.01, 0.1, and 0.2 mg/kg with five replications. The limit of quantification (LOQ) of 54 pesticides was determined as the concentration with the signal-to-noise ratio (S/N) to be 10.

4. Conclusions

In this paper, QuEChERS-UPLC-MS/MS method was developed and validated for the simultaneous detection of multi-class pesticides in proso millet. A series of optimizations were carried out in terms of chromatographic conditions, extraction, and purification. Satisfactory validation results including linearity, matrix effect, LOQs, accuracy, and precision, were obtained for all target pesticides. The method was applied for the analysis of 50 samples, which demonstrated its feasibility for monitoring pesticide residue in proso millet.

Author Contributions: Methodology, C.D.; validation, P.R. and Y.Q.; sample analysis, P.R. and C.D.; data curation, Y.Q. and C.D.; writing—original draft preparation, C.D. and Y.Y.; writing—review and editing, S.Q.; funding acquisition, C.D. All authors have read and agreed to the published version of the manuscript.

Funding: This research was funded by Research Project Supported by Shanxi Scholarship Council of China (2020-158).

Institutional Review Board Statement: Not applicable.

Informed Consent Statement: Not applicable.

Data Availability Statement: Not applicable.

Conflicts of Interest: The authors declare no conflict of interest.

Sample Availability: Not applicable.

References

1. Ren, J.; Ma, C.; Li, M.; Dang, Y.; Yu, X.; Du, S. Physicochemical, structural and functional properties of non-waxy and waxy proso millet protein. *Foods* **2023**, *12*, 1116. [CrossRef] [PubMed]
2. Zheng, N.-N.; Yun, T.-T.; Yao, Y.; Ren, G.-X.; Wang, C.-L.; Qi, W.-T. Comparative analysis of nutritional and functional components in different breeds of proso millet. *Sci. Technol. Food Ind.* **2017**, *1*, 348–359.
3. Lu, R.; Tian, Q.; Sun, N.; Qiao, Z.-J.; Zhang, L.-Z.; Niu, W.; Shan, L.; Ji, P.-S. Determination of phenolic compounds in proso millet by high performance liquid chromatography. *Food Sci.* **2014**, *35*, 152–155.
4. Nishizawa, N.; Fudamoto, Y. The elevation of plasma concentration of high-density lipoprotein chloesterol in mice fed with protein from proso millet. *Biosci. Biotechnol. Biochem.* **1995**, *59*, 333–335. [CrossRef]
5. Kumari, K.S.; Thayumanavan, B. Characterization of starches of proso, foxtail, barnyard, kodo and little millets. *Plant Foods Hum. Nutr.* **1998**, *53*, 47–56. [CrossRef] [PubMed]
6. Denery-Papini, S.; Nicolas, Y.; Popineau, Y. Efficiency and limitation of immunochemical assays for the testing of gluten-free foods. *J. Cereal Sci.* **1999**, *30*, 121–131. [CrossRef]

7. Nishizawa, N.; Sato, D.; Ito, Y.; Nagasawa, T.; Hatakeyama, Y.; Choi, M.-R.; Choi, Y.-Y.; Wei, Y.M. Effects of dietary protein of proso millet on liver injury induced by D-galactosamine in rats. *Biosci. Biotechnol. Biochem.* **2002**, *66*, 92–96. [CrossRef] [PubMed]
8. Lee, J.; Jung, M.W.; Lee, J.; Lee, J.; Shin, Y.; Kim, J.H. Dissipation of the insecticide cyantraniliprole and its metabolite IN-J9Z38 in proso millet during cultivation. *Sci. Rep.* **2019**, *9*, 11648. [CrossRef] [PubMed]
9. Song, M.-H.; Yu, J.-W.; Keum, Y.-S.; Lee, J.-H. Dynamic modeling of pesticide residue in proso millet under multiple application situations. *Environ. Pollut.* **2023**, *334*, 121993. [CrossRef]
10. Wang, F.Y.; Li, X.; Yu, S.M.; He, S.H.; Cao, D.T.; Yao, S.J.; Fang, H.; Yu, Y.L. Chemical factors affecting uptake and translocation of six pesticides in soil by maize (*Zea mays* L.). *J. Hazard. Mater.* **2021**, *405*, 124269. [CrossRef]
11. Ju, C.; Dong, S.; Zhang, H.; Yao, S.; Wang, F.; Cao, D.; Xu, S.; Fang, H.; Yu, Y. Subcellular distribution governing accumulation and translocation of pesticides in wheat (*Triticum aestivum* L.). *Chemosphere* **2020**, *248*, 126024. [CrossRef] [PubMed]
12. Yu, T.; Shen, J.-S.; Bai, H.-H.; Guo, L.; Tang, J.-J.; Jiang, Y.-B.; Xie, J.-W. A photoluminescent nanocrystal-based signaling protocol highly sensitive to nerve agents and highly toxic organophosphate pesticides. *Analyst* **2009**, *134*, 2153–2157. [CrossRef] [PubMed]
13. Anastassiades, M.; Lehotay, S.J.; Stajnbaher, D.; Schenck, F. Fast and easy multiresidue method employing acetonitrile extraction/partitioning and "dispersive solid-phase extraction" for the determination of pesticide residues in produce. *J. AOAC Int.* **2003**, *86*, 412–431. [CrossRef] [PubMed]
14. Hou, X.; Han, M.; Dai, X.H.; Yang, X.F.; Yi, S. A multi-residue method for the determination of 124 pesticides in rice by modified QuEChERS extraction and gas chromatography–tandem mass spectrometry. *Food Chem.* **2013**, *138*, 1198–1205. [CrossRef] [PubMed]
15. He, Z.; Wang, L.; Peng, Y.; Luo, M.; Wang, W.; Liu, X. Multiresidue analysis of over 200 pesticides in cereals using a QuEChERS and gas chromatography–tandem mass spectrometry-based method. *Food Chem.* **2015**, *169*, 372–380. [CrossRef] [PubMed]
16. Kolberg, D.I.; Prestes, O.D.; Adaime, M.B.; Zanella, R. Development of a fast multiresidue method for the determination of pesticides in dry samples (wheat grains, flour and bran) using QuEChERS based method and GC-MS. *Food Chem.* **2011**, *165*, 1436–1442. [CrossRef]
17. Ruan, H.; Rong, W.; Ma, Y.; Ji, W.; Liu, H.; Song, N. Determination of 34 pesticides residues in rice, proso millet and wheat with QuEChERS-online gel permeation chromatography-gas chromatography-mass spectrometry. *Chin. J. Chromatogr.* **2013**, *31*, 1211–1217. [CrossRef] [PubMed]
18. Demir, E.; Göktug, Ö.; Inam, R.; Doyduk, D. Development and characterization of iron (III) phthalocyanine modified carbon nanotube paste electrodes and application for determination of fluometuron herbicides as an electrochemical sensor. *J. Electroanal. Chem.* **2021**, *895*, 115389. [CrossRef]
19. Krone, N.; Hughes, B.A.; Lavery, G.G.; Stewart, P.M.; Arlt, W.; Shackleton, C.H. Gas chromatography/mass spectrometry (GC/MS) remains a pre-eminent discovery tool in clinical steroid investigations even in the era of fast liquid chromatography tandem mass spectrometry (LC/MS/MS). *J. Steroid Biochem. Mol. Biol.* **2010**, *121*, 496–504. [CrossRef]
20. Jiao, W.; Xiao, Y.; Qian, X.; Tong, M.; Hu, Y.; Hou, R.; Hua, R. Optimized combination of dilution and refined QuEChERS to overcome matrix effects of six types of tea for determination eight neonicotinoid insecticides by ultra performance liquid chromatography–electrospray tandem mass spectrometry. *Food Chem.* **2016**, *210*, 26–34. [CrossRef]
21. Field, J.K.; Euerby, M.R.; Haselmann, K.F.; Petersson, P. Investigation into reversed-phase chromatography peptide separation systems Part IV: Characterisation of mobile phase selectivity differences. *J. Chromatogr. A* **2021**, *1641*, 461986. [CrossRef] [PubMed]
22. McCalley, D.V. Effect of buffer on peak shape of peptides in reversed-phase high performance liquid chromatography. *J. Chromatogr. A* **2004**, *1038*, 77–84. [CrossRef] [PubMed]
23. *EN15662-2018*; Foods of Plant Origin—Determination of Pesticide Residues Using GC-MS and/or LC-MS/MS Following Acetonitrile Extration/Partitioning and Clean-Up by Dispersive SPE—QuEChERS-Method. Polski Komitet Normalizacyjny: Warsaw, Poland, 2008.
24. Chiesa, L.M.; Labella, G.F.; Panseri, S.; Britti, D.; Galbiati, F.; Villa, R.; Arioli, F. Accelerated solvent extraction by using an 'in-line' clean-up approach for multiresidue analysis of pesticides in organic honey. *Food Addit. Contam. Part A Chem. Anal. Control. Expo. Risk Assess.* **2017**, *34*, 809–818. [CrossRef] [PubMed]
25. Han, L.; Sapozhnikova, Y.; Lehotay, S.J. Method validation for 243 pesticides and environmental contaminants in meats and poultry by tandem mass spectrometry coupled to low-pressure gas chromatography and ultrahigh-performance liquid chromatography. *Food Control* **2016**, *66*, 270–282. [CrossRef]
26. Li, P.P.; Cheng, J.; Le, Y. Determination of 46 pesticide residues in sulfur-containing vegetables using solid-phase extraction and gas chromatography-triple quadrupole mass spectrometry. *J. Instrum. Anal.* **2015**, *34*, 421–427.
27. Wang, S.W.; Wang, X.N.; He, Q.; Lin, H.D.; Chang, H.; Sun, H.B.; Liu, Y.P. Simultaneous Determination of Seven Pesticides and Metabolite Residues in Litchi and Longan through High-Performance Liquid Chromatography-Tandem Mass Spectrometry with Modified QuEChERS. *Molecules* **2022**, *27*, 5737. [CrossRef] [PubMed]
28. Ajibola, A.S.; Tislera, S.; Zwiener, C. Simultaneous determination of multiclass antibiotics in sewage sludge based on QuEChERS extraction and liquid chromatography-tandem mass spectrometry. *Anal. Methods* **2020**, *12*, 576–586. [CrossRef]
29. Hou, X.; Qiao, T.; Zhao, Y.; Liu, D. Dissipation and safety evaluation of afidopyropen and its metabolite residues in supervised cotton field. *Ecotox. Environ. Safe* **2019**, *180*, 227–233. [CrossRef] [PubMed]

30. *NY/T 788-2018*; Guideline for the Testing of Pesticide Residues in Crops. Ministry of Agriculture and Rural Affairs of the People's Republic of China: Beijing, China, 2018. Available online: https://www.sdtdata.com/fx/fmoa/tsLibCard/169332.html (accessed on 20 July 2023).
31. *SANTE/11312/2021*; Guidance Document on Analytical Quality Control and Method Validation Procedures for Pesticide Residues and Analysis in Food and Feed. European Commission: Brussels, Belgium, 2017. Available online: https://www.accredia.it/en/documento/guidance-sante-11312-2021-analytical-quality-control-and-method-validation-procedures-for-pesticide-residues-analysis-in-food-and-feed/ (accessed on 29 August 2023).
32. Chen, X.X.; He, S.; Gao, Y.M.; Ma, Y.C.; Hu, J.Y.; Liu, X.L. Dissipation behavior, residue distribution and dietary risk assessment of field-incurred boscalid and pyraclostrobin in grape and grape field soil via MWCNTs-based QuEChERS using an RRLC-QqQ-MS/MS technique. *Food Chem.* **2019**, *274*, 291–297. [CrossRef] [PubMed]
33. Morais, E.H.d.C.; Collins, C.H.; Jardim, I.C.S.F. Pesticide determination in sweet peppers using QuEChERS and LC–MS/MS. *Food Chem.* **2018**, *249*, 77–83. [CrossRef]
34. Pareja, L.; Cesio, V.; Heinzen, H.; Fernández-Alba, A.R. Evaluation of various QuEChERS based methods for the analysis of herbicides and other commonly used pesticides in polished rice by LC–MS/MS. *Talanta* **2011**, *83*, 1613–1622. [CrossRef]
35. *GB 2763-2021*; National Food Safety Standard-Maximum Residue Limits for Pesticides in Food. Ministry of Agriculture and Rural Affairs of the People's Republic of China: Beijing, China, 2021. Available online: https://www.sdtdata.com/fx/fmoa/tsLibCard/183688.html (accessed on 28 May 2023).

molecules

MDPI

Article

A Green Bridge: Enhancing a Multi-Pesticide Test for Food by Phase-Transfer Sample Treatment Coupled with LC/MS

Shaoming Jin [1,†], Yi Shen [2,†], Tongtong Liu [1], Ruiqiang Liang [1], Xiao Ning [1,*] and Jin Cao [1,*]

[1] Key Laboratory of Food Quality and Safety for State Market Regulation, National Institute of Food and Drug Control, Beijing 100050, China; myjackyming@126.com (S.J.)
[2] Center for Food Evaluation, State Administration for Market Regulation, Beijing 100070, China
* Correspondence: nx200730079@163.com (X.N.); caojin@nifdc.org.cn (J.C.); Tel.: +86-010-6709-5070 (X.N. & J.C.)
† These authors contributed equally to this work.

Abstract: The preparation and treatment of the sample has become an important part of the determination process, which directly affects the accuracy of detection. The preparation of the sample for final detection is actually a process of separation and transfer of the target to be tested from the sample matrix. The phase-transfer process of analysis and detection is the process of transferring the target substance to be measured from a complex multiphase system to a simple homogeneous system. This study shows a new phase-transfer process for food sample pretreatment in the determination of carbamate pesticides. Edible gum, xanthan gum, carrageenan, and gelatin were selected for purification testing from the perspective of eco-friendliness and safety. Phase-transfer purification process research was carried out on spinach and other foods. Compared with the commonly used QuEChERS method, the LC/MS results indicate that the straightforward carrageenan treatment process can significantly diminish the detection matrix effect and yield similarly superior detection parameters. The phase-transfer purification method with carrageenan has similar sensitivity and systematic error. The limits of detection and limits of quantitation of each pesticide compound in six plant sample substrates were 0.02–0.36 μg/kg and 0.06–1.9 μg/kg, respectively, which were lower than the residue limits here and abroad. Supplemental recoveries in six blank samples at 5, 20, and 100 μg/kg with the phase-transfer process method were better than those for the QuEChERS method. Positive determination results of actual samples using carrageenan phase-transfer purification proved that this method can be used for related detection from a practical point of view.

Keywords: phase-transfer purification; QuEChERS; carrageenan; carbamate pesticides

Citation: Jin, S.; Shen, Y.; Liu, T.; Liang, R.; Ning, X.; Cao, J. A Green Bridge: Enhancing a Multi-Pesticide Test for Food by Phase-Transfer Sample Treatment Coupled with LC/MS. *Molecules* **2023**, *28*, 6756. https://doi.org/10.3390/molecules28196756

Academic Editor: Miguel Ángel González-Curbelo

Received: 21 July 2023
Revised: 19 September 2023
Accepted: 20 September 2023
Published: 22 September 2023

1. Introduction

Samples for analysis and detection often have a complex matrix, which coexists with the target substance to be measured, and the co-existing matrix substance will interfere with the detection of the target substance during the determination process, resulting in deviation of the result. Therefore, the preparation and treatment of the sample has become an important part of the detection, and it directly affects the accuracy of the determination. This effect is particularly serious when there are multiple detection targets in the sample and there are certain concentration differences among the multiple detection targets. From the point of view of accurate measurement, the matrix interference co-existing with the target to be detected in the sample is an important aspect of the late detection deviation, that is, the deviation of the detection signal [1–3]. The purpose of sample pretreatment is also to remove the relevant coexisting matrix in the fully extracted sample extract so that the detection environment of the final target matches the uncomplicated environment of the corresponding reference substance and the traceable value of the target substance in the sample can be obtained by comparison [4,5]. In recent years, many QuEChERS methods have been used for the detection of multi-pesticide residues, solvent extraction, dehydration, desorption with matrix, salting out, and removal of matrix by adsorption. With the increase in applications and reference standards, a variety of related formulations for adjusting the extraction and purification ratio have been widely used [6–10]. However, at present, this simple and fast method still has some problems. First, it is not a green treatment method, and there is a large amount of wasted reagents and consumables in the treatment. Second, for the sample matrix, not all sample formulations have the corresponding universality, and there is selectivity for relevant substrates, such as in food samples. Even if it is only for plant foods and the same target substance to be detected, for the proportion of chlorophyll, oil, protein, and water, it is necessary to adjust the extraction or purification formula.

How do we effectively extract, separate, and transfer the substance to be tested into a simple substrate environment through the coexistence of the matrix and the substance to be tested, and in the determination process, the standard substance with which it is compared in the same or similar detection environment? The approximate response results can be obtained so that the accuracy of the test results can be guaranteed and the reference with the standard material can be realized. The so-called phase-transfer process of analysis and detection is the process of transferring the target substance to be measured from a complex multiphase system to a simple homogeneous system. This is done to realize the premise assumptions of various detection methods in the process of establishing the detection principle in the detection. In order to achieve the removal of ineffective macro substances, the food industry should concentrate on obtaining nutrients as much as possible and then adopt corresponding homogenization or homogenization means to obtain relatively uniform food products [11,12]. For example, the use of flocculants [13], coagulants [14], edible glue [15], etc., in the process suggests that in the detection of food products, the corresponding green flocculation, edible glue enrichment, and other means can also be used for the phase transfer or purification of substances in the sample. For example, polyacrylamide is a commonly used synthetic polymer, which is often used in the sugar industry for flocculation to remove impurities in the virgin liquid and purify sugar [16]. Microbial flocculant common glycoproteins, mucopolysaccharides, proteins, cellulose, and other substances are used for the purification of drinking water and the removal of colored substances [17]. Edible gum is a kind of food additive. In meat processing, polysaccharide gum is commonly used to fix fat and starch to form homogeneous products. The use of xanthan gum jelly or candy, pigment, and polar substances, which can be evenly dispersed in the product, shows no phase separation [18]. Gellan gum, agar, carrageenan, and so forth in preserves, icing, jelly gel, yogurt, and other products play the role of stabilizer, dispersant, and thickening agent [19–22].

As organic chlorine pesticides are banned and the number of insect varieties resistant to organophosphorus insecticides is increasing, the amount of carbamate pesticides is increasing year by year, resulting in the residues of such pesticides in the environment and crops. Countries have formulated maximum residue limit (MRL) standards for carbamate pesticides [23]. Because of the structural characteristics of carbamate pesticides, the amino group is directly connected with the carbonyl group of carbamate. Hence, the polarity is strong and the thermal stability is poor, and it needs to be derived before it can be determined by gas chromatography (GC) [24]. Liquid chromatography–triple quadrupole tandem mass spectrometry is the preferred method for trace analysis because of its high sensitivity and good anti-interference ability [25,26]. In this study, on the basis of screening, common inorganic and organic flocculants, as well as edible gum, xanthan gum, carrageenan, and gelatin, were selected for purification testing from the perspectives of greenness and safety, as well as their physical and chemical properties. The test optimization was carried out on spinach and other foods. It was determined that in the detection of carbamate pesticide residues, carrageenan can be simply used to purify the sample extract solution. Compared with the commonly used QuEChERS method, the results show that the simple process of carrageenan treatment can effectively reduce the matrix effect of detection and obtain better detection parameters. It can be used as a method for the detection of carbamate pesticide residues in plant foods.

2. Results and Discussion

2.1. Selection of Food Substrates and Target Substances

The food matrix is complex, the pesticide residue is low, and the interference factors are many. The main reason is that the residual components are not easy to separate, enrich, and purify, and so the detection of relevant pesticides is not accurate. The extraction of target substances and the selection and optimization of substrate purification methods in sample pretreatment have become the key points of pesticide multi-residue detection. In 2003, Anastassiades and Lehotay proposed the original QuEChERS method [27]. Current QuEChERS methods include the original method, the American Society of Analytical Chemists standard method (AOAC 2007.01) [28], and the European Committee for Standardization standard method (EN 15662). According to the nature of the EN 15662 method, plant foods are divided into acid samples, high-water-content samples (water content $\geq$ 80%), low-water-content samples (water content < 80%), dry samples (grains), pigmentary samples, etc. According to the classification system, two pretreatment methods were used to treat the substrates of different plant foods. Specifically, there are six representative substrates, including acid samples (lemon, pH 3), high-water-content samples (apple and cabbage, 85–90% water content), low-water-content samples (banana, 70% water content), grain (rice, $\leq$10% water content), and pigment samples (spinach and chlorophyll). At the same time, according to the Maximum Residue Limits of Pesticides in Food (GB 2763-2021) [29], a total of 30 kinds of common carbamate pesticides and their active metabolites were selected with MRL standards.

2.2. Optimization of LC–MS/MS Conditions

In the positive ion mode of the electrospray ion source, a single standard solution of 30 kinds of carbamate pesticides was scanned to obtain stable parent ions, and the breakage voltage was optimized by SIM scanning mode. Two characteristic ion pairs with high response value were then selected for each compound as quantitative and qualitative ion pairs to further optimize the collision energy. The optimization results are shown in Table 1.

Table 1. LC–MS/MS parameters of the 30 carbamate pesticides.

Pesticides	Relative Retention Time **	Transitions (*m*/*z*)	Dewell Time (mS)	Fragmentor (V)	CE (eV)
aminocarb	0.06	209.3/152.1 *, 209.3/137.1	100	70	9, 25
propamocarb	0.06	189.2/102.1 *, 189.2/144.1	100	70	13, 9
aldicarb sulfoxide	0.11	229.1/166.1 *, 229.1/109.1	60	70	5, 13
aldicarb sulfone	0.13	245.1/166.1 *, 245.1/109.1	60	60	13, 17
oxamyl	0.13	242.1/72.1 *, 242.1/121.2	60	55	17, 9
thiofanox sulfoxide	0.26	257.1/200.0 *, 257.1/137.2	40	65	5, 13
pirimicarb	0.31	239.3/72.1 *, 239.3/182.1	40	65	20, 13
thiofanox sulfone	0.31	273.1/216.1 *, 273.1/137.1	40	65	9, 21
3-hydroxycarbofuran	0.38	238.4/163.1 *, 238.4/181.1	40	100	9, 5
dioxacarb	0.38	224.2/167.1 *, 224.2/123.1	40	40	5, 13
aldicarb	0.63	213.1/89.1 *, 213.1/116.1	200	75	13, 9
metolcarb	0.73	166.2/109.1 *, 166.2/94.1	200	30	9, 35

Table 1. *Cont.*

Pesticides	Relative Retention Time **	Transitions (*m/z*)	Dewell Time (mS)	Fragmentor (V)	CE (eV)
propoxur	0.82	210.2/111.0 *, 210.2/168.3	60	30	9, 4
carbofuran	0.85	222.3/165.1 *, 222.3/123.1	60	70	9, 21
bendiocarb	0.85	224.1/167.1 *, 224.1/109.1	60	70	5, 17
carbaryl	0.95	202.1/145.0 *, 202.1/127.3	60	40	5, 33
ethiofencarb	1.00	226.1/107.1 *, 226.1/164.1	60	50	5, 9
thiofanox	1.02	241.1/184.1 *, 241.1/57.2	60	60	5, 17
thiocarb	1.06	377.0/64.1 *, 377.0/113.0	60	120	13, 9
isoproarb	1.13	194.1/137.1 *, 194.1/95.1	60	60	5, 9
2,3,5-trimethacarb	1.13	137.1/122.0 *, 137.1/107.2	60	130	17, 25
fenobucarb	1.27	208.1/95.1 *, 208.1/152.1	40	60	9, 4
diethofencarb	1.29	268.2/226.1 *, 268.2/152.2	40	55	21, 5
methiocarb	1.31	226.1/169.1 *, 226.1/121.1	40	55	5, 17
promecarb	1.33	208.1/109.1 *, 208.1/105.1	40	50	5, 13
fenoxycarb	1.53	302.3/88.1 *, 302.3/116.1	100	100	17, 5
indoxacarb	1.70	528.1/150.2 *, 528.1/293.2	100	150	21, 9
benfuracarb	1.75	411.2/195.0 *, 411.2/252.1	100	90	21, 9
furathiocarb	1.80	383.2/252.1 *, 383.2/167.1	100	90	5, 25
carbosuifan	2.06	381.6/118.1 *, 381.6/160.1	200	130	13, 9

* Quantitative ion; ** relative retention time with reference to ethiofencarb.

At the same time, an Agilent Zorbax Eclipse XDB C18 column was compared with the Japanese Shiseido Type MG III column. They were subjected to testing using identical liquid phases, 0.1% (*v*/*v*) formic acid aqueous solution, and 0.1% (*v*/*v*) formic acid methanol solution. It was found that the separation degree and peak shape of aminocarb and propamocarb, which peaked first, were better when the latter was used, and the baseline drift was smaller. Therefore, a Shiseido Type MG III column was selected. In addition, the mobile phase systems with different compositions, NH_4Ac (5 mmol/L)–acetonitrile, NH_4Ac (5 mmol/L)–methanol, 0.1% (*v*/*v*) formic acid aqueous solution–0.1% (*v*/*v*) formic acid acetonitrile solution, and 0.1% (*v*/*v*) formic acid aqueous solution–0.1% (*v*/*v*) formic acid methanol solution, were compared. The results showed that when using acetonitrile, the peak shape of methiocarb was very poor, and the separation effect of 30 pesticides was not as good as that when the organic phase was methanol. The mobile-phase system of 0.1% (*v*/*v*) formic acid aqueous solution and 0.1% (*v*/*v*) formic acid methanol solution was more convenient to prepare and had good separation effect, and so we selected it as the mobile phase. Figure 1 shows the ion chromatogram for the extraction of 30 carbamate pesticide compounds in 100 μg/L mixed standard solution.

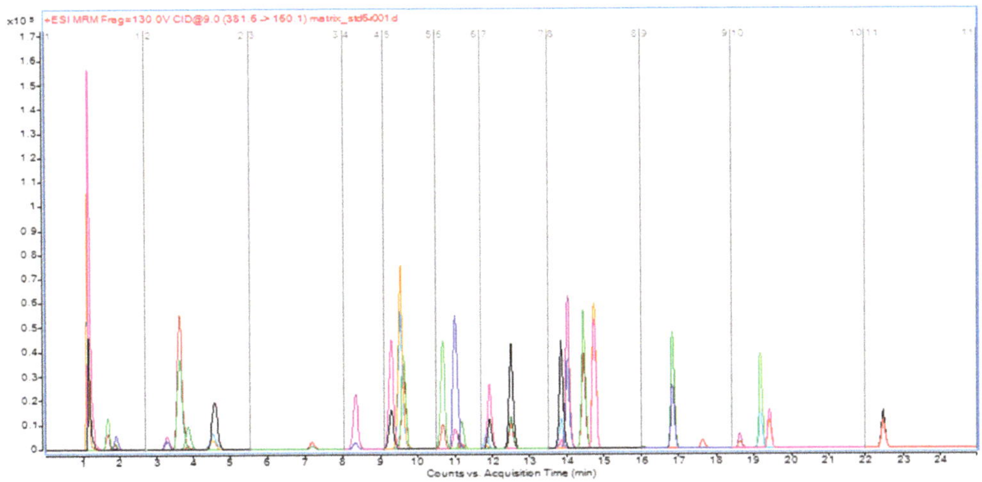

Figure 1. Extracted ion chromatogram of 30 carbamate pesticides.

2.3. *Optimization of Pre-Treatment Methods*

2.3.1. Optimization of Pre-Treatment Method 1

Selection of Purification Materials

Common flocculants include inorganic and organic chemicals, such as aluminum sulfate, magnesium sulfate, polyphosphate, and polyacrylamide. Because of certain safety problems or non-green factors, polysaccharide edible gum is mainly considered. In this study, xanthan gum, carrageenan, and gelatin were tested and selected. Xanthan gum is a water-soluble gum, which is the most characteristic of several microbial polysaccharides. It is also the largest and most widely used microbial polysaccharide in the world. Xanthan gum has good thickening, pseudoplastic rheology, water solubility, suspension, emulsion stability, acid and alkali resistance, salt resistance, temperature resistance, excellent compatibility, and other properties. It is widely used in the food industry [30]. Secondly, carrageenan also has good solubility. There are generally seven types, and the commonly used is κ-carrageenan. It can be dissolved in hot water and in cold water. Carrageenan and carrageenan sodium salt can also be dissolved, but the potassium salt and calcium salt of carrageenan can only absorb water and expand and cannot be dissolved. The gel formed by carrageenan is thermally reversible; that is, when heated, it condenses and melts into a solution, and when the solution is cooled, it forms a gel [31]. Edible gelatin consists of white or light-yellow transparent to translucent brittle flakes, particles, or powder that are lustrous, odorless, tasteless, and insoluble in cold water, ethyl ether, ethanol, chloroform, soluble in hot water, glycerin, acetic acid, salicylic acid, phthalic acid, urea, thiourea, thiocyanate, potassium bromide, and other solutions [32,33]. The relative density is 1.3–1.4, which can slowly absorb 5–10 times of cold water and expand and soften. When it absorbs more than twice the water, it is heated to 40 °C and melted into sol, and it forms a soft and elastic gel after cooling.

In the process of the experiment, the shrinking state of the gel after the addition of different acid and base was compared, and it was found that after the addition of NaOH, the water solution into the gel would leak. That is, the distribution of the target to be tested during the extraction process in the gel and the water solution is uneven, while the gelatin naturally formed in the gelatin process. Aqueous solutions also produce an unstable distribution of the situation. The state of xanthan gum strongly depends on the concentration, and the effective working concentration range used for phase transfer is too narrow, so carrageenan was finally used for treatment.

Choice of Carrageenan Quantity

After the aqueous extraction reagent is added to the sample homogenate, the amount of glue added will directly affect the co-existing matrix (mainly protein, carbohydrate, and other macro substances in the sample) and the distribution of pesticide to be detected in the solution in the later stage. According to the characteristics and solubility of carbamate components, with 20% acetonitrile water as the extraction solvent, 0.1, 0.3, 0.5, 0.7, 0.9, 1.1, 1.3, and 1.5 g of carrageenan were added separately in a water bath at 50 °C in a volume of 20 mL. After cooling to room temperature, it can be seen that under the above added amount, a gel can eventually form, and it can still become a solution by reheating. But under 0.3 g, good phase separation cannot be obtained by centrifugation. Above 1.1 g, there is expansion, indicating that it may cause uneven distribution of solute of the gel phase and water phase. Then, in the range of 0.3–1.1 g, the purification test was carried out with spinach homogenate and apple homogenate. It can be seen that in the process of gel formation, pigments and some fine fiber substances gathered in the gel, and a light-colored and transparent solution could be obtained after centrifugation. We added oxamyl and ethiofencarb to form an overall concentration of 20 μg/L in spinach and apple substrates with a corresponding volume of 20 mL. After optimization, carrageenan was added to 1.0 g and 0.5 g, and the concentrations of the two substances in spinach purification solution were 19.2 and 19.4 μg/L, respectively, with decreasing rates of 4% and 3%, and 19.5 and 19.6 μg/L in the apple purification solution, respectively, indicating that the gel phase mainly removed the matrix components. The substance to be detected is always a relatively homogeneous state in the whole system. Similarly, similar measurements were made for other substrates, and the overall concentration decreased to less than 10% in the above range, as shown in Table 2. It indicates that the overall recovery rate loss is within the allowable range. The amount of optimization for final processing is described in Section 3.2.

Table 2. Recovery of oxamyl and ethiofencarb in six types of samples with different carrageenan purification, %.

Substrates	Pesticide	Carrageenan Addition/g in 20 mL				
		0.3	0.5	0.7	0.9	1.1
Banana	Oxamyl	95.2	97.6	97.1	96.3	95.9
	Ethiofencarb	94.2	96.1	96.3	95.4	95.4
Lemon	Oxamyl	95.3	97.4	97.1	97.1	96.5
	Ethiofencarb	96.4	98.2	97.7	97.5	96.4
Apple	Oxamyl	95.4	97.5	97.2	97.1	96.5
	Ethiofencarb	94.3	98	97.9	96.4	96.2
Spinach	Oxamyl	91.4	94	94.1	95.2	96.3
	Ethiofencarb	92	93.1	95.7	96.4	96.8
Cabbage	Oxamyl	96.7	98.2	98	97.4	97.1
	Ethiofencarb	95.7	97.8	97.1	96.4	96.4
Rice	Oxamyl	97.9	97.2	96.5	96.5	95.9
	Ethiofencarb	98.1	97.2	97.1	96.8	96.1

2.3.2. Optimization of Pre-Treatment Method 2

In the experiment, the purification agents PSA and C18 and their ratio were optimized in the QuEChERS step, and the adsorption effect of purification reagent on 30 kinds of carbamate pesticides was investigated. PSA, PSA + C18, and C18 adsorbents were added to 30 kinds of pesticide mixed standard solutions with 10 μg/L concentration, and their concentrations were determined. The results show that C18 adsorbents had strong adsorption effects on carboxypropyl sulfide and carboxybutylsulfide, and their concentrations decreased by 90% and 85%, respectively. Therefore, the use of PSA as an adsorbent was eventually determined for plant foods. For samples containing part of the fat, such as rice, the fat was removed by freezing. At the same time, the influence of the dosage of PSA adsorbent on the purification effect of the blank sample matrix extract was investigated. Finally, it was determined that when the dosage of PSA reached 25 g/L, the influence of the interference peak could be significantly reduced. The purification effect was not significantly improved when the dosage continued to increase. For example, the recovery rate of six kinds of pesticides, such as methanocarb, antiaphid, dioxocarb, indocarb, prothiocarb, and furameocarb, decreased (more than 10%). For dark vegetables, the amount of GCB was investigated. It is clear that at the amount of 40.5 g/L, the influence of pigment was eliminated, and the recovery rate was not greatly affected. For acidic foods such as lemons, the amount of alkali added is optimized to reduce the influence of some carbamate substances, such as dioxocarb and ethylthiobenzocarb, by adding 400 μL of 5 mol/L NaOH solution to stabilize the relevant pesticide substances. At the same time, the alkaline solution is also used in the gumming treatment. See Section 3.2 of Materials and Methods for details.

2.4. Matrix Effect

The matrix effect will affect the repeatability, sensitivity, and accuracy of the analysis method. It is more prominent when the electrospray source is used, which mainly shows the ion inhibition effect on the target compound [34,35]. The matrix effect is the ratio of the slope of the matrix matching calibration curve to slope of the solvent standard calibration curve. The closer the ratio is to 1, the smaller is the matrix effect, and vice versa. The matrix effects of the six samples are shown in Table 3. The results showed that spinach had the largest matrix effect, followed by lemon. The basic effect of methanocarb and carbosultiocarb was the most obvious. The influence of matrix effect is often reduced by preparing a matrix matching standard curve, adding analytical protective agent and salting out. The method of matrix matching standard curve is used in this study.

Table 3. Matrix effects of 30 pesticides in six types of samples compared by using two treatment methods.

Pesticide	Banana		Lemon		Apple		Spinach		Cabbage		Rice	
	M1 *	M2 **	M1 *	M2 **	M1 *	M2 **	M1 *	M2 **	M1 *	M2 **	M1 *	M2 **
aminocarb	0.8	0.6	0.7	0.3	0.8	0.7	0.7	0.5	0.8	0.7	0.9	0.9
propamocarb	1.1	0.9	0.8	0.7	0.9	0.9	0.8	0.7	0.9	0.9	0.8	0.8
aldicarb sulfoxide	1.1	1.1	0.9	0.7	0.9	1.1	0.7	0.7	0.8	0.7	0.9	0.9
aldicarb sulfone	1.1	1.1	0.9	0.8	0.9	0.9	0.7	0.6	0.9	0.9	0.9	0.9
oxamyl	1.1	1.1	0.9	0.9	0.9	0.9	0.7	0.7	0.9	0.9	0.9	0.9
thiofanox sulfoxide	1.1	1.1	0.9	0.8	0.9	0.9	0.7	0.8	0.9	0.9	0.9	0.9
pirimicarb	1.1	1.2	0.8	0.6	0.9	0.9	0.8	0.7	0.9	0.9	0.9	0.9
thiofanox sulfone	1.1	1.1	0.8	0.7	0.9	0.9	0.8	0.7	0.9	0.9	0.9	0.9
3-hydroxycarbofuran	1.1	1.1	0.9	0.8	0.9	0.9	0.8	0.7	0.7	0.6	0.9	0.9
dioxacarb	1.1	1.1	0.7	0.7	0.9	0.9	0.7	0.6	0.9	0.9	0.8	0.8
aldicarb	1.1	1.1	0.9	0.8	0.9	0.9	0.7	0.6	0.9	0.9	0.8	0.6
metolcarb	1.1	1.1	0.8	0.8	0.9	0.9	0.7	0.6	0.9	0.9	0.9	0.9
propoxur	1.1	1.1	0.8	0.7	0.9	0.9	0.8	0.5	0.9	0.9	0.9	0.9
carbofuran	1.1	1.2	0.8	0.7	0.9	0.9	0.7	0.5	0.9	0.9	0.7	0.7
bendiocarb	1.1	1.1	0.8	0.8	0.9	0.9	0.7	0.5	0.9	0.9	0.8	0.8
carbaryl	1.1	1.2	0.8	0.9	0.9	0.9	0.7	0.5	0.9	0.9	0.8	0.8
ethiofencarb	1.1	1.1	0.8	0.7	0.9	0.9	0.7	0.4	0.9	0.9	0.8	0.7
thiofanox	1.1	1.2	0.8	0.7	0.9	0.9	0.6	0.5	0.9	0.9	0.7	0.7
thiocarb	1.2	1.3	0.9	0.8	0.9	0.8	0.8	0.8	0.9	0.9	0.9	0.9
isoproarb	1.1	1.1	0.8	0.8	0.9	0.9	0.7	0.6	0.9	0.9	0.9	0.9
2,3,5-trimethacarb	0.9	1.1	0.8	0.6	0.9	0.8	0.7	0.6	0.9	0.9	0.7	0.7
fenobucarb	1.1	1.1	0.7	0.5	0.8	0.8	0.5	0.3	0.7	0.6	0.8	0.6
diethofencarb	1.1	1.1	0.9	0.7	0.9	0.9	0.9	0.9	0.8	0.6	0.8	0.8
methiocarb	1.1	1.1	0.9	0.7	0.9	0.9	0.5	0.3	0.7	0.7	0.8	0.8
promecarb	1.1	1.1	0.8	0.7	0.9	0.9	0.6	0.4	0.7	0.7	0.9	0.9
fenoxycarb	1.1	1.1	0.8	0.7	0.9	0.9	0.6	0.3	0.9	0.9	0.8	0.8
indoxacarb	1.1	0.9	1.1	1.2	0.9	0.9	0.6	0.5	0.9	0.9	0.7	0.6
benfuracarb	1.1	0.9	0.9	0.7	0.9	0.9	0.8	0.7	0.7	0.6	0.6	0.5

Table 3. *Cont.*

Pesticide	Banana		Lemon		Apple		Spinach		Cabbage		Rice	
	M1 *	M2 **	M1 *	M2 **	M1 *	M2 **	M1 *	M2 **	M1 *	M2 **	M1 *	M2 **
furathiocarb	0.9	0.9	0.8	0.8	0.9	0.9	0.7	0.6	0.9	0.9	0.8	0.8
carbosuifan	0.9	0.9	0.8	0.6	0.9	0.9	0.7	0.6	0.8	0.8	0.9	0.9

* Sample treatment with carrageenan purification; ** sample treatment with QuEChERS.

2.5. Methodological Evaluation

2.5.1. Standard Curve and Detection Limit

According to the MRLs of Pesticides in Food (GB 2763-2021), the maximum residue range of 30 target carbamate pesticides (50–5000 μg/kg) is limited. Three concentration levels, one order of magnitude higher, one order of magnitude lower, and the same order of magnitude lower than the MRLs, were selected for investigation. Combined with the response values of each pesticide and the saturation effect of the instrument, the matrix matching mixed standard solutions of 1, 2, 5, 10, 20, 50, 100, and 200 μg/L were prepared. The linear relationship of the pesticide compounds was good, in the range of 1–100 μg/L (equivalent to 1–100 μg/kg). In the range of 1–200 μg/L, the linearity of 17 pesticide compounds, such as aminocarb and propamocarb, was poor, and better correlation coefficients could be obtained by fitting the quadratic equation, which may be due to the saturation effect of the detector or ion source [13]. Considering that the quantitative accuracy of the quadratic curve equation is not high and that the chromatographic peak shape will deteriorate when the sample concentration is too high, the maximum mass concentration of the linear curve is set at 100 μg/L. The samples with higher pesticide residues need to be diluted for detection. The linear correlation coefficients of the matrix matching curves obtained by the two treatment methods for six plant samples are all greater than 0.998. The limit of detection (LOD) and limit of quantization (LOQ) of each pesticide compound in six plant sample substrates were determined by 3 times signal-to-noise ratio and 10 times signal-to-noise ratio, which were 0.02–0.36 μg/kg and 0.06–1.9 μg/kg, respectively, which were lower than the requirements of residue limits here and abroad. The slope and intercept of the linear curve and the recovery rate at three concentration levels were compared by using the two treatment methods. The comparison values are shown in Table 4.

Table 4. Comparison of regression and recovery rates between treatment methods 1 and 2 *.

Pesticide	Rate of Linear Slope	Rate of Intercept	Comparison of Recovery		
			5 μg/kg	20 μg/kg	100 μg/kg
aminocarb	1.2	0.5	1.05	1.01	1.03
propamocarb	1.1	0.4	1.04	1.04	1.01
aldicarb sulfoxide	1.2	0.5	1.05	1.03	1.01
aldicarb sulfone	1.2	0.3	1.03	1.08	1.02
oxamyl	1.1	0.6	1.06	1.01	1.02
thiofanox sulfoxide	1.1	0.7	1.07	1.03	1.03
pirimicarb	1.3	0.5	1.05	1.05	1.04
thiofanox sulfone	1.2	0.4	1.04	1.02	1.01
3-hydroxycarbofuran	1.1	0.4	1.04	1.03	1.02
dioxacarb	1.2	0.3	1.03	1.04	1.01
aldicarb	1.1	0.4	1.04	1.01	1.01
metolcarb	1.1	0.3	1.03	1.02	1.07
propoxur	1.2	0.3	1.03	1.02	1.05
carbofuran	1.3	0.3	1.03	1.06	1.01
bendiocarb	1.2	0.4	1.02	1.04	1.01
carbaryl	1.2	0.5	1.04	1.05	1.04
ethiofencarb	1.2	0.3	1.03	1.02	1.01
thiofanox	1.2	0.3	1.03	1.01	1.06
thiocarb	1.1	0.4	1.02	1.04	1.01

Table 4. *Cont.*

Pesticide	Rate of Linear Slope	Rate of Intercept	Comparison of Recovery		
			5 μg/kg	20 μg/kg	100 μg/kg
isoproarb	1	0.2	1.02	1.02	1.04
2,3,5-trimethacarb	1	0.2	1.02	1.03	1.05
fenobucarb	1.1	0.4	1.04	1.01	1.02
diethofencarb	1.2	0.3	1.03	1.01	1.01
methiocarb	1.1	0.3	1.03	1.04	1.03
promecarb	1.2	0.3	1.03	1.02	1.04
fenoxycarb	1.1	0.3	1.03	1.01	1.01
indoxacarb	1.1	0.5	1.05	1.04	1.05
benfuracarb	1.1	0.5	1.05	1.03	1.02
furathiocarb	1.3	0.5	1.05	1.01	1.01
carbosuifan	1.2	0.5	1.05	1.03	1.03

* Treatment methods 1 and 2 refer to carrageenan purification and QuEChERS.

The results show that phase-transfer purification method 1 with carrageenan has better sensitivity and smaller systematic error.

2.5.2. Precision and Recovery Rate

For the six blank samples of banana, lemon, apple, Chinese cabbage, spinach, and rice, three levels of supplemental recovery tests were carried out. The spiking standard solutions were added into the homogenized sample solution before the purification process. The supplemental levels were 5, 20, and 100 μg/kg. Six parallel experiments were conducted for each level, and the matrix matching standard curve was quantitative. The results are shown in Table 5.

Table 5. The recoveries of low, medium, and high levels for six matrixes and RSD ranges.

Matrix	Supplemental Level			RSD Range
	5 μg/kg	20 μg/kg	100 μg/kg	
banana	59.84–127.06%	60.08–114.06%	56.13–110.05%	1.1–15%
lemon	62.05–91.61%	69.33–98.94%	71.83–107.02%	0.47–9.3%
apple	74.78–106.3%	60.51–125.2%	75.40–117.6%,	3.1–16%
cabbage	68.62–103.0%	69.81–99.31%	67.62–112.3%	1.2–13%
spinach	71.11–115.3%	77.45–125.3%	57.36–103.2%	3.2–14%
rice	85.70–106.0%	67.31–108.3%	70.70–97.61%	0.70–8.5%

The corresponding ratio of carrageenan purification is shown in Table 4. It shows that there is a better recovery rate at the three levels.

2.6. Determination of Actual Samples

After the method was established, seven batches of spinach, eight batches of apples, five batches of cabbage, five batches of bananas, three batches of rice, and three batches of lemons were determined by treatment methods 1 and 2. It was found that there were residues in the samples of spinach, apple, and Chinese cabbage, but none of them exceeded the national MRLs.

3. Materials and Methods

3.1. Instruments and Reagents

We used a 1290 Infinity liquid chromatograph with a 6460 Triple quadrupole Series Mass Spectrometer (Agilent, Santa Clara, CA, USA), a Milli-Q Pure water meter (Millipore Company, Burlington, MA, USA), a CF 16RX II centrifuge (Hitachi Corporation, Tokyo, Japan), an XHF-D high speed disperser (Ningbo Xinzhi Biotechnology Co., Ltd., Ningbo,

China), an XP205 analytical balance (1 in 100,000), and an AL204 analytical balance (1 in 10,000) (Mettler, Greifensee, Switzerland).

Ethylenediamine-*n*-propyl silane (PSA) adsorbent: 40–60 μm particle size (Tianjin Bonaigel Technology Co., Ltd., Tianjin, China). Graphitized carbon black (GCB) and C18 adsorbent: 40 μm particle size (Agilent Corporation). Sodium citrate and NaCl were highly pure, and anhydrous $MgSO_4$ was analytically pure (Sinophosphoric Chemical Reagents Co., Ltd., Ningbo, China). Disodium hydrogen citrate was analytically pure (Tokyo Chemical Industry Co., Ltd., Tokyo, Japan). Xanthan gum, carrageenan, and gelatin were purchased from Sinopharm Group Chemical Reagent Co., Ltd., Shanghai, China. Methanol and acetonitrile were chromatographically pure (Thermo Fisher Company, Waltham, MA, USA). Formic acid was analytically pure (Fluka Corporation, Everett, WA, USA). High-purity water was used in the experiment. Metolcarb, propoxur, carbofuran, bendiocarb, carbaryl, ethiofencarb, thiofanox, thiocarb, isoproarb, 2,3,5-trimethacarb, fenobucarb, diethofencarb, methiocarb, promecarb, fenoxycarb, indoxacarb, benfuracarb, furathiocarb, and carbosuifan pesticide control products were purchased from Dr. Ehrenstorfer Company in Germany. Aminocarb, propamocarb, aldicarb sulfoxide, aldicarb sulfone, oxamyl, thiofanox sulfoxide, pirimicarb, thiofanox sulfone, 3-hydroxycarbofuran, dioxacarb, and aldicarb were purchased from Sigma-Aldrich Company in the United States. Detailed information on these 30 pesticides are shown in Table 6.

Table 6. Detailed information on the 30 pesticides.

Pesticide	IUPAC Name	Formula
aminocarb	[4-(dimethylamino)-3-methylphenyl] N-methylcarbamate	$C_{11}H_{16}N_2O_2$
propamocarb	propyl N-[3-(dimethylamino)propyl]carbamate	$C_9H_{20}N_2O_2$
aldicarb sulfoxide	[(E)-(2-methyl-2-methylsulfinylpropylidene)amino] N-methylcarbamate	$C_7H_{14}N_2O_3S$
aldicarb sulfone	[(E)-(2-methyl-2-methylsulfonylpropylidene)amino] N-methylcarbamate	$C_7H_{14}N_2O_4S$
oxamyl	methyl (1Z)-2-(dimethylamino)-N-(methylcarbamoyloxy)-2-oxoethanimidothioate	$C_7H_{13}N_3O_3S$
thiofanox sulfoxide	[(Z)-(3,3-dimethyl-1-methylsulfinylbutan-2-ylidene)amino] N-methylcarbamate	$C_9H_{18}N_2O_3S$
pirimicarb	[2-(dimethylamino)-5,6-dimethylpyrimidin-4-yl] N,N-dimethylcarbamate	$C_{11}H_{18}N_4O_2$
thiofanox sulfone	[(3,3-dimethyl-1-methylsulfonylbutan-2-ylidene)amino] N-methylcarbamate	$C_9H_{18}N_2O_4S$
3-hydroxycarbofuran	(3-hydroxy-2,2-dimethyl-3H-1-benzofuran-7-yl) N-methylcarbamate	$C_{12}H_{15}NO_4$
dioxacarb	[2-(1,3-dioxolan-2-yl)phenyl] N-methylcarbamate	$C_{11}H_{13}NO_4$
aldicarb	[(E)-(2-methyl-2-methylsulfonylpropylidene)amino] N-methylcarbamate	$C_7H_{14}N_2O_4S$
metolcarb	(3-methylphenyl) N-methylcarbamate	$C_9H_{11}NO_2$
propoxur	(2-propan-2-yloxyphenyl) N-methylcarbamate	$C_{11}H_{15}NO_3$
carbofuran	(2,2-dimethyl-3H-1-benzofuran-7-yl) N-methylcarbamate	$C_{12}H_{15}NO_3$
bendiocarb	(2,2-dimethyl-1,3-benzodioxol-4-yl) N-methylcarbamate	$C_{11}H_{13}NO_4$
carbaryl	naphthalen-1-yl N-methylcarbamate	$C_{12}H_{11}NO_2$
ethiofencarb	[2-(ethylsulfanylmethyl)phenyl] N-methylcarbamate	$C_{11}H_{15}NO_2S$
thiofanox	[(3,3-dimethyl-1-methylsulfanylbutan-2-ylidene)amino] N-methylcarbamate	$C_9H_{18}N_2O_2S$
thiocarb	N,N-diethylcarbamodithioate	$C_5H_{16}NNaO_3S_2$
isoproarb	(2-propan-2-ylphenyl) N-methylcarbamate	$C_{11}H_{15}NO_2$
2,3,5-trimethacarb	(2,3,5-trimethylphenyl) N-methylcarbamate	$C_{11}H_{15}NO_2$
fenobucarb	(2-butan-2-ylphenyl) N-methylcarbamate	$C_{12}H_{17}NO_2$
diethofencarb	propan-2-yl N-(3,4-diethoxyphenyl)carbamate	$C_{14}H_{21}NO_4$
methiocarb	(3,5-dimethyl-4-methylsulfanylphenyl) N-methylcarbamate	$C_{11}H_{15}NO_2S$
promecarb	(3-methyl-5-propan-2-ylphenyl) N-methylcarbamate	$C_{11}H_{15}NO_2S$
fenoxycarb	ethyl N-[2-(4-phenoxyphenoxy)ethyl]carbamate	$C_{17}H_{19}NO_4$
indoxacarb	methyl 7-chloro-2-[methoxycarbonyl-[4-(trifluoromethoxy)phenyl]carbamoyl]-3,5-dihydroindeno [1,2-e][1,3,4]oxadiazine-4a-carboxylate	$C_{22}H_{17}ClF_3N_3O_7$
benfuracarb	ethyl 3-[[(2,2-dimethyl-3H-1-benzofuran-7-yl)oxycarbonyl-methylamino]sulfanyl-propan-2-ylamino]propanoate	$C_{20}H_{30}N_2O_5S$
furathiocarb	(2,2-dimethyl-3H-1-benzofuran-7-yl) N-[butoxycarbonyl(methyl)amino]sulfanyl-N-methylcarbamate	$C_{18}H_{26}N_2O_5S$
carbosuifan	(2,2-dimethyl-3H-1-benzofuran-7-yl) N-(dibutylamino)sulfanyl-N-methylcarbamate	$C_{20}H_{32}N_2O_3S$

The standard reserve solution of each pesticide with a mass concentration of 1 g/L was prepared with acetonitrile, and the mixed standard solution of 30 kinds of aminomethylate pesticides with a mass concentration of 5 mg/L was prepared with acetonitrile and stored at −18 °C.

Matrix matching standard solution: 2 mL of 7 blank sample extracts were separately transferred into a 15 mL centrifuge tube and gently dried with nitrogen (N_2), and 2 mL of mixed standard solutions with mass concentrations of 1, 2, 5, 10, 20, 50, and 100 μg/L were added to prepare a series of matrix matching control solutions (ready to use). The blank sample was obtained through filtration using a 0.22 μm microporous membrane for analysis. See pre-treatment methods 1 and 2 below.

3.2. *Sample Pretreatment Method*

Pre-treatment method 1: Different types of samples were homogenized or ground according to their characteristics during pre-treatment. Apple, lemon, and cabbage were cut into pieces and then packaged and stored at −18 °C, respectively. Spinach was chopped, subpackaged, and then stored at −18 °C. Banana was chopped, subpackaged, and stored at −18 °C. Prior to the determination, homogenization was carried out immediately after removal from the refrigerator. All samples were not peeled when cut or chopped. Rice samples were subpackaged after crushing and stored at −18 °C. About 10.00 g of homogenized sample or crushed rice was weighed into a 50 mL plugged centrifuge tube, 20 mL 20% acetonitrile aqueous solution was added, and carrageenan of different weights was added. The weight of carrageenan added was related to the characteristics of the samples. For apple, cabbage, lemon, and rice, the additive amount was 0.3 g; for spinach, the additive amount was 1.0 g; and for banana, the additive amount was 0.6 g. After ultrasonic treatment at 50 °C, it was cooled to room temperature for 5 min and centrifuged at 8000 rpm for 10 min, and the supernatant was taken. It was filtered for detection by using a 0.22 μm microporous filter membrane.

Pre-treatment method 2: The homogenized or ground sample procedure was the same as in Method 1. About 10.00 g of homogenized sample or crushed rice was weighed into a 50 mL plugged centrifuge tube, and 20 mL of different solvents were added according to their characteristics. The mixture was shaken vigorously for 1 min. For apple, cabbage, and spinach, the added solvent was acetonitrile. For banana, the added solvent was 10 mL of acetonitrile and 8 mL of ice water. For crushed rice, the added solvent was 10 mL of acetonitrile and 20 mL of ice water. After that, we added 4 g of anhydrous $MgSO_4$, 1 g of NaCl, 1 g of sodium citrate, and 0.5 g of disodium hydrogen citrate, shook it vigorously for 1 min, and centrifuged it at 8000 rpm for 5 min. It should be noted that lemon samples had special circumstances where an additional 0.4 mL of 5 mol/L NaOH solution was needed to be added before shaking. We took 6 mL of the upper solution and poured it into a 15 mL plugged centrifuge tube pre-filled with 150 mg PSA and 900 mg of anhydrous $MgSO_4$. For spinach, 80 mg of GCB needed to be added to absorb the pigment. We swirled it for 30 s and centrifuged it at 5 000 rpm for 2 min. We took 4 mL of the supernatant and added 40 μL of 5% (v/v) formic acid acetonitrile solution. It was filtered for analysis by using a 0.22 μm microporous filter membrane.

3.3. *LC–MS/MS Conditions*

LC conditions: Shiseido Type MG III column (150 mm × 2.0 mm, 5 μm), column temperature, 35 °C; injection volume, 2 μL; flow rate, 0.3 mL/min. Mobile phase: phase A is 0.1% (v/v) formic acid aqueous solution, and phase B is 0.1% (v/v) formic acid methanol solution. Gradient elution procedure: 0–3 min, 35% B; 3–20 min, 35% B–90% B; 20–25 min, 90% B. The running time was 25 min.

MS conditions: Electrospray ion (ESI) source in positive ion mode. Drying temperature, 330 °C; flow rate, 8 L/min; atomizing gas pressure, 30 psi; sheath temperature, 250 °C, flow rate, 11 L/min; capillary voltage, 3500 V. Scanning mode: segmented multi-response monitoring mode.

4. Conclusions

In this study, six representative plant sample substrates were selected to optimize a more green and environmentally friendly phase-transfer purification method, which was used for actual sample detection. Compared with the commonly used QuEChERS method, the results showed a more sensitive and accurate result, which also confirmed that materials closer to food can be used for relatively green inspection in food inspection. The limitations of this new purification process should also be taken into account. The target analyte of this study is carbamate pesticides, and the adsorption effect of carrageenan on them can be negligible. For other types of analytes, when using carrageenan for sample pretreatment, it is also necessary to consider the adsorption effect on the analyte, especially for polar compounds. However, this study does offer an avenue for the investigation of the corresponding low-carbonization approach.

Author Contributions: Conceptualization, S.J. and Y.S.; data curation, S.J. and X.N.; formal analysis, X.N. and R.L.; investigation, S.J. and T.L.; methodology, X.N. and Y.S.; software, Y.S. and R.L.; supervision, S.J. and J.C.; visualization, S.J. and Y.S.; resources, R.L.; writing—original draft, S.J. and J.C.; writing—review and editing, S.J., Y.S., T.L., and J.C. All authors have read and agreed to the published version of the manuscript.

Funding: This work was supported by the National Key Research and Development Program of China, grant number 2021YFC2401103.

Institutional Review Board Statement: Not applicable.

Informed Consent Statement: Not applicable.

Data Availability Statement: All data are included in this manuscript.

Acknowledgments: This research was supported by the Key Laboratory of Food Quality and Safety for State Market Regulation. We thank Bo Ding (Center for Food and Drug Inspection and Detection, Yantai) for providing the resources to complete this study.

Conflicts of Interest: The authors declare no conflict of interest.

Sample Availability: The plant food samples in the research were purchased from the supermarket and were detected as positive by testing. All processes are repeatable.

References

1. Yang, X.; Xiong, X.; Cao, J.; Luan, B.; Liu, Y.; Liu, G.; Zhang, L. Matrix precipitation: A general strategy to eliminate matrix interference for pharmaceutical toxic impurities analysis. *J. Chromatogr. A* **2015**, *1379*, 16–23. [CrossRef] [PubMed]
2. Zhang, S.; He, Z.; Zeng, M.; Chen, J. Impact of matrix species and mass spectrometry on matrix effects in multi-residue pesticide analysis based on quechers-lc-ms. *Foods* **2023**, *12*, 1226. [CrossRef] [PubMed]
3. Ning, X.; Wang, L.; Jin, S.; Fu, X.; Sun, X.; Cao, J. A strategy for sample preparation: Using egg white gel to promote the determination of aflatoxin m1 content in milk samples. *Molecules* **2022**, *27*, 5039. [CrossRef] [PubMed]
4. Gross, M.S.; Woodward, E.E.; Hladik, M.L. Evaluation of elisa for the analysis of imidacloprid in biological matrices: Cross-reactivities, matrix interferences, and comparison to lc-ms/ms. *Chemosphere* **2022**, *286*, 131746. [CrossRef] [PubMed]
5. Zhou, J.; Lin, X.; Zhao, L.; Huang, K.; Yang, Q.; Yu, H.; Xiong, X. Headspace single drop microextraction based visual colorimetry for highly sensitive, selective and matrix interference-resistant determination of sulfur dioxide in food samples. *Food Chem.* **2023**, *426*, 136659. [CrossRef] [PubMed]
6. Makni, Y.; Diallo, T.; Areskoug, F.; Guérin, T.; Parinet, J. Optimisation and implementation of quechers-based sample preparation for identification and semi-quantification of 694 targeted contaminants in honey, jam, jelly, and syrup by uhplc-q/tof high-resolution mass spectrometry. *Food Chem.* **2023**, *425*, 136448. [CrossRef]
7. Sulaiman, U.; Shah, F.; Khan, R.A. Quechers sample preparation integrated to dispersive liquid-liquid microextraction based on solidified floating organic droplet for spectrometric determination of sudan dyes: A synergistic approach. *Food Chem. Toxicol.* **2022**, *159*, 112742. [CrossRef]
8. Iwasaki, Y.; Matsumoto, H.; Okumura, M.; Inoue, H.; Kaji, Y.; Ando, C.; Kamei, J. Determination of neurotransmitters in mouse brain using miniaturized and tableted quechers for the sample preparation. *J. Pharm. Biomed. Anal.* **2022**, *217*, 114809. [CrossRef]
9. Kardani, F.; Jelyani, A.Z.; Dahanzadeh, M.; Rashedinia, M. Modified quechers sample preparation method and applied for the analysis of 30 pesticide residues in iran tea in 2020–2021 using gas chromatography-tandem mass spectrometry. *Appl. Food Res.* **2022**, *2*, 100197. [CrossRef]

10. Zhu, X.; Chen, S.; Xu, Y. Determination of non-glucosidic cyanogen in chinese liquor-fermentation ingredients using quechers sample preparation and spectrophotometric method. *Food Control* **2022**, *140*, 109101. [CrossRef]
11. Dallagi, H.; Jha, P.K.; Faille, C.; Le-Bail, A.; Rawson, A.; Benezech, T. Removal of biocontamination in the food industry using physical methods; An overview. *Food Control* **2023**, *148*, 109645. [CrossRef]
12. Sridhar, A.; Vaishampayan, V.; Senthil Kumar, P.; Ponnuchamy, M.; Kapoor, A. Extraction techniques in food industry: Insights into process parameters and their optimization. *Food Chem. Toxicol.* **2022**, *166*, 113207. [CrossRef] [PubMed]
13. Yang, J.; Zhang, X.; Lu, Q.; Wang, L.; Hu, X.; Zhang, H. Preparation, flocculation and application in sugar refining of eco-friendly dextran-polylysine complex flocculant. *Sep. Purif. Technol.* **2023**, *306*, 122673. [CrossRef]
14. Balbinoti, J.R.; Dos Santos Junior, R.E.; de Sousa, L.B.F.; de Jesus Bassetti, F.; Balbinoti, T.C.V.; de Matos Jorge, L.M.; Jorge, R.M.M. Plant-based coagulants for food industry wastewater treatment. *J. Water Process. Eng.* **2023**, *52*, 103525. [CrossRef]
15. Miazaki, J.B.; Dos Santos, A.R.; de Freitas, C.F.; Stafussa, A.P.; Mikcha, J.M.G.; Bergamasco, R.D.C.; Tonon, L.A.C.; Madrona, G.S.; Caetano, W.; Da Silva, L.H.; et al. Edible coatings and application of photodynamics in ricotta cheese preservation. *LWT* **2022**, *165*, 113697. [CrossRef]
16. Chen, X.; Zhang, L.; Xu, W.; Wang, C.; Li, H.; Xiong, L.; Zhang, H.; Chen, X. Synthesis of polyacrylamide/polystyrene interpenetrating polymer networks and the effect of textural properties on adsorption performance of fermentation inhibitors from sugarcane bagasse hydrolysate. *Bioresour. Technol.* **2020**, *318*, 124053. [CrossRef]
17. Laad, M.; Ghule, B. Removal of toxic contaminants from drinking water using biosensors: A systematic review. *Groundwater Sustain. Dev.* **2023**, *20*, 100888. [CrossRef]
18. Habibi, H.; Khosravi-Darani, K. Effective variables on production and structure of xanthan gum and its food applications: A review. *Biocatal. Agric. Biotechnol.* **2017**, *10*, 130–140. [CrossRef]
19. Ta, L.P.; Bujna, E.; Antal, O.; Ladányi, M.; Juhász, R.; Szécsi, A.; Kun, S.; Sudheer, S.; Gupta, V.K.; Nguyen, Q.D. Effects of various polysaccharides (alginate, carrageenan, gums, chitosan) and their combination with prebiotic saccharides (resistant starch, lactosucrose, lactulose) on the encapsulation of probiotic bacteria lactobacillus casei 01 strain. *Int. J. Biol. Macromol.* **2021**, *183*, 1136–1144. [CrossRef]
20. Eghbaljoo, H.; Sani, I.K.; Sani, M.A.; Rahati, S.; Mansouri, E.; Molaee-Aghaee, E.; Fatourehchi, N.; Kadi, A.; Arab, A.; Sarabandi, K.; et al. Advances in plant gum polysaccharides; Sources, techno-functional properties, and applications in the food industry—A review. *Int. J. Biol. Macromol.* **2022**, *222*, 2327–2340. [CrossRef]
21. Li, X.; Shen, Y.; Hu, F.; Zhang, X.; Thakur, K.; Rengasamy, K.R.R.; Khan, M.R.; Busquets, R.; Wei, Z. Fortification of polysaccharide-based packaging films and coatings with essential oils: A review of their preparation and use in meat preservation. *Int J. Biol. Macromol.* **2023**, *242*, 124767. [CrossRef] [PubMed]
22. Gomes, D.; Batista-Silva, J.P.; Sousa, A.; Passarinha, L.A. Progress and opportunities in gellan gum-based materials: A review of preparation, characterization and emerging applications. *Carbohydr. Polym.* **2023**, *311*, 120782. [CrossRef] [PubMed]
23. Ongono, J.S.; Béranger, R.; Baghdadli, A.; Mortamais, M. Pesticides used in europe and autism spectrum disorder risk: Can novel exposure hypotheses be formulated beyond organophosphates, organochlorines, pyrethroids and carbamates?—A systematic review. *Environ. Res.* **2020**, *187*, 109646. [CrossRef] [PubMed]
24. Wei, J.; Wei, B.; Yang, W.; He, C.; Su, H.; Wan, J.; Li, P.; Wang, Y. Trace determination of carbamate pesticides in medicinal plants by a fluorescent technique. *Food Chem. Toxicol.* **2018**, *119*, 430–437. [CrossRef]
25. Yang, B.; Wang, S.; Ma, W.; Li, G.; Tu, M.; Ma, Z.; Zhang, Q.; Li, H.; Li, X. Simultaneous determination of neonicotinoid and carbamate pesticides in freeze-dried cabbage by modified quechers and ultra-performance liquid chromatography–tandem mass spectrometry. *Foods* **2023**, *12*, 699. [CrossRef]
26. Osaili, T.M.; Al-Natour, M.Q.; Al-Abboodi, A.R.; Alkarasneh, A.Y.; El Darra, N.; Khazaal, S.; Holley, R. Detection and risk associated with organochlorine, organophosphorus, pyrethroid and carbamate pesticide residues in chicken muscle and organ meats in jordan. *Food Control* **2023**, *144*, 109355. [CrossRef]
27. Lehotay, S.J. Quechers sample preparation approach for mass spectrometric analysis of pesticide residues in foods. In *Mass Spectrometry in Food Safety: Methods and Protocols*; Zweigenbaum, J., Ed.; Humana Press: Totowa, NJ, USA, 2011; pp. 65–91.
28. *AOAC Official Method 2007.01*; Pesticide Residues in Foods by Acetonitrile Extraction and Partitioning with Magnesium Sulfate. AOAC International: Rockville, MD, USA, 2007.
29. *GB 2763-2021*; National Food Safety Standard—Maximum Residue Limits for Pesticides in Food. USDA: Washington, DC, USA, 2021.
30. Nsengiyumva, E.M.; Heitz, M.P.; Alexandridis, P. Thermal hysteresis phenomena in aqueous xanthan gum solutions. *Food Hydrocoll.* **2023**, *144*, 108973. [CrossRef]
31. Udo, T.; Mummaleti, G.; Mohan, A.; Singh, R.K.; Kong, F. Current and emerging applications of carrageenan in the food industry. *Food Res. Int.* **2023**, *173*, 113369. [CrossRef]
32. Tan, Y.; Zi, Y.; Peng, J.; Shi, C.; Zheng, Y.; Zhong, J. Gelatin as a bioactive nanodelivery system for functional food applications. *Food Chem.* **2023**, *423*, 136265. [CrossRef]
33. Portanguen, S.; Dumoulin, C.; Duconseille, A.; Meurillon, M.; Sicard, J.; Théron, L.; Chambon, C.; Sayd, T.; Mirade, P.; Astruc, T. Impact of water content and bloom index on gelatin glycation. *Food Hydrocoll.* **2023**, *145*, 109096. [CrossRef]

34. Kwon, H.; Lehotay, S.J.; Geis-Asteggiante, L. Variability of matrix effects in liquid and gas chromatography—Mass spectrometry analysis of pesticide residues after quechers sample preparation of different food crops. *J. Chromatogr. A* **2012**, *1270*, 235–245. [CrossRef] [PubMed]
35. Kmellár, B.; Fodor, P.; Pareja, L.; Ferrer, C.; Martínez-Uroz, M.A.; Valverde, A.; Fernandez-Alba, A.R. Validation and uncertainty study of a comprehensive list of 160 pesticide residues in multi-class vegetables by liquid chromatography—Tandem mass spectrometry. *J. Chromatogr. A* **2008**, *1215*, 37–50. [CrossRef] [PubMed]

molecules

MDPI

Article

Terminal Residue and Dietary Risk Assessment of Atrazine and Isoxaflutole in Corn Using High-Performance Liquid Chromatography–Tandem Mass Spectrometry

Junli Cao [1], Tao Pei [1,2], Yonghui Wang [1,2], Shu Qin [1], Yanli Qi [1], Pengcheng Ren [1] and Jindong Li [1,*]

1 Shanxi Center for Testing of Functional Agro-Products, Shanxi Agricultural University, No. 79, Longcheng Street, Taiyuan 030031, China; caojunli@sxau.edu.cn (J.C.); w1292692554@163.com (T.P.); 15212426571@163.com (Y.W.); qinshu@sxau.edu.cn (S.Q.); qiyanli0412@126.com (Y.Q.); zoe1946@126.com (P.R.)
2 College of Plant Protection, Shanxi Agricultural University, No. 81, Longcheng Street, Taiyuan 030031, China
* Correspondence: lijindong@sxau.edu.cn; Tel.: +86-0351-7965708; Fax: +86-0351-7639301

Abstract: Isoxaflutole and atrazine are representative pesticides for weed control in corn fields. Formulations containing these two pesticides have been registered in China, and their residues may threaten food safety and human health. In this study, a method for simultaneous determination of isoxaflutole, atrazine, and their metabolites in fresh corn, corn kernels, and corn straw was established based on modified QuEChERS pre-treatment and high-performance liquid chromatography–tandem mass spectrometry (HPLC–MS/MS). The linearity of seven compounds was good ($R^2 \geq 0.9912$), and the matrix effect was 48.5–77.1%. At four spiked levels of 0.01, 0.02, 0.05, and 0.5 mg kg^{-1}, all compounds' average recovery was 76% to 116%, with relative standard deviation (RSD) less than 18.9%. Field experiments were conducted in Liaoning, Heilongjiang, Inner Mongolia, Shanxi, Beijing, and Yunnan provinces to study the terminal residues. The terminal residues of all compounds were below the LOQ (0.01 mg kg^{-1}) in fresh corn and corn kernels, and atrazine residues in corn straw ranged from <0.05 mg kg^{-1} to 0.17 mg kg^{-1} Finally, a dietary risk assessment was conducted based on residues from field trials, food consumption, and acceptable daily intake (ADI). For all populations, the chronic dietary risk probability (RQ_c) of atrazine was between 0.0185% and 0.0739%, while that of isoxaflutole was 0.0074–0.0296%, much lower than 100%. The results may provide scientific guidance for using isoxaflutole and atrazine in corn field ecosystems.

Keywords: isoxaflutole; atrazine; terminal residues; dietary risk assessment

Citation: Cao, J.; Pei, T.; Wang, Y.; Qin, S.; Qi, Y.; Ren, P.; Li, J. Terminal Residue and Dietary Risk Assessment of Atrazine and Isoxaflutole in Corn Using High-Performance Liquid Chromatography–Tandem Mass Spectrometry. *Molecules* **2023**, *28*, 7225. https://doi.org/10.3390/molecules28207225

Academic Editor: Miguel Ángel González-Curbelo

Received: 19 September 2023
Revised: 17 October 2023
Accepted: 21 October 2023
Published: 23 October 2023

1. Introduction

Corn (*Zea mays* L.), also known as maize, is one of the world's major grains along with rice and wheat, rich in dietary fiber, vitamins, minerals, and other nutrients [1]. Fresh corn and corn kernels are essential to the human diet, and straw can feed livestock and poultry. China is a major corn production and consumption country, with a planting area of 43,355,859 hectares in 2021 (https://www.fao.org/faostat/en/#compare, accessed on 5 May 2023). However, weeds are seriously harmful, resulting in a 30% to 40% reduction in corn yield [2]. For weed control, pesticides are currently the most effective means to reduce crop yield losses and are indispensable for agricultural production [3,4]. However, the widespread and repeated application of pesticides around the world has resulted in pesticide residues and potential risks to humans through the food chain, which has become an issue of general concern in food and the environment [5,6]. More seriously, recent studies have reported that pesticide metabolites are frequently detected and may have higher adverse effects on non-target organisms than parent compounds [7,8]. Therefore, more attention should be paid to residue analysis and dietary intake risk of pesticide and their metabolites, which should be the basis for the comprehensive evaluation of pesticides.

As far as we know, isoxaflutole (IFT) and atrazine (ATR) have been registered as a mixture in China for controlling annual weeds in corn fields. IFT belongs to the isoxazole pre-emergence herbicides [9]. In plants or the environment, the oxazolium ring of IFT rapidly breaks down to produce diketo-nitrile metabolites (IFT-DKN) that inhibit 4-hydroxyphenylpyruvate dioxygenase (HPPD), which is essential for the biosynthesis of tocopherol and plastoquinone [10,11]. ATR belongs to the triazine herbicides and was first developed by the Swiss company Geigy in 1958 [12]. As a typical photosynthesis inhibitor, ATR occupies a vital position in the herbicide market and is widely used in crops to control annual grasses and broadleaf weeds. ATR forms four primary metabolites: deethylatrazine (DEA), deisopropyl atrazine (DIA), deethyl deisopropyl atrazine (DEDIA), and hydroxyl atrazine (HA) in environments and plants. Their chemical structures are shown in Figure 1. The residue and risk assessment of isoxazolone in plants was defined as the sum of IFT and IFT-DKN. In contrast, ATR was defined as the sum of ATR and its four main metabolites. Herbicides remain in the soil and plants after application in cornfields, potentially impacting consumers' health through the food chain. In particular, atrazine has been considered a world-recognized endocrine disruptor due to its adverse effects on the biological endocrine system, central nervous system, immune system, and human reproductive development [13–15], seriously threatening the ecological environment and consumer health [16]. Therefore, it is necessary to comprehensively study the residues of pesticides and evaluate the dietary risk to guide the safe use of IFT and ATR.

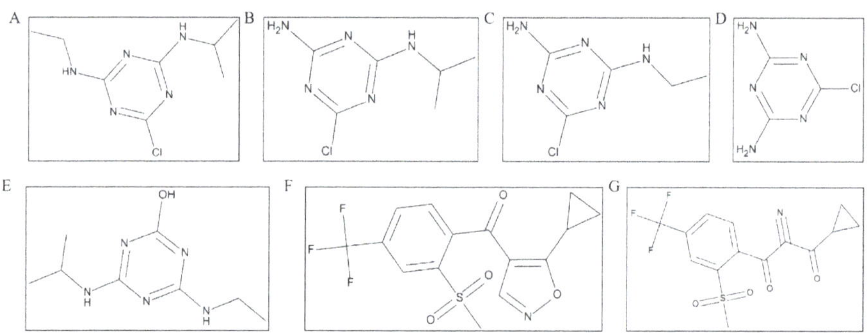

Figure 1. The chemical structural formula of ATR (**A**), DEA (**B**), DIA (**C**), DACT (**D**), HA (**E**), IFT (**F**) and IFT-DKN (**G**).

As far as we know, there have been some reports on the residue analysis of ATR, but there are few studies on IFT and its metabolites. The primary detection methods for ATR in environmental and food samples are chromatography, as well as Raman spectroscopy [17] and enzyme-linked immunosorbent assay [18]. Yuan et al. established a gas chromatography–mass spectrometry (GC–MS) method to detect apples, grapes, and tea and a gas chromatography–nitrogen and phosphorus detector (GC–NPD) to quantify ATR residues in soil [19]. Fu et al. established a solid-phase extraction and ultra-high performance liquid chromatography–tandem mass spectrometry (SPE–UPLC–MS/MS) method for detecting ATR in water [20]. Tandon et al. showed a technique for determining ATR in corn kernels, straw, and soil by liquid chromatography with a UV detector [21]. It has been reported that IFT and IFT-DKN in environmental and food samples were determined by chromatography. Lin et al. [6] developed high-performance liquid chromatography–UV (HPLC–UV) and high-performance liquid chromatography–tandem mass spectrometry (HPLC–MS/MS) method for the analysis of IFT and its two metabolites, IFT-DKN and benzoic acid metabolite (BA) in soil, water and plant samples [22,23]. Lan et al. used modified QuEChERS pre-treatment and high-performance liquid chromatography–tandem mass spectrometry to determine isoxazole and its metabolites in corn and straw, and the half-lives of IFT in Shandong and Anhui provinces were 36.4 and 42.1 days, respectively [24].

IFT can effectively kill atrazine-resistant weeds, so the mixture of IFT and ATR has broad application prospects. However, there is no report on the simultaneous detection of IFT, ATR, and their metabolites in fresh corn, corn kernels, and corn straw. Moreover, the terminal residues and dietary risks of IFT and ATR remain to be studied.

In our study, a modified QuEChERS pre-treatment and ultra-high performance liquid chromatography–tandem mass spectrometry (HPLC-MS/MS) methods were developed for the detection of seven target compounds (including two active ingredients and five metabolites) to study terminal residues of, and safety risk assessment for, IFT and ATR. A supervised field experiment was conducted on cornfield under good agricultural practice (GAP) conditions to study the end residues of IFT and ATR in fresh corn, corn kernel, and corn straw. We also assessed the dietary risk of IFT and ATR residues based on pesticide residue data, food consumption, and toxicology data. This work will provide reasonable recommendations for the safe use of IFT and ATR.

2. Results

2.1. *Optimization of UPLC–MS/MS Analysis*

Firstly, the standard solutions of 0.1 mg L^{-1} IFT, ATR and their metabolites were scanned in ESI positive ion and negative ion modes, respectively. Considering that a strong response can only be obtained in the negative mode, the negative ion mode was selected for the subsequent analysis of IFT-DKN. On the contrary, the other six compounds can only obtain strong effective peaks in the positive ion mode. Next, the standard solution was scanned in the *m/z* range of 50–400, and the strongest peak found by IFT-DKN was observed at *m/z* = 358.07, corresponding to (M − H). The strongest peaks for the other six compounds corresponded to (M + H). Next, the cone voltage, collision energy and fragmentation were determined using the instrument's optimization software. The optimized mass spectrometry parameters are shown in Table 1.

Table 1. Instrument parameters of ATR, IFT and their metabolites in multiple reaction monitoring (MRM) mode.

Compounds	Retention Time (min)	Precursor Ion (m z^{-1})	Product Ion (m z^{-1})	Cone Voltage (CV, V)	Collision Energy (CE, eV)
ATR	4.04	216.27	174.16 * 96.14	20	28 21
DEA	3.60	188.23	78.9 146.15 *	30	30 19
DIA	3.44	174.0	96.0 * 131.9	65	20 15
DACT	1.44	146.0	79.0 * 103.99	70	19 17
HA	3.30	198.0	156.0 * 113.9	30	17 21
IFT	3.96	360.0	251.19 * 219.9	30	18 40
IFT-DKN	3.85	358.07	78.77 * 278.0	−20	−17 −14

* represents quantitative ion.

To obtain optimal chromatographic peak shapes during HPLC–MS/MS analysis, different mobile phase compositions were tested. Three mobile phases, including (A) water/methanol (B) 0.1% water/methanol and (C) 0.2% water/methanol solution, were tested. The results show that adding 0.2% formic acid aqueous solution can effectively improve the sensitivity. Therefore, 0.2% formic acid water/methanol was used as the optimal mobile phase, and good peak shapes were obtained for seven compounds (Figure 2).

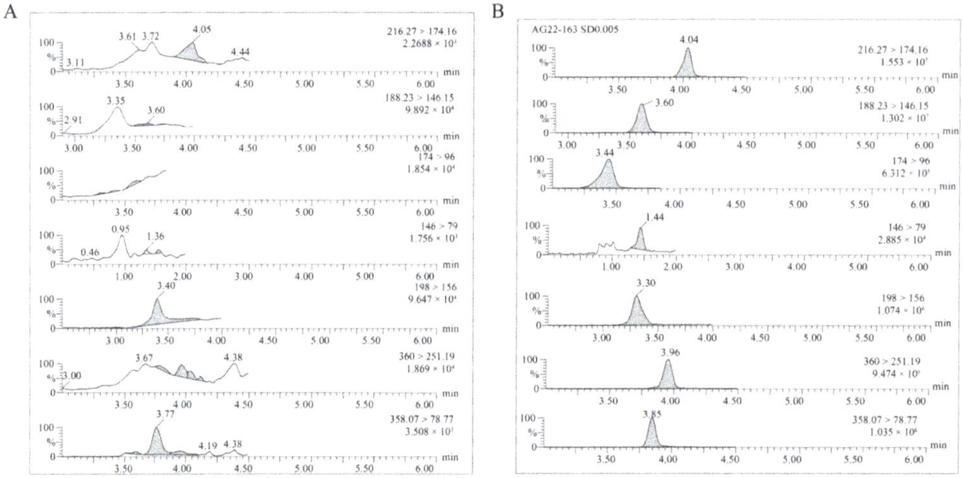

Figure 2. Typical chromatograms of ATR, IFT and their metabolites. (**A**) Blank fresh corn; (**B**) blank fresh corn spiked with 0.005 mg kg^{-1} mixed standard solution.

2.2. *QuEChERS Pre-Treatment*

The QuEChERS method was developed by Professor Anastassiades in 2003 and includes three steps: extraction, salting out, and purification [25]. It has been widely used to detect single and multiple pesticide residues [26,27]. Acetone, methanol, and ethyl acetate are common solvents used to extract pesticides from different matrices. However, acetone and methanol are miscible with water and difficult to separate, making it impossible to completely transfer pesticides to the organic phase [28]. Ethyl acetate is partially compatible with water, which is not conducive to extracting polar pesticides. As a QuEChERS solvent, acetonitrile has good extraction efficiency for most pesticides. The extract contains fewer interfering substances and is easily separated from water [29]. Therefore, acetonitrile was selected as the extraction solvent in this study. C18 and GCB are commonly used adsorbents in the QuEChERS method. C18 easily adsorbs non-polar substances, such as fats, sterols, and volatile oils; GCB is a regular polyhedron with a uniform graphitized surface and has excellent adsorption performance for samples with high pigment content. Cui et al. used C18 and GCB as purification materials for the QuEChERS method to analyze tembotrione and its metabolite in corn, corn oil, and animal-source foods [30]. Similarly, Zhong et al. used the Quechers method using C18 and GCB as purifiers to extract flumetsulam and florasulam from fresh corn, corn kernels, straw, and soil [31]. Therefore, C18 and GCB were selected as purifying agents.

2.3. *Method Validation*

According to OECD guidelines [32], the accuracy, precision, linearity, matrix effect (ME) and limit of quantitation (LOQ) of the method were verified.

The linear equation and ME of ATR, IFT and their metabolites are shown in Table 2. In the concentration range of 0.0025–0.25 mg kg^{-1}, all compounds had good linearity, with the coefficient of determination (R^2) greater than 0.9912. The chromatograms of blank fresh corn and fresh corn spiked with standard mixed solution (0.005 mg kg^{-1}) are shown in Figure 2.

Table 2. Linear equations, correlation coefficients and matrix effects of all compounds in 0.0025–0.25 mg kg^{-1}.

Compound	Matrix	Equation	Determination Coefficient (R^2)	ME (%)
ATR	Acetonitrile	$y = 1.92473 \times 10^8x + 171{,}768$	0.9961	-
	Fresh corn	$y = 1.78484 \times 10^8x + 151{,}436$	0.9972	−7.3
	Corn kernels	$y = 1.72312 \times 10^8x - 36{,}266.5$	0.9990	−10.5
	Corn straw	$y = 1.45725 \times 10^8x + 64{,}352.9$	0.9991	−24.3
DEA	Acetonitrile	$y = 1.69395 \times 10^8x + 222{,}823$	0.9912	-
	Fresh corn	$y = 1.53907 \times 10^8x + 183{,}749$	0.9945	−9.1
	Corn kernels	$y = 1.48905 \times 10^8x - 28{,}122.8$	0.9963	−12.1
	Corn straw	$y = 8.71969 \times 10^7x + 8952.86$	0.9996	−48.5
DIA	Acetonitrile	$y = 1.22314 \times 10^7x + 12{,}671.5$	0.9965	-
	Fresh corn	$y = 1.17502 \times 10^7x + 7327.07$	0.9981	−3.9
	Corn kernels	$y = 1.54207 \times 10^7x + 565.797$	0.9986	26.1
	Corn straw	$y = 8.11043 \times 10^6x - 1082.61$	0.9999	−33.7
DACT	Acetonitrile	$y = 92{,}470.9x + 15.5547$	0.9994	-
	Fresh corn	$y = 98{,}205.3x + 21.2445$	0.9977	6.2
	Corn kernels	$y = 150{,}640x + 103.791$	0.9915	62.9
	Corn straw	$y = 99{,}527x - 4.04451$	0.9991	7.6
HA	Acetonitrile	$y = 2.51197 \times 10^7x - 13{,}594.8$	0.9995	-
	Fresh corn	$y = 2.8312 \times 10^7x + 40{,}610.5$	0.9933	12.7
	Corn kernels	$y = 2.74648 \times 10^7x + 43{,}150.3$	0.9912	9.3
	Corn straw	$y = 3.17866 \times 10^7x + 49{,}750.2$	0.9906	26.5
IFT	Acetonitrile	$y = 1.31221 \times 10^8x + 97{,}738.5$	0.9940	-
	Fresh corn	$y = 1.16984 \times 10^8x + 144{,}660$	0.9944	−10.8
	Corn kernels	$y = 1.02943 \times 10^8x - 71{,}139.6$	0.9917	−21.5
	Corn straw	$y = 7.02378 \times 10^7x - 3031.07$	0.9960	−46.5
IFT-DKN	Acetonitrile	$y = 1.96818 \times 10^7x - 18{,}509.5$	0.9962	-
	Fresh corn	$y = 2.69316 \times 10^7x - 5800.49$	0.9994	36.8
	Corn kernels	$y = 1.97881 \times 10^7x - 13{,}759.9$	0.9967	0.5
	Corn straw	$y = 3.48496 \times 10^7x + 26{,}353$	0.9971	77.1

Matrix effect (ME) is an inherent aspect of the ESI source caused by impurities in the sample, which interferes with the quantitative accuracy of target compounds [33]. The ME of ATR and its four metabolites in fresh corn range from −9.1% to 12.7%; the ME of ATR, DEA, DIA and IFT in corn straw were −24.3%, −48.5%, −33.7% and −46.5%, respectively; and the ME of seven compounds in corn kernels vary greatly, with the lowest IFT-DKN being only 0.5% and the highest DACT being 62.9%. Since the ME of all compounds cannot be ignored, matrix-matched standard curves were used for quantification.

To evaluate the accuracy and precision, the mixed standards of ATR, IFT and their metabolites were spiked to the blank extracts of fresh corn, corn kernels and corn straw at 0.01, 0.02, 0.05 and 0.5 mg kg^{-1} levels, respectively. The results are shown in Figure 3. The average recoveries of ATR in fresh corn, corn kernels and corn straw were 94–101%, 90–99% and 93–113%, respectively, and the average recoveries of DEA in the three substrates were 96–99%, 95–103% and 92–106%, respectively. The average recoveries of DIA were 94–102%, 95–99% and 95–110%, respectively. The recoveries of DACT were 76–92%, 85–99% and 93–102%, respectively. HA were between 78 and 101%, 88 and 108% and 78 and 86%. The average recovery of IFT in fresh corn, corn kernels and corn straw is 81–113%, 77–90% and 88–116%, respectively. The average recovery of IFT-DKN in fresh corn was 92–104%, in corn kernels 89–109%, and in corn straw 95–103%. For fresh corn, corn kernels and corn straw, the LOQ of these compounds was all 0.01 mg kg^{-1}. The recoveries of seven compounds in fresh corn, corn grain and straw were in the range of 76 to 116%, with the relative standard deviations of all compounds ranging from 0.8% to 18.9%, which meets the requirements of OECD guidelines (SANTE 11312/2021) and the guideline for the testing of pesticide residues in crops (NY/T 788–2018).

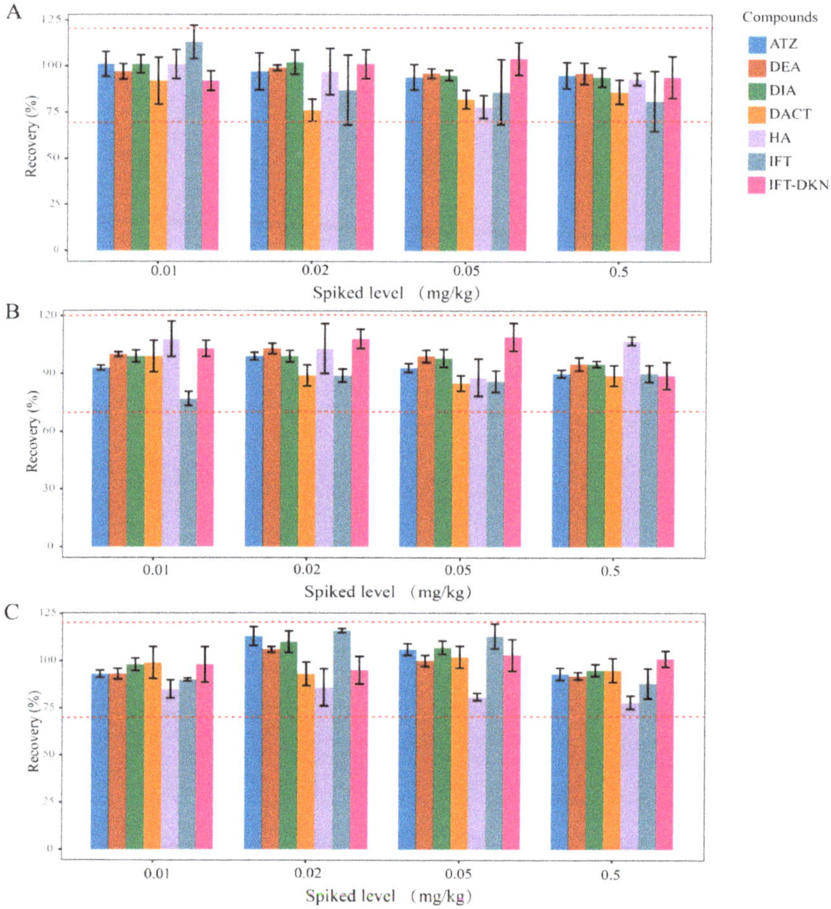

Figure 3. The recoveries and relative standard deviations of ATR, IFT and their metabolites in fresh corn (**A**), corn kernels (**B**) and corn straw (**C**) samples (n = 5).

In conclusion, the method established in this study was reliable, sensitive, has a short sample pre-treatment time and low cost, and can be used to quantify ATR, IFT and their metabolites in fresh corn, corn kernels, and corn straw.

2.4. Terminal Residue

The terminal residues of pesticides in food affect food safety and therefore arouse great concern among consumers [34]. The mixture of IFT and ATR has a broad application prospect because of their well-known complementary activity [35,36]. However, the terminal residues of these two compounds and their metabolites in corn fields have not been reported. According to the recommended dosage (active ingredient 874.5 g hm^{-2}), 53% isoxaflutole·atrazine suspending agent was sprayed during the corn three-leaf stage, and the final residues of total ATR and IFT in 6 provinces were detected, as shown in Table 3.

Table 3. Terminal residues of ATR and IFT in fresh corn, corn kernels and corn straw samples.

Location	Total ATR Residue (mg kg^{-1}) *			Total IFT residue (mg kg^{-1}) *		
	Fresh Corn	Corn Kernels	Corn Straw	Fresh Corn	Corn Kernels	Corn Straw
Liaoning	<0.05	<0.05	0.067 ± 0.0014	<0.02	<0.02	<0.02
Heilongjiang	<0.05	<0.05	<0.05	<0.02	<0.02	<0.02
Neimenggu	<0.05	<0.05	0.11 ± 0.085	<0.02	<0.02	<0.02
Shanxi	<0.05	<0.05	0.071 ± 0.0014	<0.02	<0.02	<0.02
Beijing	<0.05	<0.05	0.135 ± 0.0071	<0.02	<0.02	<0.02
Yunnan	<0.05	<0.05	<0.05	<0.02	<0.02	<0.02

* Total ATR and total IFT represent the sum of ATR and total IFT and their metabolites, respectively.

The residues of ATR, IFT and their metabolites in fresh corn and corn kernels collected at harvest were all below the LOQ (0.01 mg kg^{-1}), and he terminal residue of ATR in corn straw was in the range of <0.05 to 0.17 mg kg^{-1}. This herbicide was sprayed at the seedling stage of corn [37], which may be the main reason for the low residues of atrazine, isoxazotrione and their metabolites. Tandon et al. [21] reported that, when atrazine was sprayed before emergence, the atrazine in corn kernels, soil and straw during harvest was less than 0.005 mg kg^{-1}. According to the research results of Su et al. [33], the herbicide tembotrione residues in 10 field corn kernel samples were all less than 0.02 mg kg^{-1}. Field experiments in four provinces of China showed that the residues of glyphosate and glyphosate in corn grains were lower than 0.09 mg kg^{-1} [38]. Similarly, Zhong et al. [31] reported that the concentration levels of flumetsulam and florasulam in fresh corn and corn kernels during harvest were lower than 0.005 mg kg^{-1}. These results demonstrate that herbicides sprayed during the seedling stage or pre-emergence generally have low residue levels in corn kernels. In China, MRL values of IFT in corn and fresh corn have been established to be 0.02* mg kg^{-1}, and ATR in corn to be 0.05 mg kg^{-1} [39]. The MRL values of ATR in fresh corn developed by the United States, Japan, and Australia were 0.1, 0.2, and 0.1* mg kg^{-1}, respectively. It is safe to spray 53% isoxaflutole/atrazine suspending agent according to the recommended dosage in corn field during harvest.

2.5. Chronic Dietary Risk Assessment

The presence of pesticide residues in food may pose risk to humans. Dietary risk assessment is a necessary means to quantify the risk of pesticides in food and guide the safe use of pesticides [40,41]. The dietary risk assessment was calculated based on consumers' toxicological data, residue levels, and dietary intake. The ADI of ATR and IFT were all 0.02 mg kg^{-1} bw. Considering the risk maximization principle [34], ATR and IFT risk assessment used 0.05 mg kg^{-1} and 0.02 mg kg^{-1} as residual values to calculate NEDI and RQc, respectively. The results of intake risk are shown in Table 4. ATR and IFT are mainly used for weeding in corn fields in China. The chronic dietary risk was assessed based on the corn consumption of men and women of different age groups in the *2010–2013 Monitoring Report on Nutrition and Health Status of Chinese Residents* and the body weight provided in the 2014 National Physical Monitoring Bulletin. For all populations, the range of NEDI of ATR was 3.71×10^{-6}–1.48×10^{-5} mg kg^{-1} bw day^{-1}, and that of IFT was 1.48×10^{-6}–5.91×10^{-6} mg kg^{-1} bw day^{-1}. The RQc of ATR and IFT were in the range of 0.0074% to 0.0739%, much lower than 100%. These results suggest that the chronic risk associated with ingesting ATR and IFT through maize was acceptable. In addition, the chronic risk of ATR was higher than that of IFT. Regarding chronic dietary risks in different populations, consistent with previous studies [37], the results indicated that the intake risk for children (2–3 years old) was the highest, with increase of age the risk decreased gradually, and the intake risk was the lowest between 30 and 44 years old; for different genders, the risk for women was generally higher than that for men. These results were consistent with previous reports highlighting differences in the risk of dietary pesticide intake by age and sex [42,43], suggesting that certain populations were more susceptible to atrazine, isoxazotrione, and their metabolism through the corn, and subject to health risks.

Table 4. Chronic dietary risk assessment for ATR and IFT in a representative population.

Gender	Age (Years)	Average bw (kg)	Fi (kg)	ATR			IFT		
				STMR (mg kg^{-1})	NEDI (mg kg^{-1} bw Day^{-1})	RQ$_c$ (%)	STMR (mg kg^{-1})	NEDI (mg kg^{-1} bw Day^{-1})	RQ$_c$ (%)
Male	2–3	16.6	0.0047	0.05	1.42×10^{-5}	0.0708	0.02	5.66×10^{-6}	0.0283
	4–6	20.6	0.004		9.71×10^{-6}	0.0485		3.88×10^{-6}	0.0194
	7–10	31.8	0.0046		7.23×10^{-6}	0.0362		2.89×10^{-6}	0.0145
	11–13	46.8	0.0051		5.45×10^{-6}	0.0272		2.18×10^{-6}	0.0109
	14–17	59.1	0.0063		5.33×10^{-6}	0.0266		2.13×10^{-6}	0.0107
	18–19	63.4	0.0051		4.02×10^{-6}	0.0201		1.61×10^{-6}	0.0080
	20–29	68.8	0.0051		3.71×10^{-6}	0.0185		1.48×10^{-6}	0.0074
	30–44	71.4	0.0057		3.99×10^{-6}	0.0200		1.60×10^{-6}	0.0080
	45–59	70.3	0.0071		5.05×10^{-6}	0.0252		2.02×10^{-6}	0.0101
	60–69	67.1	0.0092		6.86×10^{-6}	0.0343		2.74×10^{-6}	0.0137
Female	2–3	15.9	0.0047		1.48×10^{-5}	0.0739		5.91×10^{-6}	0.0296
	4–6	19.6	0.0042		1.07×10^{-5}	0.0536		4.29×10^{-6}	0.0214
	7–10	29.8	0.0048		8.05×10^{-6}	0.0403		3.22×10^{-6}	0.0161
	11–13	44.4	0.0058		6.53×10^{-6}	0.0327		2.61×10^{-6}	0.0131
	14–17	51.6	0.0046		4.46×10^{-6}	0.0223		1.78×10^{-6}	0.0089
	18–19	52.7	0.0069		6.55×10^{-6}	0.0327		2.62×10^{-6}	0.0131
	20–29	54.6	0.0069		6.32×10^{-6}	0.0316		2.53×10^{-6}	0.0126
	30–44	57.9	0.0064		5.53×10^{-6}	0.0276		2.21×10^{-6}	0.0111
	45–59	59.9	0.008		6.68×10^{-6}	0.0334		2.67×10^{-6}	0.0134
	60–69	59.5	0.0099		8.32×10^{-6}	0.0416		3.33×10^{-6}	0.0166

3. Materials and Methods

3.1. Chemicals and Reagents

Certified standard IFT (purity 99.84%), ATR (99.37%), DEA (purity 99.06%), DIA (purity 99.02%), DACT (purity 99.14%) and HA (purity 97.24%) were provided by DrEhrenstorfer Ltd. (Augsburg, Germany). IFT-DKN (99.7% purity) was purchased from Beijing Qincheng Yixin Technology Development Co., Ltd. (Beijing, China). HPLC-grade acetonitrile, methanol, and formic acid were purchased from Thermo Fisher Technology Co., Ltd. (Shanghai, China). Tedia Company, Inc., Fairfield, OH, USA, provided analytical acetonitrile. Analytical anhydrous magnesium sulfate and sodium chloride were purchased from Sinopharm Chemical Reagent Co., Ltd. (Shanghai, China). Graphitized carbon black (GCB, 38–120 μm) and octadecyl silane (C18, 50 μm) were purchased from Tianjin Bona Eijer Technology Co., Ltd. (Tianjin, China). PTFE film needle filter (0.22 μm) was purchased from Tianjin Bonaigel Technology Co., Ltd. (Tianjin, China). Purified water was provided by Guangzhou Watsons Food & Beverage Co., Ltd. (Guangzhou, China).

3.2. Preparation of Solvent Standard and Matrix Matching Standard Solution

The standard solutions of ATR, DEA, DIA, DACT, and HA of 1000 mg L^{-1} were dissolved to 10 mL with methanol in a brown volumetric flask. The acetonitrile diluted 1000 mg L^{-1} standard solution of 0.0010 g IFT, and IFT-DKN was obtained similarly. The above standard solutions of 1 mL were measured in a 10 mL volumetric flask to prepare the mixed standard solutions of ATR, IFT, and their metabolites of 100 mg L^{-1}. The mixed standard solutions were diluted with acetonitrile, fresh corn, corn kernels, and corn straw to prepare different concentrations (0.0025, 0.005, 0.0125, 0.025, 0.125, 0.25 mg L^{-1}) and matrix-matching standard solutions. The prepared solution is stored in a refrigerator at 4 °C.

3.3. *Field Experiment Design and Sampling*

According to the guidelines for the Detection of Pesticide residues in crops (NY/T 788, 2018) issued by the Ministry of Agriculture and villages of the People's Republic of China, field experiments were conducted from May to October 2022 in Shenyang City, Liaoning Province, Harbin City, Heilongjiang Province, Hohhot City, Inner Mongolia, Jinzhong City, Shanxi Province, Changping District, Beijing and Maile City, Honghe Prefecture, Yunnan Province to study the residues of ATR, IFT and their metabolites in corn field. The soil pH value of these plots is between 6.5 and 8.66; the organic matter content is less than 2.9%; the average temperature during the experiment period is between 18.9 and 25.7 °C, and the rainfall is less than 367.4 mm. The commercial pesticide product 53% isoxaflutole·atrazine suspension consists of the active ingredients atrazine (48%) and isoxaflutole (5%). According to Good Agricultural Practice (GAP), 53% isoxaflutole·atrazine suspending agent were sprayed once at the three-leaf stage of corn with an active ingredient of 874.5 g hm^{-2}. Each treatment was carried out in a 100 m^2 plot, repeated twice, and separated by a buffer zone of 0.5 m. At the same time, the same volume of water was sprayed on the control plot.

During harvest, 12 (more than 2 kg) fresh corn, corn kernels, and corn straw samples with normal growth and disease-free were picked randomly from each plot. The fresh corn samples without bracts and filaments were divided into three equal-length segments. The upper, middle, and lower segments were taken respectively, and then the corn segments were crushed and thoroughly mixed. The corn kernel samples without bracts and filaments were threshed, and the grains were well mixed. The corn straw samples in the field were divided into three equal-length segments, and four upper, middle, and lower segments were chopped and thoroughly mixed. Fresh corn, corn kernels, and corn straw were divided into two samples of not less than 200 g; one was the experimental sample, and the other was the backup sample. The collected samples were packed in a sealed pocket, labeled, and stored in cold storage at less than −18 °C for further analysis.

3.4. *Residue Determination*

3.4.1. Extraction and Purification Procedures

Homogenized fresh corn (5.00 g), corn kernels (5.00 g) and corn straw (2.50 g) were weighed into 50 mL centrifuge tubes, respectively. Add 2.5 mL purified water and 10 mL acetonitrile to the centrifuge tube. Cover and shake vigorously for 10 min. Then, add 2 g anhydrous magnesium sulfate and 2 g sodium chloride into the test tube and shock again for 5 min. Centrifuge at 4000 rpm for 5 min. Then, 1.5 mL of the supernatant was transferred to a 2 mL purification tube containing different adsorbent materials. The purification tubes of fresh corn and corn kernels were filled with 150 mg anhydrous magnesium sulfate and 75 mg C18. In comparison, the corn straw purification tubes were filled with 150 mg anhydrous magnesium sulfate, 50 mg C18, and 3.75 mg GCB. The purified tube was swirled for 1 min and centrifuged at 4000 rpm for 5 min. The extracts were filtered through a 0.22 μm syringe and then transferred to an automatic sampler for HPLC-MS/MS analysis.

3.4.2. Instrumental Parameters

The compounds were analyzed by high performance liquid chromatography tandem mass spectrometry (HPLCI-ClassXEVOTQ-XS, Waters, Milford, CT, USA). The chromatographic separation was performed on an ACQUITYUPLC ®BEHC18 column (2.1 × 100 mm, 1.7 μm). The mobile phase was (A) 0.2% formic acid aqueous solution and (B) methanol. The flow rate was 0.3 mL min^{-1}. The temperature of column box was 40 °C. The injection volume was 5 μL. The conditions of gradient elution were as follows: 10% B (0–2 min), increased to 70% B (2–2.5 min), kept at 70% B (2.5–5.5 min), decreased to 10% B (5.5–5.6 min), and maintained (5.6–6 min). Except IFT-DKN (negative), other compounds were monitored by mass spectrometry with positive ionization mode. Data analysis was performed using MassLynx 4.0 software.

3.4.3. Method Validation

The method's accuracy, precision, linearity, ME and LOQ were validated. Recovery experiments were performed by spiking ATR, IFT and their metabolites to blank samples at concentrations of 0.01, 0.02, 0.05, and 0.5 mg kg^{-1}, respectively, and each was repeated five times. Solvent or matrix-matched standard solutions with concentrations ranging from 0.0025 to 0.25 mg kg^{-1} were serially diluted. The limit of quantitation (LOQ) was determined as the lowest spiked level.

Matrix effects (ME) were assessed by comparing the slope of the matrix-matched standard to the acetonitrile standard. The calculation method was:

$$ME(\%) = \frac{(Smatrix - Ssolvent)}{Ssolvent} \times 100\% \quad (1)$$

where, S_{matrix} and $S_{solvent}$ were the slopes of the matrix matching standard and solvent standard, respectively. ME can be ignored if the ME value was within the range of −20–20%. Otherwise, a matrix enhancing or attenuating effect was exhibited.

3.5. Terminal Residue and Dietary Risk Assessment

3.5.1. Residue Definition

According to the residue definition of risk assessment (JMPR, 2013), the total residue of IFT was calculated according to the following equation:

$$C_{IFT} = C_{IFT} + C_{IFT\text{-}DKN} \quad (2)$$

According to the Pesticide Registration Residue Test Residues and the Catalog of Residues for Dietary Risk Assessment in Foods of Plant Origin, the total residues of ATR were estimated with the following equation:

$$C_{ATR} = C_{ATR} + C_{DEA} \times 1.15 + C_{DIA} \times 1.24 + C_{DACT} \times 1.48 + C_{HA} \times 1.09 \quad (3)$$

where C_{IFT}, $C_{IFT\text{-}DKN}$ C_{ATR}, C_{DEA}, C_{DIA}, C_{DACT} and C_{HA} are the concentrations of IFT, IFT-DKN, ATR, DEA, DIA, DACT and HA. 1.15, 1.24, 1.48 and 1.09 are the molecular weight ratios of DEA, DIA, DACT and HA to ATR, respectively. If the residues of IFT, ATR and their metabolites were lower than their limited limits (LOQ), the value of LOQ was considered directly, and the C_{sum} was calculated directly based on the sum of LOQ values of all compounds.

3.5.2. Dietary Risk Assessment

The risk quotient (RQ_c) was used to assess further the risk of chronic dietary intake of IFT and ATR and was calculated according to the following equation.

$$NEDI = \sum(STMR \times F_i)/bw \quad (4)$$

$$RQ_c = NEDI/ADI \times 100\% \quad (5)$$

where NEDI (mg kg^{-1} bw day^{-1}) is the national estimated daily intake, and STMR (mg kg^{-1}) is the standard median residual value. F_i (kg) represents the consumption of a given food by a specific population. bw is the average body weight, kg. The ADI of IFT and ATR are both 0.02 mg kg^{-1} bw day^{-1}.

4. Conclusions

We established an improved QuEChERS pre-treatment and ultra-high performance liquid chromatography–tandem mass spectrometry method to detect ATR, IFT and metabolites. The technique has good linearity, accuracy, and precision. Under GAP conditions, samples of fresh maize, corn kernels, and corn straw after pesticide application were collected in six provinces in China. The results showed that the residues of ATR, IFT, and their metabolites in fresh corn and corn kernels were all lower than LOQ. IFT was also not detected in corn straw, but ATR residue, less than 0.135 mg kg^{-1}, was detected. All consumers' risk quotients (RQ_c) were below 100%, indicating that the chronic risk of ATR and IFT is acceptable. It should be noted that children (2–3 years old) and women have a relatively higher risk of chronic diseases than other groups.

Author Contributions: Data curation, J.C.; Funding acquisition, J.C.; Methodology, T.P., Y.W., Y.Q., P.R. and J.L.; Project administration, S.Q.; Resources, S.Q. and J.L.; Software, J.L.; Supervision, S.Q.; Validation, T.P., Y.W., Y.Q. and P.R.; Visualization, J.C., T.P., Y.W., Y.Q. and P.R.; Writing—original draft, J.C.; Writing—review & editing, S.Q. and J.L. All authors have read and agreed to the published version of the manuscript.

Funding: This research was funded by the Doctor's reward funds for working in Shanxi Province (SXBYKY2022087), Startup Foundation for Doctors of Shanxi Agricultural University (2021BQ127) and Shanxi Applied Basic Research Program Science–Youth Technology Research Fund (202103021223149).

Institutional Review Board Statement: Not applicable.

Informed Consent Statement: Not applicable.

Data Availability Statement: The data presented in this study are available from the authors upon request.

Acknowledgments: We would like to thank Zhang Junwen, Su Hailing and Zhu Li'ao for their support in this field trial, and our colleague Wang Xia' technical support.

Conflicts of Interest: The authors declare no conflict of interest.

Sample Availability: Samples of the compounds are available from the authors.

References

1. Nuss, E.T.; Tanumihardjo, S.A. Maize: A Paramount Staple Crop in the Context of Global Nutrition. *Compr. Rev. Food Sci. Food Saf.* **2010**, *9*, 417–436. [CrossRef] [PubMed]
2. Jamshidi, K.; Yousefi, A.R.; Oveisi, M. Effect of cowpea (*Vigna unguiculata*) intercropping on weed biomass and maize (Zea mays) yield. *N. Z. J. Crop Hort. Sci.* **2013**, *41*, 180–188. [CrossRef]
3. Gianessi, L.P. The increasing importance of herbicides in worldwide crop production. *Pest Manag. Sci.* **2013**, *69*, 1099–1105. [CrossRef] [PubMed]
4. Popp, J.; Pető, K.; Nagy, J. Pesticide productivity and food security. A review. *Agron. Sustain. Dev.* **2013**, *33*, 243–255. [CrossRef]
5. Carvalho, F.P. Pesticides, environment, and food safety. *Food Energy Secur.* **2017**, *6*, 48–60. [CrossRef]
6. Kim, K.-H.; Kabir, E.; Jahan, S.A. Exposure to pesticides and the associated human health effects. *Sci. Total Environ.* **2017**, *575*, 525–535. [CrossRef]
7. Wu, Y.; An, Q.; Hao, X.; Li, D.; Zhou, C.; Zhang, J.; Wei, X.; Pan, C. Dissipative behavior, residual pattern, and risk assessment of four pesticides and their metabolites during tea cultivation, processing and infusion. *Pest Manag. Sci.* **2022**, *78*, 3019–3029. [CrossRef]
8. Gao, J.; Wang, F.; Jiang, W.; Miao, J.; Wang, P.; Zhou, Z.; Liu, D. A full evaluation of chiral phenylpyrazole pesticide flufiprole and the metabolites to non-target organism in paddy field. *Environ. Pollut.* **2020**, *264*, 114808. [CrossRef]
9. Pallett, K.E.; Cramp, S.M.; Little, J.P.; Veerasekaran, P.; Crudace, A.J.; Slater, A.E. Isoxaflutole: The background to its discovery and the basis of its herbicidal properties. *Pest Manag. Sci.* **2001**, *57*, 133–142. [CrossRef]
10. Beltran, E.; Fenet, H.; Cooper, J.-F.; Coste, C.-M. Fate of isoxaflutole in soil under controlled conditions. *J. Agric. Food Chem.* **2003**, *51*, 146–151. [CrossRef]
11. Papiernik, S.K.; Yates, S.R.; Koskinen, W.C.; Barber, B. Processes affecting the dissipation of the herbicide isoxaflutole and its diketonitrile metabolite in agricultural soils under field conditions. *J. Agric. Food Chem.* **2007**, *55*, 8630–8639. [CrossRef]
12. Muller, G.; LeBaron, H.; McFarland, J.; Burnside, O. History of the discovery and development of triazine herbicides. *Triazine Herbic.* **2008**, *50*, 13–29.
13. Jin, Y.; Zhang, X.; Shu, L.; Chen, L.; Sun, L.; Qian, H.; Liu, W.; Fu, Z. Oxidative stress response and gene expression with atrazine exposure in adult female zebrafish (*Danio rerio*). *Chemosphere* **2010**, *78*, 846–852. [CrossRef] [PubMed]
14. Song, Y.; Zhu, L.S.; Wang, J.; Wang, J.H.; Liu, W.; Xie, H. DNA damage and effects on antioxidative enzymes in earthworm (*Eisenia foetida*) induced by atrazine. *Soil. Biol. Biochem.* **2009**, *41*, 905–909. [CrossRef]
15. Stoker, T.E.; Laws, S.C.; Guidici, D.L.; Cooper, R.L. The effect of atrazine on puberty in male wistar rats: An evaluation in the protocol for the assessment of pubertal development and thyroid function. *Toxicol. Sci.* **2000**, *58*, 50–59. [CrossRef] [PubMed]
16. Gammon, D.W.; Aldous, C.N.; Carr, W.C., Jr.; Sanborn, J.R.; Pfeifer, K.F. A risk assessment of atrazine use in California: Human health and ecological aspects. *Pest Manag. Sci.* **2005**, *61*, 331–355. [CrossRef] [PubMed]

17. Ding, Y.; Hao, B.; Zhang, N.; Lv, H.; Zhao, B.; Tian, Y. Rapid determination of thiram and atrazine pesticide residues in fruit and aqueous system based on surface-enhanced Raman scattering. *Spectrochim. Acta Part A* **2023**, *285*, 121873. [CrossRef]
18. Sun, W.; Huang, G.; Sai, N.; Xu, B. Carboxylated hapten coated directly coated ELISA for the detection of atrazine residue in water. *J. Hyg. Res.* **2017**, *46*, 109–119.
19. Yuan, L.; Chai, Y.; Li, C.; Liu, R.; Chen, Z.; Li, L.; Li, W.; He, Y. Dissipation, residue, dietary, and ecological risk assessment of atrazine in apples, grapes, tea, and their soil. *Environ. Sci. Pollut. Res.* **2021**, *28*, 35064–35072. [CrossRef]
20. Fu, H.; Wang, Z.; Zhang, T.; Du, M.; Lv, B.; Yang, S.; Chen, R. Simultaneous detection of topramezon, atrazine, and atrazine metabolite residues in environmental water by large-volume SPE-UPLC-MS/MS. *Int. J. Environ. Anal. Chem.* **2022**. [CrossRef]
21. Tandon, S.; Singh, A. Field dissipation kinetics of atrazine in soil and post-harvest residues in winter maize crop under subtropical conditions. *Chem. Ecol.* **2015**, *31*, 273–284. [CrossRef]
22. Lin, C.H.; Lerch, R.N.; Garrett, H.E.; Li, Y.X.; George, M.F. Improved HPLC-MS/MS method for determination of isoxaflutole (balance) and its metabolites in soils and forage plants. *J. Agric. Food Chem.* **2007**, *55*, 3805–3815. [CrossRef] [PubMed]
23. Lin, C.-H.; Lerch, R.N.; Thurman, E.M.; Garrett, H.E.; George, M.F. Determination of isoxaflutole (balance) and its metabolites in water using solid phase extraction followed by high-performance liquid chromatography with ultraviolet or mass spectrometry. *J. Agric. Food Chem.* **2002**, *50*, 5816–5824. [CrossRef] [PubMed]
24. Lan, F.; Li, Q.; Sun, L.; Yu, W.; Jiang, W.; Wang, Z.; Liu, C. Simultaneous determination of isoxaflutole and its two metabolites in corn under field conditions by LC-MS/MS. *J. Sci. Food Agric.* **2022**, *102*, 3480–3486. [CrossRef] [PubMed]
25. Anastassiades, M.; Lehotay, S.J.; Štajnbaher, D.; Schenck, F.J. Fast and easy multiresidue method employing acetonitrile extraction/partitioning and "dispersive solid-phase extraction" for the determination of pesticide residues in produce. *J. AOAC Int.* **2003**, *86*, 412–431. [CrossRef] [PubMed]
26. Lehotay, S.J.; Son, K.A.; Kwon, H.; Koesukwiwat, U.; Fu, W.; Mastovska, K.; Hoh, E.; Leepipatpiboon, N. Comparison of QuEChERS sample preparation methods for the analysis of pesticide residues in fruits and vegetables. *J. Chromatogr. A* **2010**, *1217*, 2548–2560. [CrossRef]
27. Perestrelo, R.; Silva, P.; Porto-Figueira, P.; Pereira, J.A.; Silva, C.; Medina, S.; Câmara, J.S. QuEChERS-Fundamentals, relevant improvements, applications and future trends. *Anal. Chim. Acta* **2019**, *1070*, 1–28. [CrossRef]
28. Wilkowska, A.; Biziuk, M. Determination of pesticide residues in food matrices using the QuEChERS methodology. *Food Chem.* **2011**, *125*, 803–812. [CrossRef]
29. Lehotay, S.J. Determination of pesticide residues in foods by acetonitrile extraction and partitioning with magnesium sulfate: Collaborative study. *J. AOAC Int.* **2007**, *90*, 485–520. [CrossRef]
30. Cui, K.; Wu, X.; Zhu, L.; Zhang, Y.; Dai, G.; Cao, J.; Xu, J.; Dong, F.; Liu, X.; Zheng, Y. Development and establishment of a QuEChERS-based extraction method for determining tembotrione and its metabolite AE 1417268 in corn, corn oil and certain animal-origin foods by HPLC-MS/MS. *Food Addit. Contam. Part A* **2020**, *37*, 1678–1686. [CrossRef]
31. Zhong, M.; Wang, T.; Hu, J. Dissipation kinetics and residues of triazolopyrimidine herbicides flumetsulam and florasulam in corn ecosystem. *Environ. Monit. Assess.* **2015**, *187*, 390. [CrossRef] [PubMed]
32. Pihlström, T.; Fernández-Alba, A.R.; Gamón, M.; Poulsen, M.E.; Lippold, R.; Anastassiades, M. Analytical Quality Control and Method Validation Procedures for Pesticide Residues Analysis in Food and Feed. *Sante* **2017**, *11813*, 21–22.
33. Su, Y.; Wang, W.; Hu, J.; Liu, X. Dissipation behavior, residues distribution and dietary risk assessment of tembotrione and its metabolite in maize via QuEChERS using HPLC-MS/MS technique. *Ecotoxicol. Environ. Saf.* **2020**, *191*, 110187. [CrossRef] [PubMed]
34. Lin, H.; Liu, L.; Zhang, Y.; Shao, H.; Li, H.; Li, N.; Zou, P.; Lu, N.; Guo, Y. Residue behavior and dietary risk assessment of spinetoram (XDE-175-J/L) and its two metabolites in cauliflower using QuEChERS method coupled with UPLC-MS/MS. *Ecotoxicol. Environ. Saf.* **2020**, *202*, 110942. [CrossRef] [PubMed]
35. Makinde, J.; Ogunbodede, B. Evaluation of atrazine plus isoxaflutole (Atoll®) mixture for weed control in maize. *Ghana J. Agric. Sci.* **2008**, *40*, 193–198. [CrossRef]
36. Willemse, C.; Soltani, N.; David, C.H.; Jhala, A.J.; Robinson, D.E.; Sikkema, P.H. Interaction of 4-hydroxyphenylpyruvate dioxygenase (HPPD) and atrazine alternative photosystem II (PS II) inhibitors for control of multiple herbicide-resistant waterhemp (*Amaranthus tuberculatus*) in corn. *Weed Sci.* **2021**, *69*, 492–503. [CrossRef]
37. Zhao, H.; Hu, J. Total residue levels and risk assessment of flufenacet and its four metabolites in corn. *J. Food Compos. Anal.* **2022**, *106*, 104268. [CrossRef]
38. Wang, K.; Jiao, B.; Gao, H.; Pan, X.; Wu, X.; Xu, J.; Dong, F.; Zheng, Y. Residue and dietary risk assessment of glyphosate, glufosinate-ammonium, and their metabolites in maize and soybean. *J. Food Compos. Anal.* **2023**, *120*, 105298. [CrossRef]
39. Ministry of Agriculture and Rural Affairs of the People's Republic of China. *National Food Safety Standard-Maximum Residue Limits for Pesticides in Food*; China Agricultural Press: Beiing, China, 2021.
40. Chang, C.-H.; MacIntosh, D.; Lemos, B.; Zhang, Q.; Lu, C. Characterization of Daily Dietary Intake and the Health Risk of Neonicotinoid Insecticides for the U.S. Population. *J. Agric. Food Chem.* **2018**, *66*, 10097–10105. [CrossRef]
41. Bian, Y.; Feng, Y.; Zhang, A.; Qi, X.; Pan, J.; Han, J.; Ma, X.; Liang, L. Residue distribution and risk assessment of bifenazate and its metabolite in garlic plant. *Food Chem.* **2022**, *379*, 132013. [CrossRef]

42. Zhao, H.; Zhao, Y.; Hu, J. Dissipation, residues and risk assessment of pyraclostrobin and picoxystrobin in cucumber under field conditions. *J. Sci. Food Agric.* **2020**, *100*, 5145–5151. [CrossRef] [PubMed]
43. Zhou, R.; Dong, Z.; Bian, C.; Wang, L.; Wu, T.; Zhou, W.; Li, Y.; Li, B. Residue analysis, dissipation behavior, storage stability and dietary risk assessment of florpyrauxifen-benzyl in natural paddy field environment using UPLC-QTOF-MS/MS. *J. Food Compos. Anal.* **2022**, *114*, 104781. [CrossRef]

MDPI
St. Alban-Anlage 66
4052 Basel
Switzerland
www.mdpi.com

Molecules Editorial Office
E-mail: molecules@mdpi.com
www.mdpi.com/journal/molecules

www.ingramcontent.com/pod-product-compliance
Lightning Source LLC
Chambersburg PA
CBHW040757250926
51850CB00007BA/1015
* 9 7 8 3 0 3 6 5 9 4 6 2 0 *